Selected Titles in This Series

14 **Eyal Z. Goren,** Lectures on Hilbert modular varieties and modular forms, 2002
13 **Michael Baake and Robert V. Moody, Editors,** Directions in mathematical quasicrystals, 2000
12 **Masayoshi Miyanishi,** Open algebraic surfaces, 2001
11 **Spencer J. Bloch,** Higher regulators, algebraic K-theory, and zeta functions of elliptic curves, 2000
10 **James D. Lewis,** A survey of the Hodge conjecture, Second Edition, 1999
9 **Yves Meyer,** Wavelets, vibrations and scaling, 1998
8 **Ioannis Karatzas,** Lectures on the mathematics of finance, 1996
7 **John Milton,** Dynamics of small neural populations, 1996
6 **Eugene B. Dynkin,** An introduction to branching measure-valued processes, 1994
5 **Andrew Bruckner,** Differentiation of real functions, 1994
4 **David Ruelle,** Dynamical zeta functions for piecewise monotone maps of the interval, 1994
3 **V. Kumar Murty,** Introduction to Abelian varieties, 1993
2 **M. Ya. Antimirov, A. A. Kolyshkin, and Rémi Vaillancourt,** Applied integral transforms, 1993
1 **D. V. Voiculescu, K. J. Dykema, and A. Nica,** Free random variables, 1992

Lectures on Hilbert Modular Varieties and Modular Forms

Volume 14

CRM MONOGRAPH SERIES

Centre de Recherches Mathématiques
Université de Montréal

Lectures on Hilbert Modular Varieties and Modular Forms

Eyal Z. Goren

with the assistance of Marc-Hubert Nicole

The Centre de Recherches Mathématiques (CRM) of the Université de Montréal was created in 1968 to promote research in pure and applied mathematics and related disciplines. Among its activities are special theme years, summer schools, workshops, postdoctoral programs, and publishing. The CRM is supported by the Université de Montréal, the Province of Québec (FCAR), and the Natural Sciences and Engineering Research Council of Canada. It is affiliated with the Institut des Sciences Mathématiques (ISM) of Montréal, whose constituent members are Concordia University, McGill University, the Université de Montréal, the Université du Québec à Montréal, and the Ecole Polytechnique. The CRM may be reached on the Web at www.crm.umontreal.ca.

American Mathematical Society
Providence, Rhode Island USA

The production of this volume was supported in part by the Fonds pour la Formation de Chercheurs et l'Aide à la Recherche (Fonds FCAR) and the Natural Sciences and Engineering Research Council of Canada (NSERC).

2000 *Mathematics Subject Classification.* Primary 11G10, 11G18, 14G35, 11F33, 11F41.

Library of Congress Cataloging-in-Publication Data

Goren, Eyal Z. (Zvi), 1963–
Lectures on Hilbert modular varieties and modular forms / Eyal Z. Goren ; with the assistance of Marc-Hubert Nicole.
p. cm. — (CRM monograph series, ISSN 1065-8599 ; v. 14)
Includes bibliographical references and index.
ISBN 0-8218-1995-X (alk. paper)
1. Abelian varieties. 2. Forms, Modular. 3. Moduli theory. I. Title. II. Series.

QA564.G622 2001
516.3′53—dc21 2001044830

∞ The paper used in this book is acid-free and falls within the guidelines
established to ensure permanence and durability.
This volume was submitted to the American Mathematical Society
in camera ready form by the Centre de Recherches Mathématiques.
Visit the AMS home page at URL: `http://www.ams.org/`

10 9 8 7 6 5 4 3 2 1 07 06 05 04 03 02

Contents

Introduction

This book is about abelian varieties with real multiplication, their moduli spaces, and modular forms on them. Abelian varieties are commutative, connected, projective algebraic groups that generalize the notion of elliptic curve. If the ring of integers of a totally real field L of degree g acts as endomorphisms of an abelian variety of dimension g we say that this abelian variety has real multiplication by L.

Abelian varieties with real multiplication by a given field can be parameterized by a scheme $\mathfrak{M}$, whose complex points are in natural bijection with a union of quotients of the g-fold product of the complex upper half space by suitable discrete subgroups of $\mathrm{PGL}_2(L)$, and modular forms are certain sections of line bundles on $\mathfrak{M}$ that encode an amazing amount of information about Galois representations, the geometry of $\mathfrak{M}$ and special values of L-functions. This book is devoted to these topics and their applications.

This book is written primarily for the graduate student. In particular, we assume that the reader has basic knowledge in number theory and algebraic geometry, including the rudiments of the theory of elliptic curves and one-dimensional modular forms. The book doesn't strive toward being an exhaustive presentation of the subjects of Hilbert modular forms and Hilbert moduli spaces—a formidable task—but rather toward a gentle introduction of the interested graduate student, or interested scholar, into a certain circle of ideas concerning abelian varieties and modular forms and generalizing the more widely known theory of elliptic curves and one-dimensional modular forms. It attempts to present a wide range of ideas and techniques in a uniform language and consistent notation and to show how they intertwine into a tapestry of rare beauty. By no means is every important topic introduced. The scope of the manuscript (and the graduate course out of which it grew) prevented us from presenting many topics of great importance. (For example, much to our regret, we do not discuss the theory of the U operator and the canonical subgroup for p-adic modular forms). Our goal was rather to follow one possible trail through this vast terrain of Hilbert modular varieties and modular forms, and to offer a panoramic view (with emphasis on positive characteristic) of the subject.

My hope is that the reader who dives into this book, armed with some knowledge in algebraic geometry and number theory, will emerge feeling that he knows enough about the subject to actually start working in it, and that he has a wide enough base on which to expand his knowledge.

There are already several excellent books on the same subject. Nonetheless, there is much in this book that is not covered in any of the other texts. We mention two books that we feel the reader would benefit greatly from consulting, notwithstanding other excellent references. One is G. van der Geer: *Hilbert modular surfaces* [**44**] and the other is E. Freitag: *Hilbert modular forms* [**42**].

This book is an expanded version of a year long graduate course, consisting of two weekly hours, offered at McGill University as part of the CRM Theme Year 1998-99 in Number Theory and Arithmetic Geometry. The book, as it stands, could probably be covered in a year long course of three hours per week.

I was fortunate enough to have a wonderful audience that consisted of: M. Chandrasekher, I. Chen, H. Darmon, L. Dembele, C. Greither, K. Khuri-Makdisi, H. Kisilevsky, I. Longhi, S. Mohit, M.-H. Nicole, A. Rajaei, F. Saidak, and A. Schweizer. I thank them for their attention and comments, from which this manuscript, and myself, benefited greatly. B. Conrad, H. Kisilevsky, and H. Darmon made helpful comments on the final draft of this book. I would also like to thank Jeff Achter and Eitan Bachmat for numerous corrections and suggestions on an earlier draft, and Fabrizio Andreatta for a very thorough reading of the text and numerous helpful suggestions that improved the final manuscript.

It remains to thank Marc-Hubert Nicole for his assistance and the many hours he put into this work. He prepared, based on my lecture notes, the initial version of this manuscript, proof-read many parts and was helpful in numerous others ways in the preparation of this book.

Conventions and notations. We denote by $\mathbb{Z}$, $\mathbb{Q}$, $\mathbb{R}$, and $\mathbb{C}$ the integers, rational numbers, real numbers and complex numbers, respectively. We let $\mathbb{N} = \{0, 1, 2, \dots\}$ and $\mathbb{N}^+ = \{1, 2, 3, \dots\}$. The letter $\mathbb{F}$ usually stands for a field and $\mathbb{F}_q$ is "the" field of q elements. We denote an algebraic closure of a field $\mathbb{F}$ by $\overline{\mathbb{F}}$ or $\mathbb{F}^{\text{alg}}$. We shall usually use the letter L to denote a totally real number field with ring of integers $\mathcal{O}_L$ and degree g over $\mathbb{Q}$. The symbol "=" stands also for canonical isomorphism. The symbol ":=" stands for "definition."

The term "Ring" would usually stand for a commutative associative ring with 1, with obvious exceptions like "the ring of matrices" etc.

We let $\mathbb{W}(R)$ denote the "full" Witt vectors, while $W(R)$ denotes the Witt vectors with components a power of p. The ring $W(R)$ is the "usual" ring of Witt vectors (e.g., $W(\mathbb{Z}/p\mathbb{Z}) = \mathbb{Z}_p$). The prime p implicit in the notation would be clear from the context.

Throughout the book we use the language of schemes, with our canonical reference for facts as well as notations being Hartshorne [**68**]. An exception is the section on tori where we use the classical language of algebraic geometry; varieties are identified with their set of points in an algebraic closure. The reason for that is that this gives the less advanced students a chance to catch up with the necessary algebraic geometry during the first weeks of the course. In the McGill course, we had a seminar running simultaneously that gave some background on schemes and other topics.

For a morphism of schemes $X \to S$ (also denoted X/S), we let $\mathfrak{T}^*_{X/S} = \Omega_{X/S}$ denote the sheaf of relative Kähler differentials and $\mathfrak{T}_{X/S} := \mathrm{Hom}_{\mathcal{O}_X}(\mathfrak{T}^*_{X/S}, \mathcal{O}_X)$ is the tangent sheaf. If there is a section $e\colon S \to X$, usually understood from the context (e.g., the identity section when X/S is a group scheme), we write $\mathfrak{t}^*_{X/S}$, or just $\mathfrak{t}^*_X$ if we need to make notation easier, for $e^*\mathfrak{T}^*_{X/S}$. Similarly for $\mathfrak{t}_{X/S}$.

The Dieudonné theory that we use is often covariant. This is especially so in Chapter 6 where we discuss Dieudonné–Cartier theory. Since Frobenius on the p-

divisible group induces the Verschiebung morphism on the associated module, we define the Newton polygon using the Verschiebung operator on the module and not the Frobenius operator. The intrinsic notion for us is the Newton polygon of a p-divisible group and our conventions are such that one gets the same answer as in the contravariant theory. Some authors use the other convention.

Exercises. The book contains many exercises, ranging from obvious to quite difficult. Mostly they serve to master new terminology and ideas and sometimes to introduce some extra material in a more concise fashion. The harder exercises are marked with ★.

References. We refer to, say, Theorem 4.5 in Chapter 3 by "See Theorem 4.5" if the reference is made in Chapter 3, while we refer to it by "See Chapter 3, Theorem 4.5" if the reference is made in another chapter. The same scheme is used for section numbers. Thus, we refer to Chapter 3, Section 2 as "§2" if the reference is made within Chapter 3, and by "Chapter 3, §2" if the reference is made in another chapter.

CHAPTER 1

Tori and Abelian Varieties

In the first section of this chapter we recall the definition of algebraic groups and some basic facts about them, attempting to put the classes of groups that will interest us—abelian varieties and algebraic tori—in a somewhat larger perspective. We dedicate most of our discussion of algebraic groups to diagonalizable groups and, in particular, to tori (later sections treat abelian varieties).

Tori are forms of $\mathbb{G}_m^g$ for some g. There are several good reasons to discuss tori in detail. They appear as degenerations of abelian varieties, their characters are weights of Hilbert modular forms, and their finite subgroups are involved in defining level structures on abelian varieties.

We then discuss the notion of a moduli problem in general, and provide brief remarks on Kodaira–Spencer theory. Moduli problems play a central role in algebraic geometry and this book is no exception. In our context, moduli problems are classification problems arising in algebraic geometry and one seeks to "solve the problem" by giving an appropriate variety, or scheme, whose points correspond in a natural fashion to the objects being classified. The particular case that interests us is the classification of abelian varieties with real multiplication.

Next we devote our attention to elliptic curves, which are "abelian varieties with real multiplication by $\mathbb{Q}$." We recall facts (that we assume to be familiar to the reader) concerning moduli problems for elliptic curves, modular curves and modular forms. We somewhat neglect the classical approach to modular forms, assuming it to be known to the reader, and emphasize instead the geometric approach—a recurring theme in this book. We introduce suitable factors of automorphy and line bundles and interpret modular forms as sections of those line bundles. We provide some examples of modular forms that either play a role later on or connect nicely to geometry.

We next discuss briefly, yet explicitly, complex abelian varieties and their line bundles and, in particular, the Appell–Humbert Theorem and the dual abelian variety. We define the Weil pairing and conclude the chapter by calculating it for elliptic curves.

1. Algebraic groups

In this section we follow very closely Borel [**5**] (another good reference is [**125**]), where the reader may find also a moderate introduction to algebraic geometry. We use in this section the (somewhat awkward to our taste) language of "classical" algebraic geometry. Thus, we work mostly over an algebraically closed field and when we work over a general field the concepts (such as a variety, or a morphism, being defined over a subfield) are still derived from an algebraic closure of the field and the ambient projective space. In particular, a variety is always a quasi projective variety. A subvariety (and later, a subgroup) is identified with its set of geometric points which is a subset of the ambient variety, etc. Nevertheless, hints to the scheme (or modern) point of view are spread all over the text.

Let k be a field with an algebraic closure K and let k^s be the separable closure of k in K. We denote the Galois group $\mathrm{Gal}(k^s/k)$ by Γ throughout this section. We shall denote by $k[G]$ the ring of regular functions on a variety G over the field k.

DEFINITION 1.1. An *algebraic group* over k is a quasi-projective variety G over k together with k-morphisms

$$m\colon G \times G \to G, \tag{1.1}$$

$$\mathrm{inv}\colon G \to G, \tag{1.2}$$

and a k-rational identity element e, such that the following diagrams commute:

(1.3)

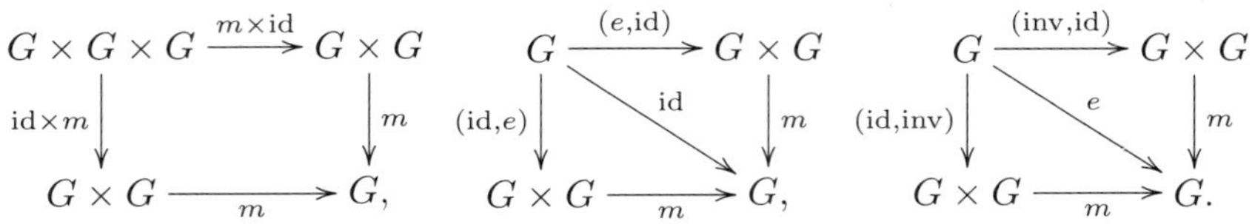

where we use e to denote also the constant map $G \to \{e\}$.

Thus, from a "classical" point of view, an algebraic group G over k is nothing else than a group $G(K)$, with identity element $e \in G(k)$, such that the group operations are algebraic and defined over k. From a "modern" point of view, i.e., scheme-theoretic, an algebraic group over k is a functor associating to each k-algebra R the group of R-rational points of G, $G(R)$, such that the natural maps $G(R) \to G(S)$ induced from morphisms of k-algebras $R \to S$ are all group homomorphisms. Group schemes are discussed in Appendix A.

An algebraic group G is called an *affine* or *linear* algebraic group if it is an affine variety or, equivalently, if it can be embedded as a group variety in $\mathrm{GL}_n(K)$ for some n; see [**5**, Prop. 1.10, p. 54]. Thus, an affine algebraic group over k is a subgroup of GL_n defined by a collection of polynomial equations with coefficients in k. Examples are provided by SL_n, O_n, SO_n, Sp_{2n}, the upper triangular matrices, the unipotent matrices. Here the reader will notice that he is already familiar with, e.g., the $\iota\delta\epsilon\alpha$ GL_n. That is, with a functor associating to any ring R a group $\mathrm{GL}_n(R)$ in a functorial way. Thus, GL_n becomes more a "group machine" than a particular group. That is precisely the scheme theoretic point of view.

If a group G is affine, the group structure may be equivalently defined by giving the maps, dual to the above, on the coordinate ring:

$$\widetilde{m}\colon k[G] \to k[G \times G] \cong k[G] \otimes_k k[G], \tag{1.4}$$

$$\widetilde{\mathrm{inv}}\colon k[G] \to k[G], \tag{1.5}$$

$$\tilde{e}\colon k[G] \to k. \tag{1.6}$$

The maps $\widetilde{m}$, $\widetilde{\mathrm{inv}}$, and $\tilde{e}$ are called co-multiplication, co-inverse and co-unit, respectively. They have the very special property of being *ring homomorphisms*. The reader is invited to reflect on that. Such a structure is called a Hopf algebra.

EXAMPLE 1.2. Let $\mathbb{G}_m/k = \mathrm{GL}_1/k$ be the multiplicative group over k. We have

$$k[\mathbb{G}_m] = k[x, x^{-1}] \cong k[x, y]/(xy - 1) \tag{1.7}$$

(the second presentation shows it is affine). The co-multiplication, co-inverse and co-unit maps are determined by:

$$\widetilde{m}(x) = x \otimes x, \quad \widetilde{\mathrm{inv}}(x) = x^{-1}, \quad \tilde{e}(x) = 1. \tag{1.8}$$

EXAMPLE 1.3. Let $\mathbb{G}_a/k = \mathbb{A}^1_k$ be the additive group over k. We have

$$k[\mathbb{G}_a] = k[x]; \tag{1.9}$$

$$\widetilde{m}(x) = x \otimes 1 + 1 \otimes x, \quad \widetilde{\mathrm{inv}}(x) = -x, \quad \tilde{e}(x) = 0. \tag{1.10}$$

EXERCISE 1.4. *Write* GL_2/k *as an affine variety. Write its co-multiplication, co-inverse and co-unit morphisms. Do the same for the upper-triangular matrices in* GL_2.

An algebraic group G is called an *abelian variety* if it is connected and projective (we will show later that an abelian variety is necessarily commutative). The reader is already familiar with elliptic curves. Except for the zero abelian variety, these are the simplest abelian varieties. Indeed, every abelian variety of dimension one is an elliptic curve. This follows, at least in the case of characteristic zero, from Theorem 6.3 below. A product of elliptic curves is an abelian variety. The Jacobian variety of a non-singular curve of genus g is an abelian variety of dimension g, and in fact every abelian variety is a homomorphic image of a Jacobian. See [**101**, §10]. See §6 for more examples.

If G is an algebraic group, then its identity component G_0 is a normal subgroup of finite index.

EXERCISE 1.5. *Prove that! What is the identity component of* $O_2(\mathbb{R})$? $O_2(\mathbb{C})$? *Find topological (or even real analytic) models for these groups.*

THEOREM 1.6 (Chevalley). *Let G be a connected algebraic group over an algebraically closed field. Then G has a* unique *maximal connected affine subgroup L and G/L is an abelian variety*:

$$0 \to L \to G \to G/L \to 0. \tag{1.11}$$

PROOF. See [**12**] and [**16**]. □

REMARK 1.7. In Theorem 1.6 abelian varieties and affine groups appear as opposite sides in the "spectrum" of algebraic groups, yet it may happen that the abelian variety G/L degenerates into the simplest kind of an affine group: a torus (see below). Such degenerations arise for example in the context of compactifications of moduli spaces of abelian varieties.

A typical example of degeneration is the family of elliptic curves over the open unit disk in $\mathbb{C}$:

$$y^2 z = x(x - \epsilon z)(x + z), \quad 0 < |\epsilon| < 1. \tag{1.12}$$

If there was a proper moduli scheme for elliptic curves (a concept we shall discuss further below, but for the time being can be thought of as a complete variety parameterizing isomorphism classes of elliptic curves) this family would have to have a limit at zero. The obvious way to complete this family is by taking the fiber above zero to be $y^2 z = x^2(x+z)$. This is no longer an elliptic curve. However, if we delete the singular point $(0 : 0 : 1)$ the remaining variety is a family of algebraic groups (with the identity point always being the point at infinity $(x : y : z) = (0 : 1 : 0)$) over $|\epsilon| < 1$. The fiber above zero being in fact isomorphic to $\mathbb{G}_m$ (see Figure 1.2

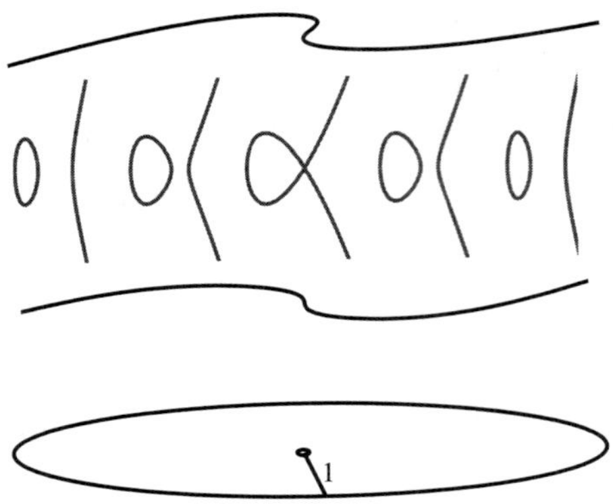

FIGURE 1.1. Degenerating family of elliptic curves.

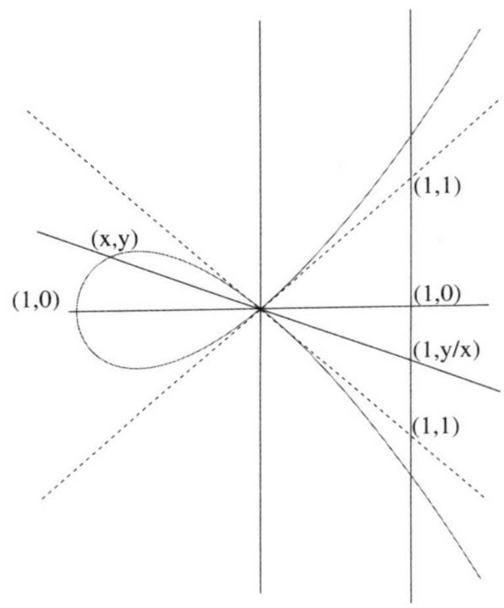

FIGURE 1.2. The nodal curve.

for easy to portray rational parameterization. To actually get an isomorphism with $\mathbb{G}_m$ use the map $(x, y) \mapsto (y-x)/(y+x)$). (We remark that the "cord-tangent" law extends to an action of the non-singular part of the fiber at zero on the whole fiber, inducing a group law on the non-singular part that makes the isomorphism with $\mathbb{G}_m$ into a group isomorphism). We obtain a family of elliptic curves degenerating to the multiplicative group $\mathbb{G}_m$. The theory of compactification of moduli spaces of abelian varieties is the content of [**39**]. In the context of elliptic curves, see [**26**] and [**34**] for a short survey.

DEFINITION 1.8. Let G, H be two algebraic groups over k. We let $\text{Hom}(G, H)$ denote the set of homomorphisms as algebraic groups from G to H defined over K. (We shall often write, e.g., $\text{Hom}(G, \text{GL}_n)$ for the more pedantic notation $\text{Hom}_K(G \otimes_k K, \text{GL}_{n/K})$). We let $\text{Hom}_{k'}(G, H)$ stand for those defined over a subfield $k \subseteq k' \subseteq K$.

Note that $\mathrm{Hom}_{k^s}(G,H)$ is a Γ-set. The action is given by

$$(^{\gamma}h)(x) = \gamma\big(h(\gamma^{-1}x)\big). \tag{1.13}$$

If we write G as a quasi-projective variety in $\mathbb{P}^n_K$, then every $x \in G(K)$ can be thought of as a $n+1$ tuple $(x_0 : \cdots : x_n)$ with $x_i \in K$. Then for $\sigma \in \Gamma$ we have $\sigma x = (\sigma x_0 : \cdots : \sigma x_n)$ (there is a unique extension of the Γ-action on $G(k^s)$ to $G(K)$) and an easy calculation shows that $^{\gamma}h$ is a morphism. However, with some basic concepts of schemes, the Galois action can be defined intrinsically:

Let X be a scheme over the field F, where F is either k^s or K. Suppose that X can be defined over k. That is, there is a scheme X_0 over k and a given identification $X = X_0 \otimes_k F$ (we often use this shorthand for $X_0 \times_{\mathrm{Spec}(k)} \mathrm{Spec}(F)$). Then for every Galois automorphism $\gamma \in \Gamma$ we can define a conjugation morphism

$$\tau_\gamma \colon X \to X, \tag{1.14}$$

as follows. Let $\phi\colon \mathrm{Spec}(F) \to \mathrm{Spec}(F)$ be the morphism induced from the field homomorphism $\gamma^{-1}\colon F \to F$. Then τ_γ is the morphism

$$X_0 \otimes_k F \xrightarrow{\mathrm{id}_{X_0} \otimes \phi} X_0 \otimes_k F. \tag{1.15}$$

We remark that if $X_0 = \mathrm{Spec}\big(k[x_1, \ldots, x_n]/(f_1, \ldots, f_m)\big)$ then

$$X = \mathrm{Spec}\big(F[x_1, \ldots, x_n]/(f_1, \ldots, f_m)\big), \tag{1.16}$$

the isomorphism of X with $X_0 \otimes_k F$ is coming from the ring isomorphism

$$F[x_1, \ldots, x_n]/(f_1, \ldots, f_m) \cong k[x_1, \ldots, x_n]/(f_1, \ldots, f_m) \otimes_k F. \tag{1.17}$$

The morphism

$$X \xrightarrow{\tau_\gamma} X, \tag{1.18}$$

is induced from

$$k[x_1, \ldots, x_n]/(f_1, \ldots, f_m) \otimes_k F \xleftarrow{a \otimes \gamma^{-1}(r) \leftarrow\!\shortmid\, a \otimes r} k[x_1, \ldots, x_n]/(f_1, \ldots, f_m) \otimes_k F. \tag{1.19}$$

Thus, the morphism τ_γ takes the ideal $(x_1 - \alpha_1, \ldots, x_n - \alpha_n)$ to the ideal $\big(x_1 - \gamma(\alpha_1), \ldots, x_n - \gamma(\alpha_n)\big)$. That is, the effect of τ_γ on points (in the naive sense) is just

$$\tau_\gamma(\alpha_1, \ldots, \alpha_n) = \big(\gamma(\alpha_1), \ldots, \gamma(\alpha_n)\big). \tag{1.20}$$

N.B.: Although τ_γ is a morphism of schemes, it is not a morphism of schemes over F.

If X_0, Y_0 are two schemes over k then Γ acts on $\mathrm{Mor}_F(X_0 \otimes_k F, Y_0 \otimes_k F)$ by the rule

$$\phi \mapsto \tau_\gamma \circ \phi \circ \tau_\gamma^{-1}. \tag{1.21}$$

In particular, getting back to the case algebraic groups G, H defined over k, we obtain (1.13) and one can prove that

$$\mathrm{Hom}_k(G,H) = \mathrm{Hom}_{k^s}(G,H)^{\Gamma}. \tag{1.22}$$

The proof is a descent argument. We refer the reader to [**111**] and [**6**] for more on Galois actions and descent. Note that if H is commutative then $\mathrm{Hom}(G,H)$ is a Γ-module.

DEFINITION 1.9. The group of *characters* of G is the group

$$X(G) = \mathrm{Hom}(G, \mathbb{G}_m). \tag{1.23}$$

Note that if $f \in X(G)$ then, in particular, f is a non-vanishing regular function on G. Thus, for abelian varieties, $X(G)$ is trivial (since any morphism from a projective variety to an affine variety is constant). However, as mentioned above, abelian varieties may degenerate into affine groups, and even into affine groups G for which $X(G)$ determines completely the group G. These are called diagonalizable groups and are discussed extensively below.

DEFINITION 1.10. The set of *co-characters* (or, multiplicative one parameter subgroups) of an algebraic group G is

$$X_*(G) = \mathrm{Hom}(\mathbb{G}_m, G). \tag{1.24}$$

There is a pairing

$$X_* \times X \to \mathrm{Hom}(\mathbb{G}_m, \mathbb{G}_m) = \mathbb{Z}. \tag{1.25}$$

EXERCISE 1.11. *Prove that $X(\mathrm{GL}_2)$ is a free abelian group of rank 1 generated by the determinant character. Find $X_*(\mathrm{GL}_2)$ and the pairing $X_* \times X \to \mathbb{Z}$ explicitly.*

LEMMA 1.12 (Independence of characters). *The subset $X(G) \subset K[G]$ is linearly independent over K.*

PROOF. This is a classical theorem due to Dedekind. See [**5**, Lemma 8.1, p. 111]. □

DEFINITION 1.13. We say that G is *diagonalizable* if $K[G] = K[X(G)]$, where $K[G]$ is the affine coordinate ring of G and $K[X(G)]$ is the linear span over K of $X(G)$. If furthermore, $X(G)_k := \mathrm{Hom}_k(G, \mathbb{G}_m)$ spans $k[G]$, we say that G is *split over* k.

THEOREM 1.14. *Assume $Y \subset X(G)_k$ spans $k[G]$. Then:*

(1) *$Y = X(G)$. In particular, $X(G)_k = X(G)$.*

(2) *$k[G] = k[X(G)]$ as Hopf algebras, where the right-hand side stands for the group algebra of the group $X(G)$ with co-multiplication induced from the diagonal map*

$$\Delta\colon X(G) \to X(G) \times X(G), \quad \Delta(f) = (f, f), \tag{1.26}$$

and co-inverse given by

$$\widetilde{\mathrm{inv}}\colon X(G) \to X(G), \quad \widetilde{\mathrm{inv}}(f) = f^{-1}. \tag{1.27}$$

(3) *If H is a closed subgroup of G, then H is diagonalizable and is split over k.*

PROOF. (1) We have $k[G] = k[Y] \subset k[X(G)_k] \subset k[G]$ and $X(G)_k$ is independent over k. Thus $Y = X(G)_k$. Applying that to $Y = X(G)_k \subset X(G)_K$ we get the rest.

(2) The only thing to verify is that the structure on both sides agree. This is straightforward. The algebra structure is provided by the definition. The co-algebra structure follows from these considerations: if we denote by $\widetilde{m}\colon k[G] \to k[G] \otimes k[G]$ the co-multiplication, then $(\widetilde{m}f)(x, y) = f(x * y)$, where $*$ is multiplication in G. If f is in $X(G)$, then $f(x * y) = f(x)f(y)$. This shows that $\widetilde{m}f = f \otimes f$, i.e., the diagonal morphism on $X(G)$ induces $\widetilde{m}f = f \otimes f$.

(3) By definition, $K[H]$ is a homomorphic image of $K[G] = K[X(G)]$ and is thus spanned by the images of $X(G)$, which are characters on H. Thus, H is diagonalizable, $K[H] = K[X(H)]$, and $K[G] \to K[H]$ is the group algebra epimorphism $\psi_*\colon K[X(G)] \to K[X(H)]$ induced by the group homomorphism $\psi\colon X(G) \to X(H)$. But then, since $X(G) = X(G)_k$, it is clear that the kernel is defined over k and $X(H) = X(H)_k$. In fact, $\operatorname{Ker}(\psi_*)$ is generated by $\{m - \mathbf{1} \mid m \in \operatorname{Ker}(\psi)\}$. □

REMARK 1.15. (1) Under the assumptions of the theorem, since G is of finite type over k, $X(G)$ is finitely generated, and therefore we have a surjective group algebra homomorphism $k[\mathbb{G}_m^r] = k[\mathbb{Z}^r] \to k[X(G)] = k[G]$ for some $r \in \mathbb{N}^+$. Thus, G is a closed subgroup of $\mathbb{G}_m^r$. One views $\mathbb{G}_m^r$ as embedded diagonally in GL_r, hence the name "diagonalizable." Conversely, Theorem 1.14 says that any closed subgroup of $\mathbb{G}_m^r$ is diagonalizable in our sense.

(2) Let G be diagonalizable and $k \subset k' \subset K$. Then G is split over k' iff $X(G) = X(G)_{k'}$.

PROPOSITION 1.16. *The contravariant functor of characters,*

$$G \mapsto X(G), \tag{1.28}$$

from the category of k-split diagonalizable groups and k-homomorphisms of algebraic groups, to the category of finitely generated $\mathbb{Z}$-modules and homomorphisms of modules is fully faithful.

PROOF. We have a natural map

$$\operatorname{Hom}_k(G, H) \to \operatorname{Hom}_{\mathbb{Z}}\big(X(H), X(G)\big). \tag{1.29}$$

The claim we are making is that it is an isomorphism. In particular, (exchanging k with K), we see that all the homomorphisms between such groups are defined over k.

One simply verifies that any closed circle (starting from any point) in the following triangle is the identity:

$$\begin{array}{ccc} & \operatorname{Hom}_k(G, H) & \\ \nearrow & & \searrow \\ \operatorname{Hom}_{\text{Hopf algebras}}(k[H], k[G]) & \longleftarrow & \operatorname{Hom}_{\mathbb{Z}}\big(X(H), X(G)\big). \end{array} \tag{1.30}$$

□

DEFINITION 1.17. An algebraic group G isomorphic over K to $\mathbb{G}_m^n$ is called an *n-dimensional torus.*

PROPOSITION 1.18. *The following are equivalent:*

(1) *G is a n-dimensional torus;*
(2) *G is a connected diagonalizable group of dimension n;*
(3) *G is diagonalizable and $X(G) \cong \mathbb{Z}^n$.*

PROOF. Let G be an n-dimensional torus, then G is a connected diagonalizable group of dimension n. As remarked before, if G is diagonalizable and $X(G) \cong \mathbb{Z}^n$ then G is a n-dimensional torus. Suppose that G is a connected diagonalizable group of dimension n. Since $\mathbb{G}_m$ is connected of dimension 1, its only connected reduced subgroups are 1 and $\mathbb{G}_m$. (In characteristic p the assumption of reduced is important. See Example 1.22). Thus every element of $X(G)$ is either trivial or surjective. Hence, $X(G)$ is a free abelian group of rank n. □

COROLLARY 1.19. *A closed connected subgroup of a torus is a torus and is a direct factor.*

PROOF. Apply Propositions 1.16 and 1.18. □

FACT 1.20. Every diagonalizable group G over k is split over k^s, thus over a finite extension. (Essentially because Γ is compact (being a profinite group) and the representation ρ associated to $X(G)$ is continuous with discrete image, hence finite; the group G is split over L, the fixed field of $\mathrm{Ker}(\rho)$, which is a finite field extension of k. See also Remark 1.24 below.)

Recall that a bilinear group homomorphism $\phi\colon A \times B \to C$ is called a *perfect pairing* if

$$\forall a \in A, \phi(a,b) = 0 \implies b = 0, \tag{1.31}$$

and

$$\forall b \in B, \phi(a,b) = 0 \implies a = 0. \tag{1.32}$$

PROPOSITION 1.21. *Let T be a torus defined over k, $\Gamma = \mathrm{Gal}(k^s/k)$.*

(a) *$X(T) = X(T)_{k^s}$, $X_* = X_*(T)_{k^s}$. Hence,*

$$X(T)_k = X(T)^\Gamma, \quad X_*(T)_k = X_*(T)^\Gamma. \tag{1.33}$$

(b) *The pairing*

$$X_*(T) \times X(T) \to \mathbb{Z} \tag{1.34}$$

is a perfect pairing of Γ-modules.

PROOF. By Proposition 1.16, for any k^s-split diagonalizable group G we have

$$\mathrm{Hom}(G,T) = \mathrm{Hom}_{k^s}(G,T). \tag{1.35}$$

In particular, $X_* = X_*(T)_{k^s}$, and hence $X_*(T)_k = X_*(T)^\Gamma$. We leave the proof of (b) to the reader as an exercise (be sure to check also the compatibility with the action of $\gamma \in \Gamma$). □

Let G be a diagonalizable k-group (but not necessarily k-split). Since G splits over k^s, we have

$$K[G] = K \otimes_{k^s} k^s[X(G)], \quad k^s[G] = k^s[X(G)]. \tag{1.36}$$

The action of Γ on an element $\sum a_\alpha \alpha \in k^s[X(G)]$ is

$$\gamma\Big(\sum a_\alpha \alpha\Big) = \sum (\gamma a_\alpha) \cdot {}^\gamma\alpha \quad \text{for } \gamma \in \Gamma. \tag{1.37}$$

We conclude that the Γ-module $X(G)$ determines $k[G] = k^s[X(G)]^\Gamma$ uniquely. If H is another diagonalizable group, then $\mathrm{Hom}(G,H) = \mathrm{Hom}_{k^s}(G,H)$, and hence the following assertions are equivalent:

(1) $h \in \mathrm{Hom}(G,H)$ is defined over k;
(2) h is Γ-equivariant;
(3) $h^*\colon k^s[H] \to k^s[G]$ is Γ-equivariant;
(4) $X(h)\colon X(H) \to X(G)$ is Γ-equivariant.

Let $\mathcal{A}$ be the category of diagonalizable groups over k with morphisms being k-homomorphisms of algebraic groups. Let $\mathcal{B}$ be the category of finitely generated continuous Γ-modules X, without p-torsion if $\operatorname{char} k = p > 0$, with Γ-equivariant homomorphisms. Here continuous means that the map

$$\Gamma \times X \to X \tag{1.38}$$

is continuous where Γ is given its profinite topology and X the discrete topology. Or still simpler, the action of Γ factors through a finite quotient (X being finitely generated).

We remark that for X an abelian group, $k[X]$ contains nilpotent elements if and only if char $k = p > 0$ and X has p-torsion. Since varieties are *reduced* by definition (i.e., the sheaf of regular functions contains no nilpotent elements), and algebraic groups are varieties (Definition 1.1), we have to exclude this case. To illustrate we give

EXAMPLE 1.22. Suppose $k = \mathbb{Z}/p\mathbb{Z}$, and $X = \mathbb{Z}/p\mathbb{Z}$. We get

$$k[X] = k[y]/(y^p - 1), \tag{1.39}$$

the group scheme $\mu_{p,k}$ of p-th roots of unity over k, and $y-1$ is clearly nilpotent. In the context of group *schemes* $\mu_{p,k}$ is a connected diagonalizable subgroup scheme of $\mathbb{G}_m$, but in the context of group *varieties* it must be excluded. See Appendix A, §3.

THEOREM 1.23. *The functor of characters*

$$X\colon \mathcal{A} \to \mathcal{B}, \quad G \mapsto X(G) \tag{1.40}$$

is an anti-equivalence of categories, between the category $\mathcal{A}$ of diagonalizable groups and the category $\mathcal{B}$ of continuous Γ-modules with no p-torsion if $\operatorname{char}(k) = p$.

PROOF. By Proposition 1.16 and the remarks just made, the only thing we need to show is essential surjectivity, that is, every object in $\mathcal{B}$ is the character module of some object in $\mathcal{A}$.

Let M be a finitely generated Γ-module. Let G be the associated affine group with affine coordinate ring $k^s[M]$, an algebra of finite type over k^s. If $m \in M$, then m defines a character of G by

$$k^s[x, x^{-1}] \to k^s[G] = k^s[M], \tag{1.41}$$

$$x \mapsto m. \tag{1.42}$$

(m being invertible in $k^s[M]$), and one easily verifies that in fact $M = X(G)$ (because $M \subset X(G)$ and it spans $k^s[G]$ as in Theorem 1.14). Hence, G is diagonalizable.

Consider now $(k^s[M])^\Gamma$. It satisfies

$$k^s \otimes (k^s[M])^\Gamma = k^s[M], \tag{1.43}$$

and hence it gives G a k-structure. The equality (1.43) is essentially the assertion that if Γ acts continuously semi-linearly on a k^s-vector space V,

$$\gamma(ax) = \gamma(a)\gamma(x) \quad (a \in k^s, x \in V), \tag{1.44}$$

there exists a k-basis to V. (We apply this to $V = k^s[M]$). This in turn follows from a generalization of Hilbert's Theorem 90:

$$H^1\big(\Gamma, \mathrm{GL}_n(k^s)\big) = 1. \tag{1.45}$$

(See [**149**, Chapter X]). One can also prove it directly (see [**5**, AG §14.2]). $\square$

REMARK 1.24. We have seen that diagonalizable groups are completely determined by their character module considered as a Γ-module. In particular, since $\mathrm{GL}_n(\mathbb{Z})$ is the automorphism group of $\mathbb{G}_m^n$, one sees that the n-dimensional tori are equivalent to continuous Galois representations $\rho\colon G \to \mathrm{GL}_n(\mathbb{Z})$ up to conjugacy. All n-dimensional tori are forms of $\mathbb{G}_m^n$ (that is, they become isomorphic to $\mathbb{G}_m^n$ over K), and in fact $H^1(\Gamma, \mathrm{Aut}(\mathbb{G}_m^n))$—the pointed set classifying forms of $\mathbb{G}_m^n$—is indeed naturally identified with the set of Galois representations ρ modulo conjugacy.

EXERCISE 1.25. (1) *Classify n-dimensional tori over the field $\mathbb{Z}/p\mathbb{Z}$. Suggestion: Do the one-dimensional and two-dimensional cases first.*

(2) *If you know some class field theory, classify one-dimensional tori over $\mathbb{Q}$.*

REMARK 1.26. One often writes $k^s[M]$ in the form

$$k^s[X^\alpha : \alpha \in M]/(X^{\alpha+\beta} - X^\alpha X^\beta, X^0 - 1). \tag{1.46}$$

We finish this section by giving some general methods to construct tori.

EXAMPLE 1.27 (Restriction of scalars). Let L/k be a separable field extension of degree d. We assume $L \subset k^s$, where k^s is our fixed separable closure. Let $\{\sigma_1, \dots, \sigma_d\}$ be the embeddings of L into k^s. This a Γ-set:

$$\begin{array}{ccc} L & \xrightarrow{\sigma_i} & k^s \\ & \underset{\tau\circ\sigma_i}{\searrow} & \downarrow{\scriptstyle \tau} \\ & & k^s. \end{array} \tag{1.47}$$

Let us write

$$\tau \circ \sigma_i = \sigma_{\tau(i)}. \tag{1.48}$$

Consider the continuous Γ-module

$$M = \bigoplus_{i=1}^{d} \sigma_i \mathbb{Z}. \tag{1.49}$$

Let G be the unique, up to isomorphism, algebraic torus with $X(G) = M$. Note that G is a d-dimensional torus defined over k. It is denoted $\mathbf{Res}_{L/k}\, \mathbb{G}_m$.

EXERCISE 1.28. *Let L/k be a quadratic extension of fields. Write $\mathbf{Res}_{L/k}\, \mathbb{G}_m$ explicitly as an affine variety over k. Write also the co-multiplication and co-inverse morphisms.*

To generalize our construction, consider an algebraic variety V/L. Define $\mathbf{Res}_{L/k}\, V$ as a functor

$$\mathbf{Res}_{L/k}\, V\colon k\text{-}\mathbf{alg} \to \mathbf{Sets}, \quad \mathbf{Res}_{L/k}\, V(B) = V(B \otimes_k L). \tag{1.50}$$

One can show that there exists an algebraic variety over k, unique up to k isomorphism, such that $\mathbf{Res}_{L/k}\, V$ is its functor of points. This variety is denoted by $\mathbf{Res}_{L/k}\, V$ as well.

To tie together the functorial and the lattice approach, note that a k^s-point of G amounts to a homomorphism $M \to k^{s\times}$, or, with the choice of the standard basis on M, simply to a d-tuple $(x_1, \dots, x_d)$ of elements of $k^{s\times}$. Therefore, for every field extension k' of k in k^s, we have

$$G(k') = \{(x_1, \dots, x_d) : \tau(x_i) = x_{\tau(i)}, \forall \tau \in \mathrm{Gal}(k^s/k')\}. \tag{1.51}$$

In particular, $G(k)$ is naturally isomorphic to $L^\times$, and in fact for every separable field extension k' of k we find a natural identification of $G(k')$ with $(k' \otimes_k L)^\times$. That shows that G *is* $\mathbf{Res}_{L/k}\,\mathbb{G}_m$.

EXAMPLE 1.29 (Tensor and Hom constructions). Let T be a d-dimensional torus over k and let $\mathcal{O}$ be the ring of integers of an algebraic number field. Let us assume that $\mathcal{O}$ is contained in the endomorphism ring of T. We remark that such examples exist in abundance, because the endomorphism ring of T over k^s is $M_d(\mathbb{Z})$.

Given a projective $\mathcal{O}$-module M of rank r, we define two new tori over k, both of dimension rd. They are denoted $T \otimes_{\mathcal{O}} M$ and $\mathrm{Hom}_{\mathcal{O}}(M, T)$. As functor of points they are defined by the following formulae: For every k-algebra B

$$(T \otimes_{\mathcal{O}} M)(B) = T(B) \otimes_{\mathcal{O}} M, \tag{1.52}$$

$$\mathrm{Hom}_{\mathcal{O}}(M, T)(B) = \mathrm{Hom}_{\mathcal{O}}\big(M, T(B)\big). \tag{1.53}$$

EXERCISE 1.30. *Verify that these are indeed the functor of points of tori over k. Prove that the character modules are given by*

$$X(T \otimes_{\mathcal{O}} M) = X(T) \otimes_{\mathcal{O}} M^\vee, \quad X\big(\mathrm{Hom}_{\mathcal{O}}(M, T)\big) = X(T) \otimes_{\mathcal{O}} M, \tag{1.54}$$

where $M^\vee = \mathrm{Hom}_{\mathcal{O}}(M, \mathcal{O})$. Note that $\mathrm{Hom}(M, T) \cong T \otimes_{\mathcal{O}} M^\vee$.

2. Moduli problems

In this section we discuss moduli problems and moduli spaces. We mainly focus on the ideas rather than on the solutions because the solutions—the moduli spaces themselves—are studied closely in later chapters. Some paragraphs in this section will be better understood only after reading on the book.

The notion of a moduli space is rather a set of examples or problems than a well-defined notion. Usually, by a moduli problem one means a certain kind of classification problem. A large part of mathematics is devoted to , or motivated by, classification problems: classifying topological spaces up to homotopy; classifying all the finite simple groups; classifying Banach spaces, etc. Usually a distinguishing feature of the classification problems called moduli problems is that they appear in the context of geometry, and that one considers families of objects as well.

In the context of algebraic geometry one may choose the most general approach and define a moduli problem as follows: Fix a base scheme S. A moduli problem over S is a contravariant functor

$$\Phi\colon \mathbf{Sch}_S \to \mathbf{Sets} \tag{2.1}$$

from the category of schemes over S to the category of sets. Assume that this functor is *representable*, i.e., that there exists a scheme $T \to S$ such that the functor of points of T, h_T, is naturally equivalent to Φ. That is, that there exists a natural isomorphism

$$h_T(R) := \mathrm{Mor}_S(R, T) \cong \Phi(R), \tag{2.2}$$

where R varies over S-schemes. One may then call T a *fine moduli scheme.* If T is equidimensional over S (and nice enough such that this makes sense), then one may call $\dim_S(T)$ the number of moduli of the moduli problem Φ.

One of the first moduli problems, as such, was described and solved by B. Riemann. It is still a problem that motivates much research in algebraic geometry.

In this case one fixes an integer $g \geq 2$ and takes Φ to be the functor associating to a scheme R the isomorphism classes of smooth curves $\mathcal{C} \to R$ of genus g (here $S = \mathrm{Spec}(\mathbb{Z})$ and hence omitted). Riemann made the statement that this problem (considered over $\mathbb{C}$) has $3g-3$ moduli. Local deformation theory as developed by Kodaira–Spencer [**86**], i.e., the theory of infinitesimal deformations (and in particular deformations over the base $k[\epsilon]$, $\epsilon^2 = 0$, k a field, called also *first order deformations*), dictates that the $k[\epsilon]$ deformations of a curve C/k are parameterized by $H^1(C, \mathfrak{T}_C)$, where $\mathfrak{T}_C$ designates the tangent sheaf of C. By the Riemann–Roch Theorem and Serre's duality

$$\dim H^1(C, \mathfrak{T}_C) = \dim H^0\big(C, \mathcal{O}(2K_C)\big) = 3g - 3. \tag{2.3}$$

This agrees with Riemann's statement because the first order deformations are the tangent space to the moduli functor at C, and if this functor was representable, the first order deformations would naturally constitute the tangent space to the moduli space at the moduli point corresponding to C. See §6 and Chapter 3, §5 for more examples.

There does not exist a fine moduli scheme for this problem. It is a general philosophy that the existence of a fine moduli space that represents a functor of isomorphism classes of certain objects,[1] usually rules out the possibility that the objects being parameterized have non-trivial automorphisms, as some curves do. However, there exists a coarse moduli space for this problem.

A *coarse moduli scheme* (for a functor $\Phi\colon \mathbf{Sch}_S \to \mathbf{Sets}$ as above) is a scheme $T \to S$ such that (i) there exists a morphism of functors $\Phi \to h_T$; (ii) for every algebraically closed field k, with $\mathrm{Spec}(k) \to S$, the morphism of functors gives a bijection $h_T(\mathrm{Spec}(k)) = \mathrm{Mor}_S(\mathrm{Spec}(k), T) \cong \Phi\big(\mathrm{Spec}(k)\big)$, and moreover (iii) every morphism $\Phi \to h_X$ from Φ to the functor of points of a S-scheme X, factor through $\Phi \to h_T$ for a unique morphism $T \to X$. Thus, loosely speaking, a coarse moduli scheme is the "best approximation" for Φ in the category of schemes.

Coarse moduli schemes may exist when fine moduli schemes do not. For example, there exists a coarse moduli scheme for the functor Φ of isomorphism classes of curves of genus g (still $g \geq 2$). We shall denote it by $\mathcal{M}_g$.

Often the existence and the "justification" for the existence of coarse moduli spaces is the following. One adds structure to the original problem, i.e., considers a functor Φ' with a natural ("forgetful") morphism $\Phi' \to \Phi$, such that in fact Φ is obtained from Φ' by a "nice" equivalence relation. For example, in the case of curves, we may take Φ' to be the moduli problem of isomorphism classes of m-pointed smooth curves. That is, a smooth curve together with m distinct marked points on it. When m is large enough, that is $m \geq N(g)$ where $N(g)$ depends only on the genus, this moduli problem is *rigid*: an automorphism of a curve of genus g fixing the m marked points is necessarily the identity. The functor Φ' is representable; there exists a fine moduli space for it, say $\mathcal{M}_{g,m}$. The space $\mathcal{M}_g$ is then a quotient of $\mathcal{M}_{g,m}$ by an equivalence relation coming from forgetting the marked points.

It follows tautologically(!) that if there exists a fine moduli scheme T for a moduli problem Φ, then there exists a universal object $\mathcal{U} \to T$. That is, there

[1]Without the proviso on the functor representing isomorphism classes it is easy to give examples where automorphisms form no obstruction. Grassmannians and formal group laws are such examples.

exists an element $\mathcal{U} \in \Phi(T)$ with the following property: Let R be a S-scheme and $E_R \colon \Phi(R) \to T(R)$ be the given bijection. Let $x \in \Phi(R)$ and $E_R(x) \colon R \to T$ the corresponding morphism (remember that a point $x \in T(R)$ is, by definition, a morphism $R \to T$). By contravariance we get an object $E_R(x)(\mathcal{U}) \in \Phi(R)$. It is equal to x. The object $\mathcal{U}$ is of course just the object corresponding to the tautological point of $T(T)$—the identity morphism $T \to T$. In the case of m-pointed curves, this boils down to saying that there exists a universal m-pointed curve $\mathcal{U} \to \mathcal{M}_{g,m}$ such that for every scheme R the isomorphism classes of m-pointed curves over R are precisely $\{f^*(\mathcal{U}) : f \in T(R)\}$.

The moduli problems we are interested in are the problems of classifying abelian varieties with certain extra structure. This extra structure is meant to rigidify the moduli problem. Some moduli schemes of elliptic curves are discussed in §3, and of abelian varieties in Chapter 2, §2.2. We do not presume to actually prove the existence of such moduli schemes. What we shall show is that for some moduli problems (e.g., elliptic curves and abelian varieties with real multiplication) there are certain quotients of a power of the complex upper half plane $\mathcal{H}$ that are "likely" to be coarse moduli schemes. In fact they are, but we shall only demonstrate a natural bijection between the points of such quotients and the value of the respective moduli problem on the scheme Spec($\mathbb{C}$). This falls short of actually proving they are coarse moduli schemes.

We want to draw the reader's attention to a phenomenon she encountered before (or so we expect): There is no universal elliptic curve.

Recall that the coarse moduli scheme of elliptic curves is given by the j-line; two elliptic curves are isomorphic over an algebraic closure if and only if they have the same j-invariant. There doesn't exist an elliptic curve over the j-line such that its fiber over every j_0 is the elliptic curve with invariant j_0. This has to do, of course, with the j-line not being quite a fine moduli scheme. There is, however, such an elliptic curve over the j-line with the points 0 and 1728 removed: e.g., $y^2 + xy = x^3 - 36x/(j - 1728) - 1/(j - 1728)$. If we ask for the moduli of elliptic curves with a 7 torsion point (this is a rigid problem) then there exists a fine moduli scheme and thus a universal elliptic curve over it. This moduli scheme is an open subvariety of $\mathbb{P}^1$. Given $t \in \mathbb{A}^1$ associate to it the elliptic curve $y^2 + (1 + t - t^2)xy + (t^2 - t^3)y = x^3 + (t^2 - t^3)x^2$. It has discriminant $t^7(t-1)^7(t^3 - 8t^2 + 5t + 1)$. The moduli scheme is then the complement of the zero locus of the discriminant. The 7 torsion point is $(0, 0)$. We have taken these examples from Silverman [**156**, Chapter III.1, Appendix C.13].

For elliptic curves, if one takes as extra structure, say, an isomorphism of the n-torsion points ($n \geq 3$) with $(\mathbb{Z}/n\mathbb{Z})^2$ then one gets a rigid moduli problem representable by a fine moduli scheme over $\mathbb{Z}$. The same level structure is not rigid (that is, objects might have automorphisms besides the identity) for abelian varieties and one needs to put some more structure to obtain nice moduli schemes. For example, on the square of an elliptic curve, the endomorphism given by the matrix $\left(\begin{smallmatrix} 1 & 3 \\ 0 & 1 \end{smallmatrix}\right)$ is an automorphism of infinite order that acts trivially on the 3 torsion points.

The usual procedure to rigidify the structure is to realize all the objects one parameterizes as subvarieties of a projective space of a fixed dimension having special properties. Thus they are typically parameterized by a subvariety of an appropriate Hilbert scheme. For curves of genus greater than 1, the divisor $3K$

(K the canonical divisor) is very ample, and thus all curves of genus g appear as subvarieties of $\mathbb{P}^{5g-6}$. See [**24**, §1]. For elliptic curves, $3 \cdot 0_E$ is a very ample divisor, and all elliptic curves appear as subvarieties of $\mathbb{P}^2$ (Weierstrass' equation). One proceeds to construct the moduli scheme as a subvariety of a quotient of the appropriate Hilbert scheme.

Giving a projective embedding into a fixed projective space requires a choice of a very ample divisor and a choice of basis for its global sections. All these are extra data and one attempts to discard it. The choice of basis is corrected by dividing the Hilbert scheme by the action of PGL_n. The choice of ample divisor is corrected by considering divisors up to algebraic equivalence. For elliptic curves, the last correction is not needed because there is a canonical choice of an ample divisor, viz., 0_E. However, for abelian varieties there is no such choice and one needs to choose an ample divisor up to algebraic equivalence. Since two divisors on an abelian variety are algebraically equivalent if and only if they define the same map from the abelian variety to its dual, one finds the usual definition of a polarization as a map from the abelian variety to its dual induced by an ample line bundle (see Definition 6.31, and Chapter 3, Definition 1.15). We note that while for elliptic curves the Néron–Severi group—the group of divisors modulo algebraic equivalence—is just $\mathbb{Z}$, for an abelian variety, though it is still a torsion free finitely generated abelian group, it may happen that it is of rank greater than 1 and thus there is no natural choice of polarization.

Thus, a typical moduli problem for abelian varieties is that of parameterizing triples (A, λ, γ) where A is an abelian variety of a given dimension (otherwise the moduli scheme would be a disjoint union according to dimension), a polarization $\lambda\colon A \to A^t$ (that is: a map from A to its dual abelian variety that comes, in a suitable sense, from an ample line bundle) which should be thought of as an equivalence class of projective embeddings, and a level structure γ (typically a choice of a point, or several points on the abelian variety that rigidifies the moduli problem). One also imposes some conditions on the polarization λ—often a bound on $\deg(\lambda)$. (The point being that up to automorphisms, every abelian variety has finitely many polarizations of a given degree ([**116**])).

3. Moduli of elliptic curves

In this section, as the title suggests, we examine the moduli schemes corresponding to pairs composed of an elliptic curve and an associated level structure. We assume the reader is familiar with the basic theory of elliptic curves.

By an elliptic curve E over a ring R, we mean a group scheme E over $\mathrm{Spec}(R)$, of relative dimension one, which is proper and smooth over $\mathrm{Spec}(R)$ with geometrically connected fibers. Equivalently, locally on $\mathrm{Spec}(R)$, the scheme E can be given by a Weierstrass equation $y^2z + a_1xyz + a_3yz^2 = x^3 + a_2x^2z + a_4xz^2 + a_6z^3$ in $\mathbb{P}^2_R$ with variables x, y, z and invertible discriminant. The zero section is given by $z = x = 0$.

DEFINITION 3.1. Let R be a ring in which n is invertible and assume that R is a $\mathbb{Z}[\zeta_n]$-module ($\zeta_n = e^{2\pi i/n}$).

(1) A *full symplectic level-n-structure* (or a $\Gamma(n)$*-structure*) on an elliptic curve E over R is an isomorphism of constant group schemes over R

$$\alpha\colon (\mathbb{Z}/n\mathbb{Z})^2 \to E[n], \tag{3.1}$$

such that under α the Weil pairing on $E[n]$ corresponds to the symplectic pairing on $(\mathbb{Z}/n\mathbb{Z})^2$ determined by $\langle(1,0),(0,1)\rangle = \zeta_n$.

(2) A $\Gamma_1(n)$-*level structure* is a choice of a point $P \in E[n](R)$ of order n.

(3) A $\Gamma_0(n)$-*level structure* is a choice of cyclic subgroup $H \subset E[n]$ of order n defined over R.

REMARK 3.2. (1) The phrase "constant group schemes over R" amounts to saying that $E[n]$ consists of n^2 disjoint sections $\operatorname{Spec}(R) \to E$.

(2) By a symplectic pairing of $(\mathbb{Z}/n\mathbb{Z})^2$ we mean a map

$$\phi\colon (\mathbb{Z}/n\mathbb{Z})^2 \times (\mathbb{Z}/n\mathbb{Z})^2 \to \mu_n(R), \tag{3.2}$$

such that ϕ is a bilinear, perfect and antisymmetric pairing.

We consider the following functors:

$$M(n)\colon R \mapsto M(n)(R) := \{\text{isomorphism classes over } R \text{ of } (E,\alpha)\}, \tag{3.3}$$

$$M_1(n)\colon R \mapsto M_1(n)(R) := \{\text{isomorphism classes over } R \text{ of } (E,P)\}, \tag{3.4}$$

$$M_0(n)\colon R \mapsto M_0(n)(R) := \{\text{isomorphism classes over } R \text{ of } (E,H)\}. \tag{3.5}$$

To clarify, two pairs (E_1,α_1) [resp. (E_1,P_1); resp. (E_1,H_1)] and (E_2,α_2) [resp. (E_2,P_2); resp. (E_2,H_2)] in $M(n)(R)$ [resp. $M_1(n)(R)$; resp. $M_0(n)(R)$] are isomorphic over R iff there exists an isomorphism $f\colon E_1 \to E_2$, defined over R, such that $f \circ \alpha_1 = \alpha_2$ [resp. $f(P_1) = P_2$; resp. $f(H_1) = H_2$]. Thus, we always have $(E,P) \cong \big(E, h(P)\big)$, for $h \in \operatorname{Aut}_R(E)$. For most elliptic curves h could only be ± 1, but for some curves there are more possibilities.

We have the following natural transformations of functors:

$$M(n) \to M_1(n), \tag{3.6}$$

given by

$$(E,\alpha) \mapsto \big(E, \alpha(1,0)\big), \tag{3.7}$$

and

$$M_1(n) \to M_0(n), \tag{3.8}$$

given by

$$(E,P) \mapsto (E, \langle P\rangle), \tag{3.9}$$

where $\langle P\rangle$ denotes the cyclic group generated by P. Note that the functors $M_1(n)$ and $M_0(n)$, as well as the morphism $M_1(n) \to M_0(n)$, are defined in fact over $\mathbb{Z}$, but the morphisms $M(n) \to M_1(n)$ and $M(n) \to M_0(n)$ are only defined over $\mathbb{Z}[\zeta_n]$.

Consider now elliptic curves over $\mathbb{C}$. We first remark that the underlying complex manifold of a complex elliptic curve is a one-dimensional complex torus and vice versa, every one-dimensional torus is obtained from a complex elliptic curve, unique up to isomorphism. Let $\mathcal{H}$ be the complex upper half plane,

$$\mathcal{H} = \{z \in \mathbb{C} : \operatorname{Im}(z) > 0\}. \tag{3.10}$$

Let $\tau \in \mathcal{H}$, and let $\mathcal{L}_\tau$ be the lattice corresponding to τ, i.e.,

$$\mathcal{L}_\tau = \mathbb{Z} + \mathbb{Z}\cdot\tau. \tag{3.11}$$

Let $\mathcal{E}_\tau$ be the elliptic curve,

$$\mathcal{E}_\tau = \mathbb{C}/\mathcal{L}_\tau. \tag{3.12}$$

Later, in Example 6.44, we shall see that every elliptic curve is isomorphic to $\mathcal{E}_\tau$ for a suitable $\tau \in \mathcal{H}$, and we shall prove that under the Weil pairing

$$\langle \cdot, \cdot \rangle \colon \mathcal{E}_\tau[n] \times \mathcal{E}_\tau[n] \to \mu_n, \tag{3.13}$$

we have

$$\langle 1/n, \tau/n \rangle = \exp(2\pi i/n) = \zeta_n. \tag{3.14}$$

Granted this, we get a symplectic level-n-structure

$$\alpha \colon (\mathbb{Z}/n\mathbb{Z})^2 \to \mathcal{E}_\tau[n], \tag{3.15}$$

by setting $\alpha(1,0) = 1/n$ and $\alpha(0,1) = \tau/n$. We also get

- $\Gamma_1(n)$-level structure by taking P_τ to be $\alpha(1,0)$ on $\mathcal{E}_\tau$.
- $\Gamma_0(n)$-level structure by taking the subgroup $\{a/n \mid a = 0, \ldots, n-1\}$ on $\mathcal{E}_\tau$, which we denote by H_τ.

Let $\Gamma(n)$, $\Gamma_1(n)$, $\Gamma_0(n)$ be the following congruence subgroups of $\mathrm{SL}_2(\mathbb{Z})$:

$$\Gamma(n) = \left\{ \begin{pmatrix} a & b \\ c & d \end{pmatrix} : \begin{pmatrix} a & b \\ c & d \end{pmatrix} \equiv \begin{pmatrix} 1 & 0 \\ 0 & 1 \end{pmatrix} \bmod n \right\}, \tag{3.16}$$

$$\Gamma_1(n) = \left\{ \begin{pmatrix} a & b \\ c & d \end{pmatrix} : \begin{pmatrix} a & b \\ c & d \end{pmatrix} \equiv \begin{pmatrix} 1 & * \\ 0 & 1 \end{pmatrix} \bmod n \right\}, \tag{3.17}$$

$$\Gamma_0(n) = \left\{ \begin{pmatrix} a & b \\ c & d \end{pmatrix} : \begin{pmatrix} a & b \\ c & d \end{pmatrix} \equiv \begin{pmatrix} * & * \\ 0 & * \end{pmatrix} \bmod n \right\}. \tag{3.18}$$

Let $\mu = \left(\begin{smallmatrix} a & b \\ c & d \end{smallmatrix}\right) \in \mathrm{SL}_2(\mathbb{Z})$, $\tau' = \mu\tau$. An easy calculation yields that τ and τ' define isomorphic couples in the following settings:

$$(\mathcal{E}_\tau, \alpha_\tau) \cong (\mathcal{E}_{\tau'}, \alpha_{\tau'}) \iff \mu \in \Gamma(n), \tag{3.19}$$

$$(\mathcal{E}_\tau, P_\tau) \cong (\mathcal{E}_{\tau'}, P_{\tau'}) \iff \mu \in \Gamma_1(n), \tag{3.20}$$

$$(\mathcal{E}_\tau, H_\tau) \cong (\mathcal{E}_{\tau'}, H_{\tau'}) \iff \mu \in \Gamma_0(n). \tag{3.21}$$

Moreover, we have the following diagram:

$$\begin{array}{ccc}
\Gamma(n)\backslash\mathcal{H} & \xrightarrow[\alpha(1,0)=1/n,\ \alpha(0,1)=\tau/n]{\tau\mapsto\mathbb{C}/(\mathbb{Z}+\mathbb{Z}\tau)} & M(n)(\mathbb{C}) \\
\downarrow & & \downarrow \\
\Gamma_1(n)\backslash\mathcal{H} & \xrightarrow[P=1/n]{\tau\mapsto\mathbb{C}/(\mathbb{Z}+\mathbb{Z}\tau)} & M_1(n)(\mathbb{C}) \\
\downarrow & & \downarrow \\
\Gamma_0(n)\backslash\mathcal{H} & \xrightarrow[H=\langle 1/n\rangle]{\tau\mapsto\mathbb{C}/(\mathbb{Z}+\mathbb{Z}\tau)} & M_0(n)(\mathbb{C}).
\end{array} \tag{3.22}$$

EXERCISE 3.3. *Take a point τ for which one of the coverings is ramified. Express that in terms of automorphisms of the elliptic curve $\mathcal{E}_\tau$.*

We recall that the open Riemann surfaces $\Gamma(n)\backslash\mathcal{H}$, $\Gamma_1(n)\backslash\mathcal{H}$ and $\Gamma_0(n)\backslash\mathcal{H}$, denoted traditionally by $Y(n)(\mathbb{C})$, $Y_1(n)(\mathbb{C})$ and $Y_0(n)(\mathbb{C})$ respectively, can be compactified canonically by adding the finitely many orbits (called cusps) of $\mathbb{P}^1(\mathbb{Q})$ under the group $\Gamma(n)$, $\Gamma_1(n)$ and $\Gamma_0(n)$ respectively. The resulting curves $\Gamma(n)\backslash\mathcal{H}^*$, $\Gamma_1(n)\backslash\mathcal{H}^*$ and $\Gamma_0(n)\backslash\mathcal{H}^*$ (where $\mathcal{H}^* = \mathcal{H} \cup \mathbb{P}^1(\mathbb{Q})$) are denoted accordingly $X(n)(\mathbb{C})$, $X_1(n)(\mathbb{C})$ and $X_0(n)(\mathbb{C})$. The Y's and the X's are isomorphic as Riemann surfaces with the complex points of algebraic curves that we denote, abusing notation, by the same letters. See, e.g., [**154**], especially §1.5.

The discussion above shows that $\Gamma(n)\backslash\mathcal{H}$, $\Gamma_1(n)\backslash\mathcal{H}$ and $\Gamma_0(n)\backslash\mathcal{H}$ are in bijection with $M(n)(\mathbb{C})$, $M_1(n)(\mathbb{C})$ and $M_0(\mathbb{C})$ respectively. In fact $Y(n)$, $Y_1(n)$ and $Y_0(n)$ are coarse moduli spaces for these functors.

For completeness we give information on indices and genera. The reader may find the proofs in [**154**, Chapter 1.6] and [**102**, Chapter 4.2]. Let Γ be a congruence subgroup of $\mathrm{SL}_2(\mathbb{Z})$. That is, $\Gamma \supset \Gamma(n)$ for some n. Let $\overline{\Gamma}$ be the image of Γ in $\mathrm{PSL}_2(\mathbb{Z})$. Let

$$\mu_\Gamma = [\mathrm{PSL}_2(\mathbb{Z}) : \overline{\Gamma}]. \tag{3.23}$$

Let X_Γ be the compactified quotient $\Gamma\backslash\mathcal{H}^*$. Let $\pi_\Gamma\colon X_\Gamma \to X(1)$ be the projection. Let v_2 be the number of points in the fiber $\pi_\Gamma^{-1}(i)$ that are not ramified (i.e., $\mathrm{Stab}_\Gamma(x) \cong \mu_4$), let v_3 be the number of points in $\pi_\Gamma^{-1}(\omega)$ ($\omega = e^{2\pi i/3}$) that are unramified, and let $v_\infty = |\pi_\Gamma^{-1}(i\infty)|$. Let g_Γ be the genus of X_Γ. Then, it follows from Hurwitz's formula that

$$g_\Gamma = 1 + \frac{\mu_\Gamma}{12} - \frac{v_2}{4} - \frac{v_3}{3} - \frac{v_\infty}{2}. \tag{3.24}$$

In particular:

(1) For $\Gamma = \Gamma(n)$ one finds

$$\mu_{\Gamma(n)} = \begin{cases} (n^3/2)\prod_{p|n}(1 - 1/p^2), & n \geq 3, \\ 6, & n = 2, \end{cases} \tag{3.25}$$

and

$$v_2 = v_3 = 0, \quad v_\infty = \mu_{\Gamma(n)}/n. \tag{3.26}$$

It follows that $g_{\Gamma(2)} = 0$ and for $n \geq 3$ we have

$$g_{\Gamma(n)} = 1 + \frac{(n-6)n^2}{24}\prod_{p|n}\left(1 - \frac{1}{p^2}\right). \tag{3.27}$$

(2) For $\Gamma = \Gamma_0(n)$ one finds that

$$\mu_{\Gamma_0(n)} = n\prod_{p|n}\left(1 + \frac{1}{p}\right), \tag{3.28}$$

$$v_2 = \begin{cases} 0, & 4 \mid n, \\ \prod_{p|n}\left(1 + \left(\frac{-4}{p}\right)\right), & \text{otherwise}, \end{cases} \tag{3.29}$$

$$v_3 = \begin{cases} 0, & 2 \mid n \text{ or } 9 \mid n, \\ \prod_{p|n}\left(1 + \left(\frac{-3}{p}\right)\right), & \text{otherwise}. \end{cases} \tag{3.30}$$

Letting φ be Euler's function, we have

$$v_\infty = \sum_{d|n}\varphi\bigl(\gcd(d, n/d)\bigr). \tag{3.31}$$

(3) For $\Gamma = \Gamma_1(n)$ one finds that the genus of $X_1(n)$ for $n \leq 4$ is zero and the number of cusps is 1, 2, 2 and 3 for $n = 1$, 2, 3 and 4, respectively.

For $n > 4$ we have the following formulae:

$$\mu_{\Gamma_1(n)} = \frac{n^2}{2}\prod_{p|n}\left(1-\frac{1}{p^2}\right), \tag{3.32}$$

$$v_\infty = \frac{1}{2}\sum_{d|n}\varphi(d)\varphi(n/d) = \frac{n}{2}\prod_{p|n}\left(1-\frac{1}{p^2}+\mathrm{val}_p(n)\left(1-\frac{1}{p}\right)^2\right). \tag{3.33}$$

Since $v_2 = v_3 = 0$ for $n \geq 4$, we obtain

$$g_{\Gamma_1(n)} = 1+\frac{n^2}{24}\prod_{p|n}\left(1-\frac{1}{p^2}\right)-\frac{n}{4}\prod_{p|n}\left(1-\frac{1}{p^2}+\mathrm{val}_p(n)\left(1-\frac{1}{p}\right)^2\right), \quad n>4. \tag{3.34}$$

4. Modular forms

In this section, we present modular forms as sections of line bundles. Though our presentation is quite self-contained, we assume that the reader is familiar with the usual definition of modular forms, and our intention is to present a re-interpretation and some interesting phenomena.

Let Γ be a discrete subgroup of $\mathrm{SL}_2(\mathbb{R})$.

DEFINITION 4.1. A holomorphic function $j\colon \Gamma\times\mathcal{H}\to\mathbb{C}^\times$ is called a *factor of automorphy* if for all $\mu_1,\mu_2\in\Gamma$ and $\tau\in\mathcal{H}$:

$$j(\mu_1\mu_2,\tau) = j(\mu_1,\mu_2\tau)j(\mu_2,\tau). \tag{4.1}$$

One calls (4.1) the *cocycle relation.* Indeed, if we view Γ as acting on the multiplicative abelian group $\mathcal{O}_\mathcal{H}^\times$ of non-vanishing holomorphic functions on $\mathcal{H}$ by the rule $(f^\gamma)(\tau) = f(\gamma\tau)$, then the association

$$\Gamma\to\mathcal{O}_\mathcal{H}^\times, \quad \mu\mapsto j(\mu,\cdot) \tag{4.2}$$

is a 1-cocycle in $Z^1(\Gamma,\mathcal{O}_\mathcal{H}^\times)$. Figure 4.1 demonstrates the compatibility of the action of Γ on $\mathcal{H}\times\mathbb{C}$.

EXAMPLE 4.2. Let k be an integer. Let

$$j_k(\mu,\tau)\colon \Gamma\times\mathcal{H}\to\mathbb{C}, \quad j_k(\mu,\tau) = (c\tau+d)^k, \tag{4.3}$$

where $\mu = \left(\begin{smallmatrix} a & b\\ c & d\end{smallmatrix}\right)\in\Gamma, \tau\in\mathcal{H}$. Then j_k is a factor of automorphy.

DEFINITION 4.3. For $\Gamma\subset\mathrm{SL}_2(\mathbb{Z})$ torsion free (i.e., Γ contains no elliptic elements), we define the Hodge bundle $\mathbb{E}^k$ as the quotient:

$$\Gamma\backslash\mathcal{H}\times\mathbb{C}, \tag{4.4}$$

with the action of Γ given by $(\tau,\alpha)\sim_\mu(\mu\tau, j_k(\mu,\tau)\alpha)$.

Using that Γ acts *properly and discontinuously* on $\mathcal{H}\times\mathbb{C}$: for any pair of compact subsets $K_1,K_2\subset\mathcal{H}\times\mathbb{C}$ the set

$$\{\gamma\in\Gamma : \gamma K_1\cap K_2\neq\varnothing\} \tag{4.5}$$

is finite, and using that Γ is torsion free, hence acts freely on $\mathcal{H}\times\mathbb{C}$, we get that the Hodge bundle $\mathbb{E}^k$ has a well-defined structure of a complex manifold and is, in fact, a line bundle over $\Gamma\backslash\mathcal{H}$.

More generally, given a factor of automorphy j for Γ we obtain (for the same reasons) a line bundle

$$\Gamma\backslash\mathcal{H}\times\mathbb{C}\to\Gamma\backslash\mathcal{H}, \tag{4.6}$$

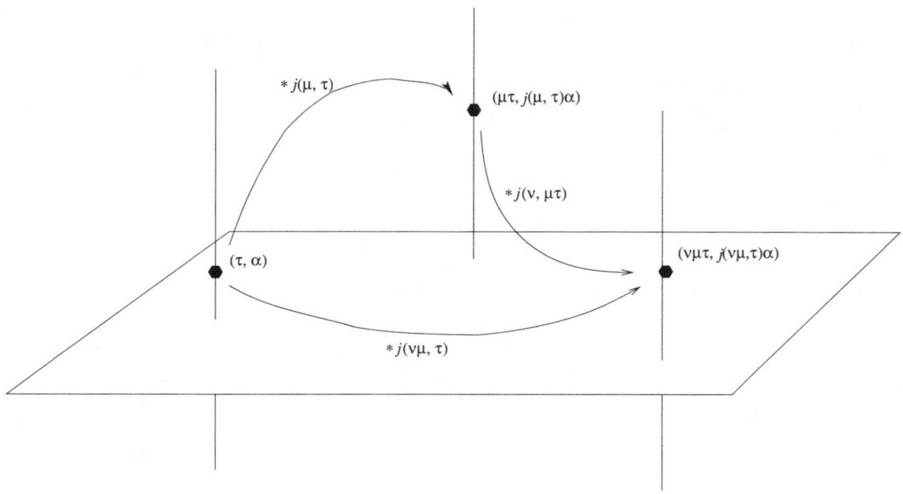

FIGURE 4.1. The cocycle relation.

where $\mu \in \Gamma$ acts by $(\tau, \alpha) \mapsto (\mu\tau, j(\mu, \tau)\alpha)$.

Conversely, given a line bundle $L \to \Gamma\backslash\mathcal{H}$, denoting by $\pi\colon \mathcal{H} \to \Gamma\backslash\mathcal{H}$ the canonical map we obtain a trivial line bundle π^*L on $\mathcal{H}$ ($\mathcal{H}$ is simply connected). Choose an isomorphism $\Psi\colon \pi^*L \to \mathcal{H} \times \mathbb{C}$. We get a canonical isomorphism

$$(\mathcal{H} \times \mathbb{C})_\tau \overset{\Psi_\tau^{-1}}{\cong} (\pi^*L)_\tau \cong (\mu^*\pi^*L)_\tau \cong (\pi^*L)_{\mu\tau} \overset{\Psi_{\mu\tau}}{\cong} (\mathcal{H} \times \mathbb{C})_{\mu\tau}, \tag{4.7}$$

which is multiplication by a non-zero scalar $j(\mu, \tau)$. One verifies that j is a factor of automorphy for Γ, and that this gives an inverse construction to the one above.

We note that $\mathbb{E}^{\otimes kk'} \cong (\mathbb{E}^{\otimes k})^{\otimes k'}$, because, in general, if L_i, $i = 1, 2$, are two line bundles defined by factors of automorphy h_i, then $L_1 \otimes L_2$ is defined by the factor of automorphy $h_1 h_2$.

DEFINITION 4.4. A *modular form* of weight k and level Γ is a global section of the line bundle $\mathbb{E}^k$ over $\Gamma\backslash\mathcal{H}$, i.e., an element of $H^0(\Gamma\backslash\mathcal{H}, \mathbb{E}^k)$. Equivalently, a function

$$f\colon \mathcal{H} \to \mathbb{C}, \tag{4.8}$$

such that for every $\mu \in \Gamma$ and $\tau \in \mathcal{H}$

$$f(\mu\tau) = j_k(\mu, \tau) f(\tau). \tag{4.9}$$

We denote the vector space of modular forms of weight k and level Γ by $\mathcal{F}(\mathbb{C}, k, \Gamma)$.

REMARK 4.5. Some care has to be taken here. A modular form in the sense above need not be holomorphic at the cusps. The Hodge bundle $\mathbb{E}$ extends to the compactified curve $\Gamma\backslash\mathcal{H}^*$ if Γ is sufficiently small[2] and a modular form of weight k and level Γ is usually taken to be a section of $\mathbb{E}^k$ over $\Gamma\backslash\mathcal{H}^*$. That is, one adds

[2] The extension is natural in the sense that it is the relative cotangent bundle at the origin of a family $\mathcal{E} \to \Gamma\backslash\mathcal{H}^*$ of generalized elliptic curves. One also has the Kodaira–Spencer isomorphism $\mathbb{E}^2 \cong \Omega^1_{(\Gamma\backslash\mathcal{H}^*)/\mathbb{C}}(\mathbf{cusps})$. On the other hand, this isomorphism could be taken as the definition of the extension of $\mathbb{E}^2$ (hence of $\mathbb{E}^{2k}$) to $\Gamma\backslash\mathcal{H}^*$.

the requirement of being holomorphic at the cusps. To distinguish we shall say "holomorphic modular forms" and use the notation $\mathcal{M}(\mathbb{C}, k, \Gamma)$.

Starting with property (4.9), one may define modular forms of weight k and level Γ, for Γ not necessarily torsion free, e.g., for $\mathrm{SL}_2(\mathbb{Z})$, as holomorphic functions

$$f\colon \mathcal{H} \to \mathbb{C}, \tag{4.10}$$

satisfying

$$f(\mu\tau) = j_k(\mu, \tau) f(\tau). \tag{4.11}$$

Again, the usual terminology requires holomorphity at infinity as well. In the case where Γ is not torsion free, these are the global sections of a sheaf on $\Gamma\backslash\mathcal{H}$ which is generally *not* invertible. The problem is with the ramification of the map $\mathcal{H} \to \Gamma\backslash\mathcal{H}$ that obstructs the descent of the Hodge bundle to the quotient. To see that indeed there *is* an obstruction take the case $\Gamma = \mathrm{SL}_2(\mathbb{Z})$ and note that

$$\begin{aligned}
&\text{from } \mu_0 = \begin{pmatrix} -1 & 0 \\ 0 & -1 \end{pmatrix} && \text{we get } f(\mu_0\tau) = f(\tau) = (-1)^k f(\tau), \\
&\text{from } \mu_1 = \begin{pmatrix} 0 & -1 \\ 1 & 0 \end{pmatrix} && \text{we get } f(\mu_1 i) = f(i) = i^k f(i), \\
&\text{from } \mu_2 = \begin{pmatrix} 0 & -1 \\ 1 & 1 \end{pmatrix} && \text{we get } f(\mu_2\omega) = f(\omega) = (-\omega^2)^k f(\omega),
\end{aligned}$$

($\omega = \exp(2\pi i/3)$) and hence every section of the sheaf of modular forms of weight k and level $\mathrm{SL}_2(\mathbb{Z})$ vanishes: (i) identically, if $2 \nmid k$; (ii) at i, if $4 \nmid k$; (iii) at ω, if $6 \nmid k$. Thus, for k odd this is the zero sheaf. But also for k even the sheaf need not be invertible. Recall the isomorphism $\mathrm{SL}_2(\mathbb{Z})\backslash\mathcal{H} \cong \mathbb{C}$. It is given in fact by the j-invariant. Since $\mathbb{C}$ is simply connected, any line bundle over it is trivial. But triviality of the line bundle implies the existence of a non-vanishing section! That is impossible in general, since $\mu_1, \mu_2 \in \mathrm{SL}_2(\mathbb{Z})$. See also the digression on factors of automorphy in §6.1

EXERCISE 4.6. *Show that $\mathbb{E}^{12}$ descends to a line bundle on $\mathrm{SL}_2(\mathbb{Z})\backslash\mathcal{H}$. How does that fit with the existence of the cusp form Δ?*

Interesting references in this context are [**104**] and [**108**].

We return to the case of a torsion free subgroup Γ. We may interpret the Hodge bundle $\mathbb{E}^1$ in the following way: Given $\tau \in \mathcal{H}$, $\gamma = \left(\begin{smallmatrix} a & b \\ c & d \end{smallmatrix}\right) \in \Gamma$, we have an isomorphism of elliptic curves (induced by multiplication):

$$\mathcal{E}_\tau = \mathbb{C}/(\mathbb{Z} + \mathbb{Z}\tau) \xleftarrow[\times(c\tau+d)=j(\gamma,\tau)]{\sim} \mathcal{E}_{\gamma\tau} = \mathbb{C}\Big/\left(\mathbb{Z} + \mathbb{Z}\left(\frac{a\tau+b}{c\tau+d}\right)\right), \tag{4.12}$$

which induces a map between the tangent spaces at the origin

$$\mathbb{C} = \mathfrak{t}_{\mathcal{E}_\tau} \xleftarrow[\times c\tau+d]{} \mathfrak{t}_{\mathcal{E}_{\gamma\tau}} = \mathbb{C}, \tag{4.13}$$

and therefore, a map between the cotangent spaces

$$\mathfrak{t}^*_{\mathcal{E}_\tau} \xrightarrow[\times c\tau+d]{\sim} \mathfrak{t}^*_{\mathcal{E}_{\gamma\tau}}. \tag{4.14}$$

Consider the projection map

$$\mathcal{E}_{\mathrm{univ},\Gamma} := \Gamma\backslash\tilde{\mathcal{E}}_{\mathrm{univ},\Gamma} \xrightarrow{\pi} \Gamma\backslash\mathcal{H}, \tag{4.15}$$

where $\tilde{\mathcal{E}}_{\text{univ},\Gamma} = \{(\tau, x) : \tau \in \mathcal{H},\ x \in \mathbb{C}/(\mathbb{Z} + \mathbb{Z}\tau)\}$, the action of Γ being given by $(\tau, x) \sim (\gamma\tau, j(\gamma, \tau)^{-1}x)$. We see that the family of elliptic curves $\tilde{\mathcal{E}}_{\text{univ},\Gamma}$ over $\mathcal{H}$ descends to a family of elliptic curves $\mathcal{E}_{\text{univ},\Gamma}$ over $\Gamma\backslash\mathcal{H}$ (which is the "universal family with level Γ") and that the relative tangent space of $\mathcal{E}_{\text{univ},\Gamma}$ over $\Gamma\backslash\mathcal{H}$ is defined by the factor of automorphy $j(\gamma, \tau)^{-1}$.

Conclusion. Let Γ be a torsion free subgroup of $\mathrm{SL}_2(\mathbb{Z})$. Let $\mathcal{E}_{\text{univ},\Gamma} \to \Gamma\backslash\mathcal{H}$ be the universal elliptic curve of level Γ. A modular form of weight k and level Γ is a section of the k-th tensor power of the relative cotangent space of $\mathcal{E}_{\text{univ},\Gamma}$ at the origin, i.e., of ${\mathfrak{t}^*_{\mathcal{E}_{\text{univ},\Gamma}}}^{\otimes k} = e^*({\mathfrak{T}^*_{\mathcal{E}_{\text{univ},\Gamma}/(\Gamma\backslash\mathcal{H})}}^{\otimes k})$, where $e\colon \Gamma\backslash\mathcal{H} \to \mathcal{E}_{\text{univ},\Gamma}$ is the identity section induced from $\tau \mapsto (\tau, 0)$. (Recall our convention: For X/S with a section $s\colon S \to X$ we let $\mathfrak{T}^*_{X/S} = \Omega_{X/S}$ denote the sheaf of relative Kähler differentials, and $\mathfrak{t}^*_{X/S}$, or just $\mathfrak{t}^*_X$ if we need to make notation easier, is $s^*\mathfrak{T}^*_{X/S}$.)

REMARK 4.7. The reader may notice how the fact that there exists a universal family is connected to the fact that the map $\mathcal{H} \to \Gamma\backslash\mathcal{H}$ is unramified and to the fact that the automorphism group of an elliptic curve with "Γ structure" is trivial. See Exercise 3.3.

Let f be a modular form of weight k and level Γ. Taking the last perspective, we therefore have a map:

$$(E, \Gamma)_R \overset{f}{\mapsto} f(E, \Gamma) \in H^0(E, \underline{\omega}^{\otimes k}). \tag{4.16}$$

Here $\pi\colon E \to \mathrm{Spec}(R)$ is an elliptic curve over a $\mathbb{C}$-algebra R and $\underline{\omega}$ is the sheaf over $\mathrm{Spec}(R)$ of relative holomorphic forms on E. That is, $\underline{\omega} := \pi_*\Omega^1_{E/R}$. It is a projective rank 1 module over R. Every such holomorphic differential form is translation invariant and therefore there is a canonical identification of $\underline{\omega}$ with the relative cotangent sheaf at zero $\mathfrak{t}^*_{E/R}$.

Suppose we choose ω_0 to be some non-vanishing differential on E, and consider:

$$(E, \Gamma, \omega_0) \mapsto f(E, \Gamma, \omega_0) := \frac{f(E, \Gamma)}{\omega_0^k} \in R. \tag{4.17}$$

Replacing ω_0 with $\lambda\omega_0$, for $\lambda \in R^\times$, we get:

$$(E, \Gamma, \lambda\omega_0) \mapsto f(E, \Gamma, \lambda\omega_0) = \frac{f(E, \Gamma)}{(\lambda\omega_0)^k} = \lambda^{-k} f(E, \Gamma, \omega_0). \tag{4.18}$$

Therefore, we may think of a modular form f (with level Γ, weight k) as a rule

$$(E, \Gamma, \omega_0)_R \overset{f}{\mapsto} f(E, \Gamma, \omega_0) \in R, \tag{4.19}$$

such that $f(E, \Gamma, \lambda\omega_0) = \lambda^{-k} f(E, \Gamma, \omega_0)$ and f depends only on the isomorphism class of (E, Γ, ω_0) and commutes with base change. The point of this gymnastics is that such a definition makes sense if we replace $\mathbb{C}$ by any base scheme B and allow R to be any B-algebra! This is the point of view taken by Katz.

REMARK 4.8. The last interpretation of modular forms is perhaps best motivated by the classical view of modular forms as certain homogeneous functions of lattices. Take for simplicity $\Gamma = \mathrm{SL}_2(\mathbb{Z})$.

To give an elliptic curve $E_{/\mathbb{C}}$ up to isomorphism is to give a lattice $\mathcal{L}_E$ in $\mathbb{C}$ up to homothety. Thus, to give an elliptic curve with a non-vanishing differential

$(E_{/\mathbb{C}}, \omega_0)$ is to give a unique lattice in $\mathbb{C}$, $\mathcal{L}_{E,\omega_0}$. Hence, a modular form f gives a function on lattices in $\mathbb{C}$. Since we have the relation

$$\mathcal{L}_{E,\lambda\omega_0} = \lambda \mathcal{L}_{E,\omega_0}, \tag{4.20}$$

the property $f(E, \lambda\omega_0) = \lambda^{-k} f(E, \omega_0)$ is translated into the property $f(\lambda\mathcal{L}) = \lambda^{-k} f(\mathcal{L})$, i.e., f is a function of lattices that is homogeneous of weight k. See also [**147**, Chapter VII], [**89**, Chapter 3, §2].

Let f be a modular form with respect to the group $\Gamma_0(N)$ for some N. Since the matrix $\left(\begin{smallmatrix} 1 & 1 \\ 0 & 1 \end{smallmatrix}\right)$ belongs to $\Gamma_0(N)$ it follows that $f(\tau+1) = f(\tau)$. Therefore, f has a Fourier expansion, called a *q-expansion* with respect to the variable $q = e^{2\pi i \tau}$;

$$f = \sum_{n \geq -N} a_n q^n. \tag{4.21}$$

A similar expansion can be obtained in every cusp with respect to a suitable parameter. The holomorphic modular forms have q-expansions with $a_n = 0$ at every cusp if $n < 0$, and the cusp forms, by definition, have q-expansions with $a_n = 0$ at every cusp if $n \leq 0$. We note that if we let $\mathbb{E}^k$ denote the extension of $\mathbb{E}^k$ mentioned above to a sheaf on $X_0(N)(\mathbb{C}) = \Gamma_0(N)\backslash\mathcal{H}^*$, then the holomorphic modular forms of weight k are just the sections $H^0(X_0(N)(\mathbb{C}), \mathbb{E}^k)$, while the cusp forms of weight k are sections of $H^0\big(X_0(N)(\mathbb{C}), \mathbb{E}^k(-\mathbf{cusps})\big)$.

5. Some examples of modular forms

5.1. Level 1. Let k be an even integer, $k \geq 4$. The *Eisenstein series of weight* k is the complex valued function on the upper half plane

$$G_k^{\mathbb{Q}}(\tau) = \sum_{(n,m)}{}' \frac{1}{(m\tau + n)^k}; \tag{5.1}$$

the sum extends over all integers m, n such that $(m, n) \neq (0, 0)$. It is well-known that this is a modular form of level one and weight k. See [**147**, Chapter VII]. We normalize $G_k^{\mathbb{Q}}$ such that the q-expansion starts with one and denote it then by $E_k^{\mathbb{Q}}$ or simply by E_k. Some examples of the q-expansion of $E_k^{\mathbb{Q}}$ at the cusp $i\infty$ are given in Table 5.1, where $\sigma_r(n) = \sum_{d|n} d^r$.

TABLE 5.1. Some q-expansions of Eisenstein series.

form	q-expansion
$E_4^{\mathbb{Q}}$	$1 + 240 \sum_{n=1}^{\infty} \sigma_3(n) q^n$
$E_6^{\mathbb{Q}}$	$1 - 504 \sum_{n=1}^{\infty} \sigma_5(n) q^n$
$E_8^{\mathbb{Q}}$	$1 + 480 \sum_{n=1}^{\infty} \sigma_7(n) q^n$
$E_{10}^{\mathbb{Q}}$	$1 - 264 \sum_{n=1}^{\infty} \sigma_9(n) q^n$
$E_{12}^{\mathbb{Q}}$	$1 + \frac{65520}{691} \sum_{n=1}^{\infty} \sigma_{11}(n) q^n$
$E_{14}^{\mathbb{Q}}$	$1 - 24 \sum_{n=1}^{\infty} \sigma_{13}(n) q^n$

TABLE 5.2. Spaces of modular forms of low weight.

weight	dimension	basis	weight	dimension	basis
0	1	1	12	2	$\Delta, E^{\mathbb{Q}}_{12}$
2	0	–	14	1	$E^{\mathbb{Q}}_{14}$
4	1	$E^{\mathbb{Q}}_4$	16	2	$\Delta E^{\mathbb{Q}}_4, E^{\mathbb{Q}}_{16}$
6	1	$E^{\mathbb{Q}}_6$	18	2	$\Delta E^{\mathbb{Q}}_6, E^{\mathbb{Q}}_{18}$
8	1	$E^{\mathbb{Q}}_8$	20	2	$\Delta E^{\mathbb{Q}}_8, E^{\mathbb{Q}}_{20}$
10	1	$E^{\mathbb{Q}}_{10}$	22	2	$\Delta E^{\mathbb{Q}}_{10}, E^{\mathbb{Q}}_{22}$

In general (loc. cit.)

$$E^{\mathbb{Q}}_k(q) = 1 + 2\zeta(1-k)^{-1}\sum_{n=1}^{\infty}\sigma_{k-1}(n)q^n. \tag{5.2}$$

It is well-known that the graded ring of holomorphic modular forms of level 1 over the complex numbers, $\bigoplus_{k=0}^{\infty}\mathcal{M}(\mathbb{C}, k, \mathrm{SL}_2(\mathbb{Z}))$, is isomorphic to the free polynomial ring in two variables $\mathbb{C}[E_4, E_6]$ of weights 4 and 6 respectively. The modular form $\Delta = (E_4^3 - E_6^2)/1728$ is a cusp form of weight 12 that induces an injection

$$\mathcal{M}(\mathbb{C}, k, \mathrm{SL}_2(\mathbb{Z})) \xrightarrow{\times\Delta} \mathcal{M}(\mathbb{C}, k+12, \mathrm{SL}_2(\mathbb{Z})), \tag{5.3}$$

whose image is of codimension one, consisting of all the cusp forms of weight $k+12$. Thus, for $k \geq 0$,

$$\dim\mathcal{M}(\mathbb{C}, k+12, \mathrm{SL}_2(\mathbb{Z})) = \dim\mathcal{M}(\mathbb{C}, k, \mathrm{SL}_2(\mathbb{Z})) + 1. \tag{5.4}$$

The dimensions for small weight are given in Table 5.2.

The modular form Δ has some important properties making it one of the key players in the theory. First, it has a product expansion

$$\Delta = q\prod_{n=1}^{\infty}(1-q^n)^{24}, \tag{5.5}$$

which shows that Δ does not vanish on the upper half plane and has integral Fourier coefficients $\sum_{n=1}^{\infty}\tau(n)q^n$. The coefficients $\tau(n)$ are the famous Ramanujan τ-function (see page 118 for a table of some values of this function). Second, it is the unique cusp form of its weight and thus an eigenfunction for all the Hecke operators. Third, it has the interpretation of being (up to a constant) the discriminant of the elliptic curve $\mathcal{E}_\tau$ associated to $\tau \in \mathcal{H}$ by means of the Weierstrass $\wp$-function (and that shows again that Δ does not vanish).

5.2. Higher level. Let f be a modular form of level 1 and weight k considered as a function on $\mathcal{H}$. Consider the function

$$\tau \mapsto g(\tau) := f(N\tau). \tag{5.6}$$

Let $\gamma = \left(\begin{smallmatrix} a & b \\ c & d\end{smallmatrix}\right)$ be a matrix in $\Gamma_0(N)$. Then $g(\gamma\tau) = f(N\gamma\tau) = f(N\gamma N^{-1}N\tau) = f(\left(\begin{smallmatrix} a & Nb \\ c/N & d\end{smallmatrix}\right)N\tau) = (c\tau+d)^k f(N\tau) = (c\tau+d)^k g(\tau)$. Thus $g(\tau)$ is a modular form of level $\Gamma_0(N)$ and weight k.

EXERCISE 5.1. *Define an action of* $(\mathbb{Z}/N\mathbb{Z})^\times$ *on modular forms of level* $\Gamma_1(N)$ *and weight* k *by* $\langle d\rangle f(\tau) = (c\tau + d)^{-k} f(\gamma\tau)$ *where* $\gamma = \left(\begin{smallmatrix} a & b \\ c & d \end{smallmatrix}\right)$ *is in* $\Gamma_0(N)$. *Prove that this is well-defined, i.e., independent of the choice of* γ *and preserves the space of modular forms of weight* k.

Let $\epsilon\colon (\mathbb{Z}/N\mathbb{Z})^\times \to \mathbb{C}^\times$ *be a Dirichlet character. A modular form* f *of level* $\Gamma_1(N)$ *and weight* k *is called a modular form of character* ϵ *if it satisfies the relation* $\langle d\rangle f = \epsilon(d) f$. *The space of modular forms of level* $\Gamma_1(N)$ *and weight* k *decomposes according to Dirichlet characters. Now, let* $M \mid N$ *and* ϵ *a Dirichlet character* mod M *that we consider also as a character* mod N. *Assume that* $aM \mid N$ *and let* f *be a modular form of level* $\Gamma_1(M)$ *and character* ϵ. *Prove that* $g(\tau) = f(a\tau)$ *is a modular form of level* $\Gamma_1(N)$ *and character* ϵ. *If* $M \neq N$ *we call* g *an* old form.

Prove that the diamond operator $\langle d\rangle$ *is a modular isomorphism in the following sense. If* τ mod $\Gamma_1(N)$ *corresponds to a pair* (E, P) *consisting of an elliptic curve* E *and a point* P *on* E *of exact order* N, *then* $\gamma\tau$ *corresponds to the pair* (E, dP), *if* $\gamma = \left(\begin{smallmatrix} a & b \\ c & d \end{smallmatrix}\right)$ *is in* $\Gamma_0(N)$.

EXERCISE 5.2. *Let* X *be a modular curve of genus* g *and having* s *cusps. Using Riemann–Roch, prove that the dimension of the weight* 2 *cusp forms on* X *is* g *and the dimension of holomorphic modular forms of weight* 2 *is* $g + s - 1$.

Modular forms obtained in this fashion are very important. Consider for example the modular function of level N given by

$$\tau \mapsto u(\tau) := \frac{\Delta(N\tau)}{\Delta(\tau)}. \tag{5.7}$$

The divisor of this function is supported at the cusps. In fact, for N prime there are two cusps c_1, c_2 on $X_0(N)$ (See Appendix B) and this function proves the Manin–Drinfeld Theorem:

THEOREM. *Every divisor* D *of degree zero supported at the cusps is torsion. That is, there exists a positive integer* n *and a function* h *whose divisor is* nD.

EXERCISE 5.3. *For* N *prime, bound the order of the divisor* $c_1 - c_2$. *In fact, one can prove that the order is exactly the numerator of* $(N-1)/12$ *but this is harder. See* [**123**, §3] *and* [**124**, §4] *for generalizations. A deep theorem of Mazur* [**94**] *asserts that, in fact, the class* $c_1 - c_2$ *generates* $J_0(N)(\mathbb{Q})_{\mathrm{tor}}$!

In general, one considers the value of this function at a point τ, i.e., one considers $\Delta(N\tau)/\Delta(\tau)$. Since this function has integral Fourier coefficients it extends to a function over the arithmetic scheme over $\mathbb{Z}[1/N]$—the proper regular model of $X_0(N)$—and still its divisor is supported at the cusps because the g.c.d. of the Fourier coefficients is 1. Thus, from the moduli interpretation we see that if the elliptic curve corresponding to τ is defined over $\mathbb{Q}^{\mathrm{alg}}$ and has everywhere potential good reduction then $u(\tau)$ is at least an S-unit, where S is the set of primes dividing N. Moreover, this unit lies in the moduli field of $\mathcal{E}_\tau$, which is equal to the residue field of τ as a point on the algebraic curve $X_0(N)$. In case τ corresponds to an elliptic curve with complex multiplication one can further find this moduli field and the Galois action via the theory of complex multiplication and Shimura's reciprocity law. In the complex multiplication setting, by a suitable normalization of this construction one obtains units called *Siegel's units.* See [**89**, Chapter 12] and [**32**, Chapter 2.2].

It is not our intention to give a systematic introduction to the theory of elliptic modular forms, i.e., to modular forms for some congruence subgroup of $\mathrm{SL}_2(\mathbb{Z})$, but rather to give some interesting examples that connect to the material to be discussed in this book. We thus restrict our attention to two constructions of modular forms: Eisenstein series and theta series. We follow Ogg [**122**], where the reader can find proofs and more details.

Eisenstein Series. Let $k \geq 3$.[3] Let N be a fixed positive integer and c, d be two integers. Generalizing the construction of the Eisenstein series of level one, one defines

$$G_k(\tau; c, d, N) = \sum_{\substack{m \equiv c \bmod N \\ n \equiv d \bmod N}}' \frac{1}{(m\tau + n)^k}. \tag{5.8}$$

It is easy to verify that $G_k(\tau; c, d, N)$ is a modular form of weight k and level $\Gamma(N)$. One can modify the construction and define the restricted Eisenstein series $G_k^*(\tau; c, d, N)$ for $(c, d, N) = 1$ as

$$G_k^*(\tau; c, d, N) = \sum_{\substack{m \equiv c \bmod N \\ n \equiv d \bmod N \\ (m,n)=1}} \frac{1}{(m\tau + n)^k}. \tag{5.9}$$

Also $G_k^*(\tau; c, d, N)$ is a modular form of weight k and level $\Gamma(N)$. The two kinds of Eisenstein series are related by explicit formulae ([**122**, IV-34]).

The restricted Eisenstein series $G_k^*(\tau; c, d, N)$ have the wonderful property that their q-expansion starts with zero at all cusps except for the cusp $-d/c$ where the q-expansion starts with 1. The restricted Eisenstein series are thus seen to generate the space of Eisenstein series.

We introduce the notation $f \mid \gamma$:

$$(f \mid \gamma)(\tau) = (ad - bc)^{k/2}(c\tau + d)^{-k} f(\gamma\tau), \quad \gamma = \begin{pmatrix} a & b \\ c & d \end{pmatrix} \in \mathrm{GL}_2(\mathbb{R})^+. \tag{5.10}$$

We have with this notation

$$G_k^*(\tau; (c, d), N) \mid \gamma = G_k^*(\tau; (c, d)\gamma, N), \quad \gamma \in \mathrm{SL}_2(\mathbb{Z}) \tag{5.11}$$

([**122**, IV-33]). Thus, from orbits of $\Gamma_0(N)$ in the set of pairs of residues (c, d) modulo N such that $(c, d, N) = 1$, we get Eisenstein series for $\Gamma_0(N)$.

Theta Series. Let r be an even integer and let

$$Q(x) = \frac{1}{2} {}^t x A x = \frac{1}{2} \sum_{i,j=1}^{r} a_{ij} x_i x_j \tag{5.12}$$

be a positive definite integral quadratic form. That is $A = (a_{ij})$ is a symmetric $r \times r$ matrix of integers whose diagonal elements are even, and A is positive definite. Note that for $v \in \mathbb{Z}^r$ the value $Q(v)$ is an integer.

One defines the *determinant* of Q, denoted D, as the determinant of A, and the *discriminant* of Q, denoted Δ, as $(-1)^{r/2} D$. The least positive integer N such

[3] For level subgroups, unlike the case of $\mathrm{SL}_2(\mathbb{Z})$, there may be non-trivial modular forms of odd weight, though there are none of negative weight.

that $\frac{1}{2}{}^t xNA^{-1}x$ is an integral quadratic form is called the *level* of Q. We put $A^* = NA^{-1}$. The integral positive definite form

$$Q^* = \frac{1}{2}{}^t xA^*x \tag{5.13}$$

is the *adjoint form* to Q. One always has $N \mid D \mid N^r$. See [**122**, Chapter VI].

The theta series associated to Q is defined as

$$\begin{aligned}\theta(\tau, Q) &= \sum_{v\in\mathbb{Z}^r} e^{2\pi i\cdot Q(v)\tau} \\ &= 1 + \sum_{n=1}^{\infty} a_Q(n)q^n, \quad q = e^{2\pi i\tau}.\end{aligned} \tag{5.14}$$

Here the coefficients $a_Q(n)$ are the representation numbers of Q, i.e., $a_Q(n)$ is the number of vectors $v \in \mathbb{Z}^r$ solving the equation $Q(x) = n$.

Let ϵ be the character $\epsilon(n) = (\Delta/n)$ (Jacobi symbol). See [**73**, p. 56]). Simple properties of that symbol show that ϵ may be considered as a Dirichlet character of conductor dividing N. Then the main theorem is

THEOREM 5.4. *The series $\theta(\tau, Q)$ is a modular form for $\Gamma_0(N)$ of character ϵ and weight $r/2$.*

(This construction can be generalized using spherical functions. See [**122**, VI-10], [**143**, Chapter 9]). It turns out that to get modular forms of level 1 the number of variables must be divisible by 8. See [**147**, Chapter VII, §6.5].

EXAMPLE 5.5. (1) Though we assumed that r is even, the case of $r = 1$ is classical and worth mentioning. Let $Q_N(x) = Nx^2$ corresponding to the matrix $A_N = 2N$. Note that $\theta(\tau, Q_N) = \theta(N\tau, Q_1)$ and

$$\theta(\tau, Q_N) = 1 + 2\sum_{n=1}^{\infty} q^{Nn^2}. \tag{5.15}$$

We note that

$$\theta(\tau, Q_1) = 1 + 2\sum_{n=1}^{\infty} q^{n^2} = \Theta\left[\begin{smallmatrix}0\\0\end{smallmatrix}\right](0, 2\tau), \tag{5.16}$$

(Riemann's theta function, [**41**, Chapter VI]). Since r is odd, the situation is a bit different from the one described above. Still we have some modularity results, e.g., $\Theta\left[\begin{smallmatrix}0\\0\end{smallmatrix}\right](0,\tau)^4$ is a modular form of level $\Gamma_1(2)$ and weight 2.

(2) Let $Q(x) = x_1^2 + x_2^2 + x_3^2 + x_4^2$ corresponding to the diagonal matrix $A = \mathrm{diag}[2, 2, 2, 2]$. It is a quadratic form in four variables, of level 4 and discriminant 16. It therefore defines a modular form of level $\Gamma_0(4)$, weight 2 and trivial character.

$$\theta(\tau, Q) = 1 + \sum_{n=1}^{\infty} a_n q^n, \tag{5.17}$$

where a_n is the number of expressions of n as a sum of four squares (negative numbers allowed). Note that

$$\theta(\tau, Q) = \theta(\tau, Q_1)^4. \tag{5.18}$$

A classical formula gives an elementary expression for the coefficients a_n:

PROPOSITION 5.6 (Jacobi). *The number a_n of representations of n as a sum of four squares is*

$$a_n = \begin{cases} 8\sum_{d|n} d, & n \text{ odd} \\ 24\sum_{\substack{d|n \\ d\equiv 1 \bmod 2}} d, & n \text{ even.} \end{cases} \tag{5.19}$$

Two strategies for proving this Proposition are as follows: (i) One proves that $\theta(\tau, Q)$ is in fact a modular form of weight 2 and a certain level Γ such that the corresponding modular curve X_Γ is of genus 0 and has two cups. There is thus a unique modular form of level 2 on X_Γ up to a scalar. One constructs a modified Eisenstein series of weight two and compares coefficients. See [**109**, Chapter 1, §15]. (ii) Use the product expansion of $\theta(\tau, Q_1)$ to derive the formula directly (loc. cit.).

Connection to Supersingular Elliptic Curves. Let p be a prime number and consider the set of isomorphism classes of supersingular elliptic curves over $\overline{\mathbb{F}_p}$. There are $\hbar = g+1$ of them, where g is equal to the genus of $X_0(p)(\mathbb{C})$:

$$g = \begin{cases} (p+1)/12 - \left(1+\left(\frac{-1}{p}\right)\right)/4 - \left(1+\left(\frac{-3}{p}\right)\right)/3 & p\neq 2, \\ 0 & p=2. \end{cases} \tag{5.20}$$

EXERCISE 5.7. *Prove formula* (5.20) *from Hurwitz's genus formula using that* $\mathrm{SL}_2(\mathbb{Z})\backslash\mathcal{H}$ *has genus zero.*

Let E be an elliptic curve over $\overline{\mathbb{F}_p}$. We recall here various equivalent ways of saying that the elliptic curve E is supersingular: (i) E has no physical p-torsion: $E[p](\overline{\mathbb{F}_p}) = \{0\}$; (ii) there exists an embedding of the group scheme α_p into E; (iii) there does not exist an embedding of μ_p into E; (iv) the Verschiebung morphism $\mathrm{Ver} = \widehat{\mathrm{Fr}_E}\colon E^{(p)} \to E$ is purely inseparable; (v) the multiplication by p map $[p]\colon E \to E$ is purely inseparable; (vi) the ring of endomorphisms of E is isomorphic to a (maximal) order of the quaternion algebra $B_{p,\infty}$—the quaternion algebra over $\mathbb{Q}$ ramified at the two places p and ∞; (vii) for n large enough Fr_E^n is an endomorphism of E and its characteristic polynomial has Newton polygon with respect to the p-adic valuation which is a straight line; (viii) the formal group of E has height 2.

See [**156**, V.3] for some of the equivalences. The others you can do once you read Chapter 6, §7 and Appendix A. We recall that furthermore the j-invariant of a supersingular elliptic curve is in $\mathbb{F}_{p^2}$.

Another characterization of supersingularity is the following. Let k be a finite field of characteristic $p>2$. Choose a Weierstrass equation for E/k, say $y^2 = f(x)$, with $f(x)$ a separable cubic polynomial. Then E is supersingular iff the coefficient of x^{p-1} in $f(x)^{(p-1)/2}$ is zero. See [**156**, Theorem 4.1]. What's interesting about this criterion is that it is really a criterion about the Newton polygon of a Jacobian of a curve. We give the case of genus two in page 218.

EXERCISE 5.8. *The curve $y^2 = x^3+1$ is non-singular for $p>3$. Using the last criterion, prove that it is supersingular iff p is inert in $\mathbb{Q}(\zeta_3)$.*

There is an interesting characteristic p method to prove formula (5.20). One considers the $\Gamma_0(p)$-level moduli problem of parameterizing elliptic curves with a given subgroup of order p. One proves that there exists a coarse moduli scheme for this moduli problem, which is a regular scheme that is a *flat* relative curve $Y_0(p)$ over $\mathbb{Z}$. Over the complex numbers $Y_0(p)(\mathbb{C}) \cong \Gamma_0(p)\backslash\mathcal{H}$.

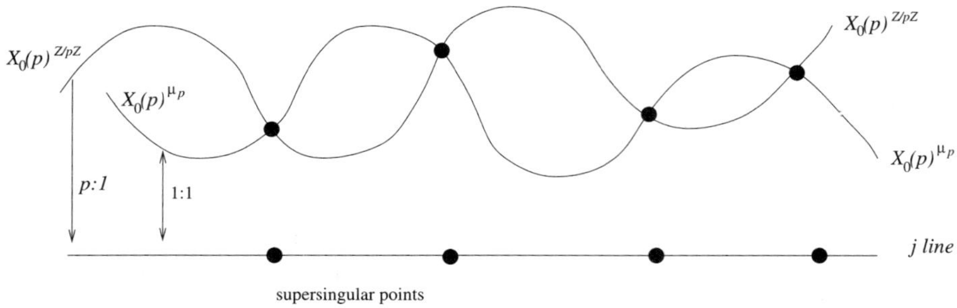

FIGURE 5.1. The modular curve $X_0(p)$ in characteristic p.

To actually have a proper morphism $X_0(p) \to \mathrm{Spec}(\mathbb{Z})$, one needs to throw in the cusps. The cusps may also be given a moduli interpretation using the concept of generalized elliptic curves. See [**26**] and Appendix B. This interpretation shows that there are two cusps lying over the unique cusp of $X_0(1)$ (the j-line). One is unramified, and the other has ramification of order p, and moreover this persists in characteristic p (loc. cit.). One proves that $X_0(p)$ is a regular integral scheme and the morphism $X_0(p) \to \mathrm{Spec}(\mathbb{Z})$ is *proper* and *flat*. Over the complex numbers $X_0(p)(\mathbb{C}) \cong \Gamma_0(p)\backslash\mathcal{H}^*$.

One goes on to prove that in characteristic p, the reduction of $X_0(p)$ consists of two components that intersect transversely precisely above the supersingular points of $X_0(p)$. See Figure 5.1.

The proof is based on the construction of two "sections"

$$X_0(1) \bmod p \rightrightarrows X_0(p) \bmod p. \tag{5.21}$$

One section is a true section and is obtained by sending an elliptic curve E to the pair $(E, \mathrm{Ker}(\mathrm{Fr}_E))$ (note that if E is ordinary $\mathrm{Ker}(\mathrm{Fr}_E)$ is a form of μ_p) and the other is obtained by sending E to $(E, \mathrm{Ker}(\mathrm{Ver}_E))$, where $\mathrm{Ver}_E\colon E \to E^{(1/p)}$ (note that if E is ordinary $\mathrm{Ker}(\mathrm{Ver}_E)$ is a form of $\mathbb{Z}/p\mathbb{Z}$). Leaving aside the problem of non-integral base schemes, there is still the problem that Ver_E is usually defined only after a base change. See Appendix A for groups schemes and the morphisms Fr, Ver.

Thus the true state of affairs is that $X_0(p) \bmod p$ is composed of two components that we denote $X_0(p)^{\mu_p}$ and $X_0(p)^{\mathbb{Z}/p\mathbb{Z}}$. Those two components intersect precisely over the supersingular points. The map

$$X_0(p)^{\mu_p} \to X_0(1) \bmod p, \tag{5.22}$$

is an isomorphism, while the map

$$X_0(p)^{\mathbb{Z}/p\mathbb{Z}} \to X_0(1) \bmod p, \tag{5.23}$$

is purely inseparable of degree p. Thus, both components have genus zero, and the total genus of $X_0(p) \bmod p$ (which, courtesy of flatness, is also the genus of *every* fiber of $X_0(p) \to \mathrm{Spec}(\mathbb{Z})$) is the genus of the intersection graph of the components. This graph consists of two vertices, with $\hbar$ edges between them, and has thus genus $\hbar - 1$.

Fix representatives for the supersingular elliptic curve $E_1, \dots, E_{\bar{h}}$. For every i and j we consider the abelian group

$$I_{i,j} = \mathrm{Hom}_{\overline{\mathbb{F}_p}}(E_i, E_j). \tag{5.24}$$

It is a left $\mathrm{End}(E_j)$ module and a right $\mathrm{End}(E_i)$ module. The function $f \mapsto \deg(f)$ is a positive definite integral quadratic form $Q_{i,j}$ on $I_{i,j}$ in four variables. One defines the theta series

$$\theta(\tau, Q_{i,j}) := \frac{1}{|\operatorname{Aut}(E_j)|} \sum_{b \in I_{i,j}} q^{\deg(b)}. \tag{5.25}$$

The main theorem is

THEOREM 5.9. *The series $\theta(\tau, Q_{i,j})$ are modular forms of weight 2 and level $\Gamma_0(p)$ (with trivial character) that span the space of modular forms of weight 2 and level $\Gamma_0(p)$.*

We refer the interested reader to [**57, 97, 133**] for more on this fascinating story. The proof that the level and character are as stated follows of course from Theorem 5.4 given above and can be found in [**133**, Theorem 2.14].

EXAMPLE 5.10. There is a unique supersingular elliptic curve E over an algebraically closed field of characteristic two. Its endomorphism ring is the ring $\mathbb{Z}[i, j, (1+i+j+k)/2]$ in the rational Hamilton quaternions $\mathbb{H}$, where following classical notation we write the Hamilton quaternions as $\mathbb{Q} \oplus \mathbb{Q}i \oplus \mathbb{Q}j \oplus \mathbb{Q}k$ with the relations

$$-1 = i^2 = j^2 = k^2, \quad ij = k = -ji. \tag{5.26}$$

This ring is the free $\mathbb{Z}$ module on $1, i, j$ and $(1+i+j+k)/2$. The norm of an element $a + bi + cj + dk$ is $a^2 + b^2 + c^2 + d^2$, and is equal to its degree as an endomorphism of E. If we take the non-maximal order $\mathbb{Z}[i, j]$ with basis $1, i, j$ and k, we get $Q(a + bi + cj + dk) = a^2 + b^2 + c^2 + d^2$ and we arrive at Example 5.5(2).

If we take the maximal order above with basis $e_1 = 1$, $e_2 = i$, $e_3 = j$ and $e_4 = (1+i+j+k)/2$ then an element $\sum_{i=1}^4 x_i e_i = (x_1 + x_4/2) + (x_2 + x_4/2)i + (x_3 + x_4/2)j + x_4/2 \cdot k$ has norm $\sum_{i=1}^4 x_i^2 + \sum_{i=1}^3 x_i x_4$. We therefore get the quadratic form $Q(x_1, \dots, x_4) = \frac{1}{2}\, {}^t x A x$, where A is the matrix

$$\begin{pmatrix} 2 & 0 & 0 & 1 \\ 0 & 2 & 0 & 1 \\ 0 & 0 & 2 & 1 \\ 1 & 1 & 1 & 2 \end{pmatrix}. \tag{5.27}$$

This is a matrix of determinant and discriminant 4 and level 2. It therefore defines a modular form $\theta(\tau, Q)$ of weight 2 and level 2 whose first Fourier coefficients are given by

$$\theta(\tau, Q) = 1/24 + q + q^2 + 4q^3 + q^4 + 6q^5 + 4q^6 + 8q^7 + q^8 + 13q^9 + 6q^{10} + \cdots. \tag{5.28}$$

EXERCISE 5.11. *Use the uniqueness of E to deduce that the coefficient of q^n is $\sum_{d|n, d \equiv 1 \bmod 2} d$.*

EXERCISE 5.12. *The situation above can be made explicit. We leave the verifications of details to the reader.*

The equation for E may be chosen as

$$y^2 + y = x^3. \tag{5.29}$$

The substitutions

$$x \mapsto u^2x + s^2, \quad y \mapsto y + su^2x + t, \tag{5.30}$$

give an automorphism of E if and only if the following equations are satisfied:

$$u^3 = 1, \quad s^4 + s = 0, \quad t^2 + t + s^6 = 0. \tag{5.31}$$

We write this automorphism as $(u, s, t)x, (u, s, t)y$. The group of automorphisms is visibly of order 24. Now,

$$(u_1, s_1, t_1)(u, s, t) = (u_1u, s_1 + u_1s, t_1 + t + u_1^2s_1s^2). \tag{5.32}$$

This shows that there is a unique element of order two. It is $(1, 0, 1)$ and we denote it by -1. The notation is justified since in its action on E it is equal to the automorphism $[-1]$, acting by $(x, y) \mapsto (x, y + 1)$. Handy formulas are

$$(u, s, t)^2 = \big(u^2, s(1 + u), us^3\big), \quad -(u, s, t) = (u, s, t + 1). \tag{5.33}$$

We deduce that there are precisely six elements of order four. They are the elements of the set $\{(1, s, t) : s^3 = 1, t^2 + t + 1 = 0\}$.

Let s_1, s_2 and s_3 be the solutions to $s^3 = 1$ and let t_1, t_2 be the solutions to $t^2 + t + 1 = 0$. We may assume that $s_1s_2^2 = t_1$. With these conventions we let

$$i := (1, s_1, t_1), \quad j := (1, s_2, t_1), \quad k := (1, s_3, t_1). \tag{5.34}$$

The usual relations are satisfied:

$$i^2 = j^2 = k^2 = -1, \quad ij = -ji = k. \tag{5.35}$$

It follows that $\{\pm 1, \pm i, \pm j, \pm k\}$ is Q_8—the group of quaternions of order 8. The cyclic group C_3 of order 3 given by $\{(u, 0, 0) : u^3 = 1\}$ acts by conjugation on Q_8, permuting cyclically the elements i, j, k, and

$$\mathrm{Aut}(E) \cong C_3 \ltimes Q_8. \tag{5.36}$$

To see that $(1+i+j+k)/2 = (1+i)(1+j)/2$ is an endomorphism, it is enough to show that $(1 + i)(1 + j)$ kills $E[2]$. Use the fact that $E[2]$ is a non-split extension of α_2 by α_2 to deduce that it is enough to show that both $(1 + i)$ and $(1 + j)$ kill the unique copy of α_2 embedded in E. To obtain this property, use the description of $\mathrm{Aut}(\alpha_p)$ given in Appendix A to show that both i and j act trivially on α_2.

6. Abelian varieties over $\mathbb{C}$

In this section we discuss abelian varieties over the complex numbers. We often refer to the excellent book by Lange and Birkenhake [**90**]. The reader may also consult [**18, 107, 115, 159**]. We found it necessary, regrettably, to omit most proofs in order to keep these lecture notes to a reasonable size. Nevertheless, we go into detail in describing abelian varieties over $\mathbb{C}$ in hope of arming the novice with some intuition when facing the much more strenuous presentation of abelian varieties (or schemes) over arbitrary fields in later sections.

We assume some familiarity with Lie groups. Specifically, with the definition of a complex Lie group as an analytic manifold with a group structure such that the group operations are analytic, and with the exponential map that we recall below.

DEFINITION 6.1. Let k be a field. An *abelian variety* A over k is a projective, connected algebraic group. A morphism $f\colon A \to B$ between two abelian varieties is called an *isogeny* if it is a surjective homomorphism with finite kernel.

We consider the case $k = \mathbb{C}$. Let V denote a g-dimensional vector space over $\mathbb{C}$ and Λ a lattice in V (i.e., a subgroup of rank $2g$ such that $\mathbb{R} \otimes \Lambda \cong V$ under the map $\alpha \otimes \lambda \mapsto \alpha\lambda$). A *complex torus* is a complex Lie group isomorphic to $X = V/\Lambda$ for some lattice Λ. Note that X is compact.

REMARK 6.2. Do not confuse a complex torus with an algebraic torus as in §1. The underlying analytic manifold of an algebraic torus over $\mathbb{C}$ is homotopic to a topological torus ($\mathbb{C}^\times$ is homotopic to the circle S^1 and a complex torus of dimension g is homeomorphic to $(S^1)^{2g}$), which may explain the terminology. However, as complex Lie groups there are no non-trivial maps between an algebraic torus and a complex torus.

THEOREM 6.3. *Any connected compact complex Lie group X is a complex torus.*

PROOF. We first prove that

- X is abelian.

Consider the commutator map

$$X \times X \to X, \qquad (x, y) \mapsto [x, y] = xyx^{-1}y^{-1}. \tag{6.1}$$

Let U be an open neighborhood of 1. Let $x \in X$. Since $[x, 1] = 1 \in U$ and $[\cdot,\cdot]$ is continuous, for every $x \in X$ we have open sets V_x and W_x, with $x \in V_x$ and $1 \in W_x$, such that $[V_x, W_x] \subset U$. Since X is compact,

$$X = \bigcup_{x\in X} V_x = \bigcup_{i=1}^{n} V_{x_i} \quad \text{(for some } x_1, \dots, x_n). \tag{6.2}$$

Let $W = \bigcap_{i=1}^n W_{x_i}$, an open neighborhood of 1. Then

$$[X, W] \subset U. \tag{6.3}$$

We may assume that there exists an embedding $U \hookrightarrow \mathbb{C}^g$. Hence, for every $w \in W$ we get an analytic function

$$X \to \mathbb{C}^g, \quad x \mapsto [x, w]. \tag{6.4}$$

This function must be constant since X is compact. Hence, for every $w \in W$, we have $[X, w] = 1$ and thus

$$[X, W] = 1. \tag{6.5}$$

The set W is open and contains the identity element, so it generates an open subgroup Q of X that has the property $[X, Q] = 1$; but in a topological group, any open subgroup is also closed. Thus, $Q = X$ is abelian.

- It follows from the theory of Lie groups that for every complex Lie group X, there exists a unique holomorphic map

$$\mathbb{C} \times V \to X; \quad (t, v) \mapsto \phi_v(t), \tag{6.6}$$

where $V = \mathfrak{t}_X$ is the tangent space at 1, such that

(1) $\phi_v(t)$ is a holomorphic homomorphism in t: $\phi_v(t_1+t_2) = \phi_v(t_1)\phi_v(t_2)$. In our case, since X is abelian, we write $\phi_v(t_1 + t_2) = \phi_v(t_1) + \phi_v(t_2)$.

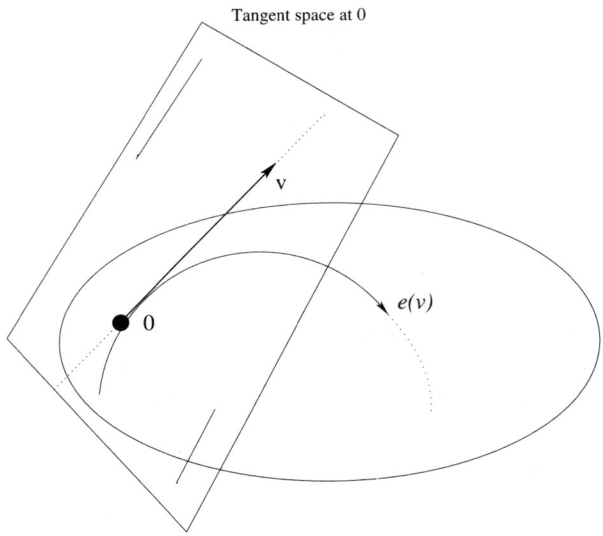

FIGURE 6.1. The exponential map.

(2) $d\phi_v$ (the induced map on tangent spaces at zero) takes the unit tangent vector to $\mathbb{C}$ at zero to v (i.e., $d\phi_v(d/dt|_0) = v$).

One defines the *exponential map*

$$e\colon V \to X \tag{6.7}$$

by

$$e(v) = \phi_v(1). \tag{6.8}$$

The two maps $\mathbb{C} \to X$ given by

$$t \mapsto \phi_{sv}(t) \quad \text{and} \quad t \mapsto \phi_v(st) \qquad (s \text{ fixed}), \tag{6.9}$$

are holomorphic homomorphisms with derivatives at zero equal to sv. Thus, by uniqueness,

$$\phi_{sv}(t) = \phi_v(st), \tag{6.10}$$

and therefore

$$e(sv) = \phi_v(s). \tag{6.11}$$

Consider the map

$$f\colon \mathbb{C} \to X \tag{6.12}$$

given by

$$f(t) = e(tx) + e(ty) = \phi_x(t) + \phi_y(t), \tag{6.13}$$

for some fixed $x, y \in X$. Then

$$f(t_1+t_2) = e\big((t_1+t_2)x\big) + e\big((t_1+t_2)y\big) \tag{6.14}$$
$$= \phi_x(t_1+t_2) + \phi_y(t_1+t_2)$$
$$= \phi_x(t_1) + \phi_x(t_2) + \phi_y(t_1) + \phi_y(t_2)$$
$$= f(t_1) + f(t_2).$$

Therefore, f is a holomorphic homomorphism with derivative at 0 equal to $x+y$ (use the chain rule). Thus, $f(t) = \phi_{x+y}(t)$. Plugging in $t = 1$, we get

$$e(x) + e(y) = e(x+y). \tag{6.15}$$

Therefore, $e\colon V \to X$ is a homomorphism.

The image of e is an open subgroup (it contains a neighborhood of 1) and since X is connected, e is surjective. Having X compact gives us that the kernel is a lattice Λ, because lattices are the only discrete subgroups of vector spaces with compact quotient. □

COROLLARY 6.4. *Every abelian variety over $\mathbb{C}$ is a complex torus, hence the group law is commutative.*

COROLLARY 6.5. *Let X be a complex torus of dimension g and let $X[n]$ be the subgroup of n torsion points of X, i.e., $X[n] := \{x : x \in X, nx = 0\}$. Then*

$$X[n] \cong \frac{1}{n}\Lambda/\Lambda \cong (\mathbb{Z}/n\mathbb{Z})^{2g}. \tag{6.16}$$

COROLLARY 6.6. *Once we fix isomorphisms $\Lambda_i \cong \mathbb{Z}^{2g_i}$, we have an isomorphism of groups*

$$\operatorname{Hom}(X_1, X_2) \cong \{M \in M_{g_2\times g_1}(\mathbb{C}) : M\Lambda_1 \subset \Lambda_2\} \hookrightarrow \{M \in M_{2g_2\times 2g_1}(\mathbb{Z})\}. \tag{6.17}$$

In particular, $\operatorname{Hom}(X_1, X_2)$ is a torsion free abelian group of rank $\leq 4g_1g_2$.

EXERCISE 6.7. (1) *Let $\Lambda = \mathbb{Z} + \mathbb{Z}\tau$ be a lattice in $\mathbb{C}$ such that the endomorphism ring of $\mathbb{C}/\Lambda$ strictly contains $\mathbb{Z}$. Prove that $\mathbb{Q}(\tau)$ is an imaginary quadratic extension of $\mathbb{Q}$. Prove the converse.*

(2) *Let A be an abelian variety of dimension g and let f be an endomorphism of A. Prove that $\mathbb{Q}(f) \subset \operatorname{End}(A) \otimes \mathbb{Q}$ is an algebra of degree $\leq 2g$. Let $\mathbb{Q}(f)$ be a field of degree r. Show that there exists an abelian variety of dimension $\leq r$ such that $\mathbb{Q}(f) \subset \operatorname{End}(A) \otimes \mathbb{Q}$. In general, one cannot improve this dimension as the example of a real quadratic field shows.*

COROLLARY 6.8. *Complex compact connected g-dimensional Lie groups are parameterized by:*

$$\operatorname{GL}_g(\mathbb{C})\backslash \operatorname{GL}_{2g}(\mathbb{R})/\operatorname{GL}_{2g}(\mathbb{Z}). \tag{6.18}$$

PROOF. Any two lattices in $\mathbb{R}^{2g} \cong \mathbb{C}^g$ are equivalent under $\operatorname{GL}_{2g}(\mathbb{R})$. The equivalence is obtained by first choosing a basis for the lattice (this gives the matrix in $\operatorname{GL}_{2g}(\mathbb{R})$). Changing the choice of basis amounts to moding out by $\operatorname{GL}_{2g}(\mathbb{Z})$. Thus, $\operatorname{GL}_{2g}(\mathbb{R})/\operatorname{GL}_{2g}(\mathbb{Z})$ is in bijection with lattices in $\mathbb{R}^{2g}$. Moding out by $\operatorname{GL}_g(\mathbb{C})$ (embedded in $\operatorname{GL}_{2g}(\mathbb{R})$ by the isomorphism $\mathbb{R}^{2g} \cong \mathbb{C}^g$) amounts to the fact that two complex tori $\mathbb{C}^g/\mathcal{L}_1$ and $\mathbb{C}^g/\mathcal{L}_2$ are isomorphic iff there exists a matrix $M \in \operatorname{GL}_g(\mathbb{C})$ such that $M\mathcal{L}_1 = \mathcal{L}_2$. □

EXERCISE 6.9. *Exhibit a bijection between the sets* $\mathrm{GL}_1(\mathbb{C})\backslash \mathrm{GL}_2(\mathbb{R})/\mathrm{GL}_2(\mathbb{Z})$ *and* $\mathrm{SL}_2(\mathbb{Z})\backslash\mathcal{H}$.

REMARK 6.10. While every g-dimensional abelian variety over $\mathbb{C}$ is a complex torus the converse does not hold if $g > 1$. In fact, the moduli of complex tori is g^2-dimensional, while that of abelian varieties is $g(g+1)/2$-dimensional. We remarked before (see §2) that the moduli of curves of genus g is $3g-3$ for $g > 1$. We sketch here the argument for abelian varieties following the same reasoning of Kodaira–Spencer we used for curves.

The Kodaira–Spencer theory dictates that the universal $k[\epsilon]$-deformation, i.e., first order deformation, of an abelian variety X is given by $H^1(X, \mathfrak{T}_X)$, where $\mathfrak{T}_X$ is the tangent sheaf to X, i.e., the sheaf of derivations. Moreover, the cup product map "evaluating the differential on a derivation"

$$H^1(X, \mathfrak{T}_X) \times H^0(X, \Omega^1_X) \to H^1(X, \mathcal{O}_X), \tag{6.19}$$

yields an isomorphism

$$H^1(X, \mathfrak{T}_X) = \mathrm{Hom}\big(H^0(X, \Omega^1_X), H^1(X, \mathcal{O}_X)\big). \tag{6.20}$$

Now, for an abelian variety one can identify $H^0(X, \Omega^1_X)$ with the cotangent space at zero $\mathfrak{t}^*_X$, and $H^1(X, \mathcal{O}_X)$ with the tangent space of the dual abelian variety $\mathfrak{t}_{X^t}$ discussed extensively below (§6.2). Thus,

$$H^1(X, \mathfrak{T}_X) = \mathfrak{t}_X \otimes \mathfrak{t}_{X^t}. \tag{6.21}$$

This shows that the first order deformations are of dimension g^2.

Moreover, if X is principally polarized (§6.2)[4] then this polarization λ gives an isomorphism λ_* from $\mathfrak{t}_X$ to $\mathfrak{t}_{X^t}$. The theory then identifies the deformations preserving the polarization with the symmetric tensors of $\mathfrak{t}_X \otimes \mathfrak{t}_{X^t}$ (under the involution induced by $v \otimes w \mapsto \lambda_*^{-1}(w) \otimes \lambda_*(v)$) giving $g(g+1)/2$ moduli.

EXERCISE 6.11. *Deduce from Corollary 6.4 that if a complex compact curve is a complex Lie group then it must be of genus 1. Prove, using just topological tools, that a closed real two-dimensional manifold is a topological group if and only if it has genus 1.*

EXERCISE 6.12. *Prove that any holomorphic differential on an abelian variety* X/k *is translation invariant (and hence non-vanishing). Conclude the identification* $H^0(X, \Omega^1_{X/k}) = \mathfrak{t}^*_X$ *used above.*

6.1. The Appell–Humbert Theorem. Recall Chow's theorem:

THEOREM 6.13. *Let P be a complete algebraic variety over $\mathbb{C}$ and $\mathcal{Z} \subset P$ a closed analytic subset of P; then there is an algebraic subvariety Z of P such that $Z(\mathbb{C}) \cong \mathcal{Z}$, as complex analytic spaces.*

PROOF. See [**56**, p. 167]. □

DEFINITION 6.14. Let X be a complete analytic variety. A line bundle $\mathcal{L} \to X$ is *ample* if $\exists k > 0$ such that $\mathcal{L}^{\otimes k}$ is *very ample*, i.e., if we have an embedding:

$$\begin{aligned} X &\hookrightarrow \mathbb{P}\big(H^0(X, \mathcal{L}^{\otimes k})\big), \\ x &\mapsto \{\text{hyperplane of sections vanishing at } x\}. \end{aligned} \tag{6.22}$$

[4]More generally, in any characteristic in fact, if X has a polarization that is a separable map.

If one chooses a basis $s_0, \ldots, s_n$ to $\Gamma(X, \mathcal{L}^{\otimes k})$, we can also write such an embedding as

$$x \mapsto \big(s_0(x) : \cdots : s_n(x)\big). \tag{6.23}$$

COROLLARY 6.15. *A complex torus is an abelian variety if and only if it has an ample line bundle* $\mathcal{L}$.

We would like to understand when is the complex torus $\mathbb{C}^g/\Lambda$ an abelian variety. Given the corollary, the main problem we are faced with is:

When does $\mathbb{C}^g/\Lambda$ *have an ample line bundle? and even more, what are all the line bundles on* $\mathbb{C}^g/\Lambda$?

The Appell–Humbert Theorem supplies a complete answer to those questions.

First, let us make a digression on *factors of automorphy* (compare with §4): Let $X = \mathbb{C}^g/\Lambda$ be a complex torus, $\pi\colon \mathbb{C}^g \to X$ the natural projection and $\mathcal{L}$ a line bundle on X. Because $\mathbb{C}^g$ is simply connected every line bundle on it is trivial, and therefore $\pi^*\mathcal{L} \cong \mathbb{C}^g \times \mathbb{C}$. Let T_λ be the translation-by-λ map on $\mathbb{C}^g$ given by $T_\lambda(x) = x + \lambda$. For every $\lambda \in \Lambda$, we have

$$T_\lambda^*\pi^*\mathcal{L} = (\pi \circ T_\lambda)^*\mathcal{L} = \pi^*\mathcal{L}. \tag{6.24}$$

But

$$\big(T_\lambda^*(\pi^*\mathcal{L})\big)_x = \pi^*\mathcal{L}_{x+\lambda}, \tag{6.25}$$

by definition of the pullback of a line bundle. This implies the existence of holomorphic non-vanishing maps, one for each $\lambda \in \Lambda$,

$$j(\lambda)\colon \mathbb{C}^g \to \mathbb{C}^\times, \tag{6.26}$$

such that

$$j(\lambda_1)(\lambda_2 + v) \cdot j(\lambda_2)(v) = j(\lambda_1 + \lambda_2)(v) \tag{6.27}$$

(it is the same compatibility requirement as in §4). We let Λ act on holomorphic functions f by ${}^\lambda f(v) = f(\lambda + v)$. We see that

$$j(\lambda_1 + \lambda_2) = {}^{\lambda_2}j(\lambda_1) \cdot j(\lambda_2). \tag{6.28}$$

It follows that

$$j \in Z^1\big(\Lambda, \mathcal{O}^\times(\mathbb{C}^g)\big), \tag{6.29}$$

where $\mathcal{O}^\times(\mathbb{C}^g)$ denotes the group of non-vanishing holomorphic functions on $\mathbb{C}^g$. Conversely, any $j \in Z^1\big(\Lambda, \mathcal{O}^\times(\mathbb{C}^g)\big)$ induces a line bundle on $\mathbb{C}^g/\Lambda$ whose isomorphism class depends only on the image of j in $H^1\big(\Lambda, \mathcal{O}^\times(\mathbb{C}^g)\big)$. One can show the following

PROPOSITION 6.16. *There exist canonical isomorphisms*:

$$\begin{aligned} (6.30)\quad \mathrm{Pic}(X) &:= \{\textit{isomorphism classes of line bundles on } X\} \\ &\cong H^1(X, \mathcal{O}_X^\times) && (\textit{Sheaf cohomology}) \\ &\cong H^1\big(\Lambda, \mathcal{O}^\times(\mathbb{C}^g)\big). && (\textit{Group cohomology}) \end{aligned}$$

PROOF. See [**90**, p. 24]. □

Thus, we are only left with the task of describing this latter group!

DEFINITION 6.17. A *Riemann form* on $\mathbb{C}^g$ *with respect to* Λ is a hermitian form $H\colon \mathbb{C}^g \times \mathbb{C}^g \to \mathbb{C}$ such that $E = \operatorname{Im} H$ (meaning $E(u,v) = \operatorname{Im}\big(H(u,v)\big)$) satisfies

$$E\colon \Lambda \times \Lambda \to \mathbb{Z}. \tag{6.31}$$

Recall that a hermitian form is $\mathbb{C}$-linear in the first variable, $\overline{\mathbb{C}}$-linear in the second variable, and $H(v,w) = \overline{H(w,v)}$.

EXERCISE 6.18. *Let τ be a complex symmetric $g \times g$ matrix such that* $\operatorname{Im}(\tau)$ *is positive definite. Define a bilinear form $H(x,y)$ on $\mathbb{C}^g \times \mathbb{C}^g$ by*

$$H(x,y) = {}^t x \big(\operatorname{Im}(\tau)\big)^{-1} \bar{y}. \tag{6.32}$$

Prove that this is a Riemann form with respect to the lattice spanned by the columns of the $2g \times g$ matrix $(I\ \tau)$. Calculate that the matrix of $E(x,y) = \operatorname{Im}\big(H(x,y)\big)$, *with respect to the basis for the lattice given by the columns of $(I\ \tau)$, is* $\left(\begin{smallmatrix} 0 & -I_g \\ I_g & 0 \end{smallmatrix}\right)$.

DEFINITION 6.19. A *semi-character* with respect to a Riemann form H is a map

$$\chi\colon \Lambda \to \mathbb{C}_1 := \{z \in \mathbb{C} : |z| = 1\} \tag{6.33}$$

such that

$$\chi(\lambda_1 + \lambda_2) = \chi(\lambda_1)\chi(\lambda_2)\exp\big(\pi i E(\lambda_1, \lambda_2)\big). \tag{6.34}$$

Note that although χ is not necessarily a character, χ^2 *is* a character.

EXERCISE 6.20. (1) *Let Λ be a lattice in $\mathbb{C}$. Find all Riemann forms and quasi-characters on Λ. In particular, deduce that a non-trivial Riemann form always exists.*
(2) *Find a lattice in $\mathbb{C}^2$ that has no non-trivial Riemann form.*

We define a group structure on the set

$$\mathcal{G} = \{(H,\chi) : H \text{ is a Riemann form with respect to } \Lambda,\ \chi \text{ is a semi-character with respect to } H\} \tag{6.35}$$

by the rule

$$(H_1, \chi_1) + (H_2, \chi_2) = (H_1 + H_2, \chi_1\chi_2). \tag{6.36}$$

An element $(H,\chi) \in \mathcal{G}$ induces a line bundle on $\mathbb{C}^g/\Lambda$ denoted $\mathcal{L}_{(H,\chi)}$, defined by the factor of automorphy $j_{(H,\chi)}$:

$$j_{(H,\chi)}(\lambda)(v) = \chi(\lambda)\exp\Big(\pi H(v,\lambda) + \frac{\pi}{2} H(\lambda,\lambda)\Big), \quad v \in \mathbb{C}^g, \lambda \in \Lambda. \tag{6.37}$$

THEOREM 6.21 (Appell–Humbert). *There exists an isomorphism of groups,*

$$\mathcal{G} \cong \operatorname{Pic}(X), \quad (H,\chi) \mapsto \mathcal{L}_{(H,\chi)}. \tag{6.38}$$

In particular,

$$\mathcal{L}_{(H_1,\chi_1)} \otimes \mathcal{L}_{(H_2,\chi_2)} \cong \mathcal{L}_{(H_1+H_2,\chi_1\chi_2)}. \tag{6.39}$$

Moreover, $\mathcal{L}_{(H,\chi)}$ is ample if and only if H is positive definite.

PROOF. See [**90**, p. 32 and Proposition 5.2, p. 86]. □

COROLLARY 6.22. *A complex torus $X = \mathbb{C}^g/\Lambda$ is an abelian variety if and only if there exists a positive definite Riemann form H with respect to Λ.*

REMARK 6.23. (1) One can show that E is alternating (i.e., skew symmetric) and satisfies the identity

$$E(ix, iy) = E(x, y). \tag{6.40}$$

Conversely, given an alternating form E satisfying Equations (6.31) and (6.40) it is easy to check that the function $H(x, y) = E(ix, y) + iE(x, y)$ is a Riemann form.

(2) One may obtain semi-characters as follows: Assume that E is non-degenerate. It is well-known (Elementary Divisors Theorem) that there exists a basis for Λ

$$x_1, \ldots, x_g, y_1, \ldots, y_g, \tag{6.41}$$

such that E is given by the matrix:

$$\begin{pmatrix} 0 & D \\ -D & 0 \end{pmatrix}, \tag{6.42}$$

where 0 stands for the $g \times g$ zero matrix and D is a $g \times g$ diagonal matrix with diagonal entries $d_i \in \mathbb{N}^+$, such that $d_1 \mid d_2 \mid \cdots \mid d_g$. Put

$$\Lambda_1 = \mathrm{Span}_{\mathbb{Z}}\{x_i : i = 1, \ldots, g\}, \quad \Lambda_2 = \mathrm{Span}_{\mathbb{Z}}\{y_i : i = 1, \ldots, g\}. \tag{6.43}$$

Clearly

$$\Lambda = \Lambda_1 \oplus \Lambda_2. \tag{6.44}$$

Then $\mathbb{C}^g = \mathbb{R}\Lambda_1 \oplus \mathbb{R}\Lambda_2$, and we write the decomposition of a vector $v \in \mathbb{C}^g$ accordingly as

$$v = v_1 + v_2. \tag{6.45}$$

Put

$$\chi(v) = \exp[\pi i \cdot E(v_1, v_2)]. \tag{6.46}$$

Then

$$\begin{aligned} \chi(v + w) &= \exp[\pi i \cdot E(v_1 + w_1, v_2 + w_2)] \\ &= \chi(v) \cdot \chi(w) \cdot \exp\big[\pi i\big(E(v_1, w_2) + E(w_1, v_2)\big)\big] \\ &= \chi(v) \cdot \chi(w) \cdot \exp\big[\pi i\big(E(v, w)\big)\big], \end{aligned} \tag{6.47}$$

because Λ_1, Λ_2 are isotropic with respect to E.

REMARK 6.24. Given $\mathcal{L} \in \mathrm{Pic}(X)$, we may ask what is the associated Riemann or alternating form?

Consider the exact sequence of sheaves on X:

$$0 \to \mathbb{Z} \to \mathcal{O} \xrightarrow{e^{2\pi i(-)}} \mathcal{O}^\times \to 0. \tag{6.48}$$

The associated cohomology sequence gives a map

$$\mathrm{Pic}(X) := H^1(X, \mathcal{O}_X^\times) \to H^2(X, \mathbb{Z}). \tag{6.49}$$

Under this map, the line bundle $\mathcal{L}$ is sent to an element $c_1(\mathcal{L}) \in H^2(X, \mathbb{Z})$ called the *first Chern class* of $\mathcal{L}$. However,

$$H^2(X, \mathbb{Z}) = \wedge^2 H^1(X, \mathbb{Z}). \tag{6.50}$$

That is, if $X = \mathbb{C}^g/\Lambda$ then $H^2(X, \mathbb{Z})$ consists of alternating $\mathbb{Z}$-valued bilinear forms on $H_1(X, \mathbb{Z}) = \Lambda$. In particular, regarding $c_1(\mathcal{L})$ as a bilinear form E, we have the relation that E is precisely $\mathrm{Im}\, H$ where H is the Riemann form associated to $\mathcal{L}$.

EXERCISE 6.25. *Let M be a free $\mathbb{Z}$ module of finite rank. Let $T = M \otimes M$ and define the antisymmetrization map $a\colon T \to T$ by extending linearly the map $x \otimes y \mapsto x \otimes y - y \otimes x$. Prove that $\operatorname{Ker}(a) = \langle x \otimes x : x \in M \rangle = T^{\mathrm{sym}}$, where T^{sym} consists of the symmetric elements of M (the elements fixed under the linear map s induced from $x \otimes y \mapsto y \otimes x$). Prove that the image of a is T^{asym}—the antisymmetric elements (elements f such that $s(f) = -f$). Remark that although $T^{\mathrm{sym}} \cap T^{\mathrm{asym}} = \{0\}$ and $T^{\mathrm{sym}} \oplus T^{\mathrm{asym}}$ is of finite index in T, this index is a positive power of 2.*

Recall that by definition, $\bigwedge^2 M = T/\langle x \otimes x : x \in M \rangle$. Conclude a canonical isomorphism

$$\bigwedge^2 M \cong (M \otimes M)^{\mathrm{asym}}. \tag{6.51}$$

Suppose now that $M = \operatorname{Hom}(N, \mathbb{Z})$. Then $M \otimes M$ is canonically identified with the bilinear forms $N \times N \to \mathbb{Z}$, and $(M \otimes M)^{\mathrm{asym}}$ corresponds to the alternating forms. Conclude that $\bigwedge^2 M$ may be identified with the alternating bilinear forms $N \times N \to \mathbb{Z}$.

6.2. The dual abelian variety.

DEFINITION 6.26. Let X be a complex torus. The Néron–Severi group of X is defined as the image of the homomorphism $H^1(X, \mathcal{O}_X^\times) \to H^2(X, \mathbb{Z})$. It is denoted by $\mathrm{NS}(X)$. We put $\mathrm{NS}^0(X) = \mathrm{NS}(X) \otimes \mathbb{Q}$.

REMARK 6.27. The Néron–Severi group is also the group of divisors on X modulo algebraic equivalence.[5] Using our knowledge on complex tori we see that for $X = V/\Lambda$, $\mathrm{NS}(X)$ can be identified with the group of Riemann forms H with respect to Λ. Since it is a subgroup of $H^2(X, \mathbb{Z}) = H^1(X, \mathbb{Z}) \wedge H^1(X, \mathbb{Z})$ (via $H \mapsto \operatorname{Im}(H)$), it is a free $\mathbb{Z}$-module of rank $\leq \binom{2g}{2}$ where $g = \dim(X)$. We let $\mathrm{NS}(X)^+$ denote the elements corresponding to positive definite Riemann forms with respect to Λ.

DEFINITION 6.28. The subgroup of $\operatorname{Pic}(X)$ given by

$$\{(0, \chi) \mid \chi\colon \Lambda \to \mathbb{C}_1 \text{ is a homomorphism}\} \tag{6.52}$$

is called $\operatorname{Pic}^0(X)$. We ease notation and write $\operatorname{Pic}^0(X) = \{\chi \mid \chi\colon \Lambda \to \mathbb{C}_1\}$.

These groups are related by an exact sequence:

$$0 \to \operatorname{Pic}^0(X) \to \operatorname{Pic}(X) \to \mathrm{NS}(X) \to 0. \tag{6.53}$$

The group $\operatorname{Pic}^0(X)$ carries a canonical complex structure, as we shall see shortly. With this canonical complex structure we call it the *dual abelian variety* and denote it by X^t or $X^\vee$. Consider the complex torus $X = V/\Lambda$, where V is a g-dimensional vector space, and Λ is a full lattice. Put

$$\Omega = \Omega(X) = \operatorname{Hom}_{\overline{\mathbb{C}}}(V, \mathbb{C}), \tag{6.54}$$

where $\overline{\mathbb{C}}$ means $\mathbb{C}$-conjugate-linear; Ω is a complex vector space of dimension g and

$$f \in \Omega \iff f(v_1) + f(v_2) = f(v_1 + v_2), \quad f(\lambda v) = \bar{\lambda} f(v). \tag{6.55}$$

The pairing

$$\Omega \times V \to \mathbb{C}, \quad (f, v) \mapsto f(v) \tag{6.56}$$

[5]This allows one to define the Néron–Severi group of every variety. It need not be torsion free.

is bi-additive, $\mathbb{C}$-linear in the first variable, and $\overline{\mathbb{C}}$-linear in the second variable. We have an isomorphism

$$\Omega \xrightarrow{\cong} \operatorname{Hom}_{\mathbb{R}}(V, \mathbb{R}) \tag{6.57}$$

given by

$$\ell \mapsto \operatorname{Im} \ell. \tag{6.58}$$

The inverse is given by $k \mapsto \ell$, where $\ell(v) = -k(iv) + ik(v)$. Thus,

$$\Omega \times V \to \mathbb{R}, \quad (\ell, v) \mapsto \operatorname{Im}\big(\ell(v)\big) \tag{6.59}$$

is a perfect $\mathbb{R}$-linear pairing.

Put

$$\hat{\Lambda} = \{\ell \in \Omega : \operatorname{Im} \ell(v) \in \mathbb{Z} \quad \forall v \in \Lambda\}. \tag{6.60}$$

It is the $\mathbb{Z}$-dual of Λ with respect to the real pairing between Ω and V. Then $\hat{\Lambda}$ is the kernel of the surjective map

$$\Omega \to \operatorname{Pic}^0(X), \quad \ell \mapsto e^{2\pi i \cdot \operatorname{Im} \ell(\cdot)}. \tag{6.61}$$

We get

$$\Omega/\hat{\Lambda} \cong \operatorname{Pic}^0(X). \tag{6.62}$$

Therefore $\operatorname{Pic}^0(X)$ has a canonical complex structure induced from that of Ω.

Before giving a theorem linking X to $X^\vee$, let us give a few definitions. Let $\mathcal{L}$ be a line bundle on X. Define

$$\phi_{\mathcal{L}}(x) = T_x^*\mathcal{L} \otimes \mathcal{L}^{-1}, \tag{6.63}$$

where

$$T_x : X \to X, \quad T_x(y) = x + y, \tag{6.64}$$

is the translation-by-x morphism and T_x^* denotes the pullback of the line bundle $\mathcal{L}$ by T_x. Recall that $\mathcal{L}_{(H,\chi)}^{-1} = \mathcal{L}_{(-H,\chi^{-1})}$. One can show:

$$T_x^*\mathcal{L}_{(H,\chi)} \cong \mathcal{L}_{(H,\chi\mu_x)}, \tag{6.65}$$

with $\mu_x(\lambda) = e^{2\pi i E(x,\lambda)}$, $\lambda \in \Lambda$. Here, we extend E to V by $\mathbb{R}$-linearity.

EXERCISE 6.29. *Prove the identity* (6.65) *by comparing the factors of automorphy of both sides.*

Thus, $\phi_{\mathcal{L}} : X \to \operatorname{Pic}^0(X)$ is given by a closed formula

$$\phi_{\mathcal{L}}(x) = e^{2\pi i E(x,-)} \in \operatorname{Hom}(\Lambda, \mathbb{C}_1). \tag{6.66}$$

COROLLARY 6.30. *The map $\phi_{\mathcal{L}}$ is surjective if and only if E is non-degenerate.*

DEFINITION 6.31. A *polarization* of X is a homomorphism $f : X \to X^\vee$ such that $f = \phi_{\mathcal{L}}$ for some ample line bundle $\mathcal{L}$.

REMARK 6.32. (1) A polarization is an *isogeny*, that is, it is a surjective homomorphism of abelian varieties with finite kernel. The *degree* of the polarization is the order of the kernel. It is its degree as a finite morphism.

(2) A polarization is an isogeny, but the converse is not true. Indeed, if ϕ is a polarization then $-\phi$ is an isogeny that is never a polarization.

EXERCISE 6.33. *Prove the last remark.*

DEFINITION 6.34. A polarization is *principal* if it is an isomorphism.

EXERCISE★ 6.35. *Find an abelian variety having no principal polarization.*

EXERCISE 6.36. *Let* $\mathcal{L} = \mathcal{L}_{(H,\chi)}$ *and* $E = \mathrm{Im}(H)$. *Prove that* $\phi_{\mathcal{L}}$ *is a principal polarization iff* $\det(E) = 1$.

We remark that for any abelian variety A, we can always find a principally polarized abelian variety B such that A is isogenous to B. This follows quite easily from the theory we have developed above and the reader may enjoy proving it. In the case of an arbitrary base field this follows from the properties of the finite Heisenberg group (see Chapter 3, §2) associated by Mumford to $(X, \mathcal{L})$.

One can parameterize abelian varieties, endowed with a rigid level structure and a polarization, by a fine moduli scheme. Let $\mathcal{H}_g$ be Siegel's upper half space

$$\{\tau \in M_g(\mathbb{C}) : \tau = \tau^t, \mathrm{Im}(\tau) \gg 0\}. \tag{6.67}$$

It turns out that the coarse moduli scheme of couples (A, λ), A a g-dimensional complex abelian variety, λ a principal polarization of A, is $\mathrm{Sp}(2g, \mathbb{Z})\backslash\mathcal{H}_g$. The action of a matrix $\left(\begin{smallmatrix} A & B \\ C & D \end{smallmatrix}\right) \in \mathrm{Sp}(2g, \mathbb{Z})$ is given by $\tau \mapsto (A\tau + B)(C\tau + D)^{-1}$. Cf. Exercise 6.18.

REMARK 6.37. The symmetry condition on the matrix τ is known as Riemann's relations.

THEOREM 6.38. *Let* X *be a complex torus,* $X = V/\Lambda$. *Let* $X^\vee$ *be the dual complex torus with its canonical complex structure.*

(1) $X^{\vee\vee}$ *is canonically isomorphic to* X.

(2) $\mathrm{Hom}(X, X^\vee)$ *consists of bilinear forms* H *on* V *such that* $\mathrm{Im}\, H(\Lambda, \Lambda) \subseteq \mathbb{Z}$ ($\mathbb{C}$-*linear in the first variable, and* $\mathbb{C}$-*conjugate-linear in the second variable*).

(3) *The map* $\phi \in \mathrm{Hom}(X, X^\vee)$ *is of the form* $\phi_{\mathcal{L}}$, *for some* $\mathcal{L}$, *iff the associated form* H *is hermitian, i.e.,* $H(x, y) = \overline{H(y, x)}$. *It is a polarization iff* H *is a positive definite Riemann form with respect to* Λ.

(4) *Let* $f\colon X \to Y$ *be a homomorphism of complex tori with* $Y = V'/\Lambda'$. *Consider the following diagram:*

$$\begin{array}{ccc} \Omega(Y) & \xrightarrow{\widetilde{f^*}} & \Omega(X) \\ \downarrow & & \downarrow \\ \mathrm{Pic}^0(Y) & \xrightarrow[f^*]{} & \mathrm{Pic}^0(X), \end{array} \tag{6.68}$$

where $f^* = f^\vee$ *is the pullback of line bundles, and* $\widetilde{f^*}$ *the lifting to the universal covering space. Then* $\widetilde{f^*}$ *is just the pull-back*

$$f^*\colon \mathrm{Hom}_{\overline{\mathbb{C}}}(V', \mathbb{C}) \to \mathrm{Hom}_{\overline{\mathbb{C}}}(V, \mathbb{C}). \tag{6.69}$$

It follows that

$$(f + g)^\vee = f^\vee + g^\vee, \quad (fg)^\vee = g^\vee f^\vee, \quad f^{\vee\vee} = f. \tag{6.70}$$

(5) *Let $\Phi\colon X \to X^\vee$ be a polarization. The Weil pairing associated to Φ is defined as:*

$$X[n] \times X[n] \xrightarrow{1\times\Phi} X[n] \times X^\vee[n] \xrightarrow{\cong} \frac{1}{n}\Lambda/\Lambda \times \frac{1}{n}\hat{\Lambda}/\hat{\Lambda} \longrightarrow \mu_n, \tag{6.71}$$

via the map

$$\frac{1}{n}\Lambda/\Lambda \times \frac{1}{n}\hat{\Lambda}/\hat{\Lambda} \to \mu_n(\mathbb{C}) \qquad (x,k) \mapsto e^{2\pi i\cdot n\cdot k(x)}, \tag{6.72}$$

where we think of $\hat{\Lambda}$ as $\{k \in \Omega = \mathrm{Hom}_{\mathbb{R}}(V,\mathbb{R}) \mid k(\lambda) \in \mathbb{Z}\ \forall \lambda \in \Lambda\}$. The Weil pairing is an alternating bilinear pairing.

EXERCISE 6.39. (1) *Complete the proof of the above theorem. At this point it is mainly unfolding the definitions. We give a few pointers though.*

First note that $v \in V$ may be viewed as an element of $\mathrm{Hom}_{\overline{\mathbb{C}}}(\Omega,\mathbb{C})$. The action of v on $f \in \Omega$ is by sending it to $f(v)$. This gives canonically $V = \mathrm{Hom}_{\overline{\mathbb{C}}}(\Omega,\mathbb{C})$. To get (1) it remains to verify that Λ is the $\mathbb{Z}$ dual of $\hat{\Lambda}$. Now, by general facts on complex tori we identify $\mathrm{Hom}(X,X^\vee)$ first with the linear maps $f\colon V \to \Omega$ such that $f(\Lambda) \subset \hat{\Lambda}$ and then use $\Omega = \mathrm{Hom}_{\overline{\mathbb{C}}}(V,\mathbb{C})$. Part (2) follows, and part (3) is obtained by calculating the maps coming from line bundles.

Part (4) results from first calculating for a homomorphism $f\colon X \to Y$ the pull back $f^*L_{(H,\chi)}$ of the line bundle $L_{(H,\chi)}$ on Y. One proves that $f^*L_{(H,\chi)} = L_{(f^*H,f^*\chi)}$, where $f^*H(x,y) = H\big(f(x),f(y)\big)$ and $f^*\chi(x) = \chi\big(f(x)\big)$.

(2) *Prove that given $f\colon X \to X^\vee$ corresponding to a bilinear form $H(x,y)$, the dual homomorphism $f^\vee\colon X = X^{\vee\vee} \to X^\vee$ corresponds to the bilinear form $\overline{H(y,x)}$.*

(3) *Prove also that the Weil pairing is perfect iff $(\deg(\Phi),n) = 1$.*

DEFINITION 6.40. Let B be a semi-simple algebra over $\mathbb{Q}$. Let Tr denote its reduced trace to $\mathbb{Q}$. We recall its definition. First, if B is a simple algebra with center K then $[B:K] = d^2$ for some integer d. If $b \in B$ then the characteristic polynomial of b over K is of the form $f(t)^d$ where $f(t) = t^d + a_1t^{d-1} + \cdots + a_0$, $a_i \in K$. We define the reduced trace from B to K by $\mathrm{Tr}_K(b) = -a_1$. For any subfield $L \subseteq K$ we define the reduced trace from B to L by $\mathrm{Tr}_L(b) = \mathrm{Tr}_{K/L}\big(\mathrm{Tr}_K(b)\big)$. The case $L = \mathbb{Q}$ is a particular case and in this case we write Tr instead of $\mathrm{Tr}_{\mathbb{Q}}$. In the general case $B = B_1 \times \cdots \times B_r$ where the B_i are simple algebras. We define $\mathrm{Tr}(b) = \mathrm{Tr}(b_1,\dots,b_r) = \sum_{i=1}^r \mathrm{Tr}(b_i)$ (where $\mathrm{Tr}(b_i)$ is the reduced trace from B_i to $\mathbb{Q}$).

By an *involution* on B we understand an isomorphism $\rho\colon B \to B$, $x \mapsto \rho(x)$ of $\mathbb{Q}$ vector spaces such that $\rho(xy) = \rho(y)\rho(x)$. An element of B is called symmetric (with respect to ρ) if $x = \rho(x)$. When ρ is understood we shall also denote it by $x \mapsto x^*$.

An involution $x \mapsto x^*$ on B is a *positive involution* if $\mathrm{Tr}(xx^*) > 0$ for all $x \neq 0$. Given a positive involution $x \mapsto x^*$ we say that an element $b \in B$ is *positive* if $\mathrm{Tr}(xbx^*) > 0$ for any $x \neq 0$.

DEFINITION 6.41. Let λ be a polarization on an abelian variety A, and put $\mathrm{End}^0(A) = \mathrm{End}(A) \otimes \mathbb{Q}$. The *Rosati involution* associated to λ is the map

$$\begin{aligned} \mathrm{End}^0(A) &\to \mathrm{End}^0(A) \\ f &\mapsto f^* := \lambda^{-1} f^\vee \lambda. \end{aligned} \tag{6.73}$$

FACT 6.42. The Rosati involution is a positive involution on $\mathrm{End}^0(A)$. The fact that it is an involution follows from Exercise 6.39 above. See [**90**, Theorem 1.8, Chapter 5].

The semi-simple rational finite-dimensional algebras with positive involution were classified by Albert. Since every abelian variety A has some polarization, it follows that the endomorphism algebras of abelian varieties all fall into this category. We follow [**153**, §1]. See also [**129**] for results concerning endomorphism rings of abelian varieties.

Every division algebra over $\mathbb{Q}$ with a positive involution belongs to one of the following four types of algebras.

Type I: Totally real algebraic number field L.

Type II: Central simple algebra B over L such that the simple components of $B \otimes \mathbb{R}$ are all isomorphic to $M_2(\mathbb{R})$.

Type III: Central simple algebra B over L such that the simple components of $B \otimes \mathbb{R}$ are all isomorphic to the Hamilton quaternions over $\mathbb{R}$.

Type IV: Central simple algebra B over a totally imaginary quadratic extension of L.

Recall that the canonical involution σ is defined by $x \mapsto \sigma(x) = x - \mathrm{Tr}(x)$. The positive involutions are respectively:

Type I: The identity.

Type II: Let $a \in B$ be an element such that a^2 is a totally negative element of L, we let $\rho(x) = a\sigma(x)a^{-1}$. Then ρ is a positive involution and every positive involution ρ is obtained that way.

Type III: The canonical involution.

Type IV: This case is more subtle. We only remark that if B is a CM field containing L then the involution is complex conjugation.

Let $\lambda \colon A \to A^t$ be a polarization. Define a map

$$\begin{aligned} \mathrm{NS}^0(A) &\hookrightarrow \mathrm{End}^0(A), \\ \phi_\lambda \colon \mathcal{L}_{(H,\chi)} &\mapsto \lambda^{-1}\phi_L. \end{aligned} \tag{6.74}$$

FACT 6.43. The set $\phi_\lambda(\mathrm{NS}^0(A))$ consists of the symmetric elements of $\mathrm{End}^0(A)$ (under the Rosati involution defined by λ). We defined $\mathrm{NS}(A)^+$ in Remark 6.27 as the set of positive definite Riemann forms, and $\mathrm{NS}^0(A)^+$ is defined as the set of positive rational multiples of positive definite Riemann forms. Theorem 6.38 allows us to regard $\mathrm{NS}(A)^+$ as the set of polarizations of A, or ample line bundles on A up to algebraic equivalence. The set $\phi_\lambda(\mathrm{NS}^0(A)^+)$ turns out to be the set of positive symmetric elements of $\mathrm{End}^0(A)$. See [**90**, Chapter V, Proposition 2.1].

EXAMPLE 6.44 (Elliptic curves). We finish this section by examining some of the results discussed above in the particular case of elliptic curves.

Fix $\tau \in \mathcal{H}$ and let

$$\Lambda = \mathbb{Z} + \mathbb{Z}\tau. \tag{6.75}$$

Let $\mathcal{E}$ be the elliptic curve $\mathbb{C}/\Lambda$. Let $E\colon \Lambda \times \Lambda \to \mathbb{Z}$ be the real alternating pairing determined by $E(\tau, 1) = 1$. We have

$$E(a\tau + b, c\tau + d) = ad - bc. \tag{6.76}$$

How does one determine the associated hermitian form H? If $H(1,1) = \alpha$, then $H(z_1, z_2) = z_1\overline{z_2}\alpha$. Recall that $H(x,y) = E(ix, y) + iE(x,y)$, so

$$\begin{aligned}\alpha = H(1,1) &= E(i,1) + iE(1,1) \\ &= E(i,1) \\ &= E\left(\frac{1}{\operatorname{Im}(\tau)} \cdot \tau - \frac{\operatorname{Re}(\tau)}{\operatorname{Im}(\tau)}, 1\right) \\ &= E\left(\frac{1}{\operatorname{Im}(\tau)} \cdot \tau, 1\right) = \frac{1}{\operatorname{Im}(\tau)} > 0,\end{aligned} \tag{6.77}$$

since $\tau \in \mathcal{H}$ (Note also that $\operatorname{Im}(\tau) =$ area of a fundamental domain of the lattice Λ). Hence, H is a positive definite Riemann form with respect to Λ (cf. Exercise 6.18). Since every other hermitian form is proportional to H by a real scalar, such a form would be a Riemann form with respect to Λ if and only if this scalar is an integer, because the associated real form must be integer valued on $\Lambda \times \Lambda$. We deduce that H generates the Néron–Severi group of $\mathcal{E}$.

Let us calculate the dual lattice $\hat{\Lambda}$. Recall the procedure of defining the dual abelian variety. It is defined as a quotient of $\Omega = \operatorname{Hom}_{\overline{\mathbb{C}}}(\mathbb{C}, \mathbb{C})$ by a lattice dual to Λ. We use the identification

$$\mathbb{C} \to \operatorname{Hom}_{\overline{\mathbb{C}}}(\mathbb{C}, \mathbb{C}), \quad c \mapsto \ell_c, \tag{6.78}$$

where $\ell_c(1) = c$, and hence, $\ell_c(a + b\tau) = \overline{a + b\tau} \cdot c$. We write a general element in $\mathbb{C}$ as $\alpha + \beta\tau$, where α and β are real numbers, and obtain

$$\begin{aligned}\hat{\Lambda} &= \{c : \operatorname{Im}(\ell_c(\Lambda)) \subseteq \mathbb{Z}\} \\ &= \{c : \operatorname{Im}(\ell_c(1)), \operatorname{Im}(\ell_c(\tau)) \in \mathbb{Z}\} \\ &= \{c : \operatorname{Im}(c) \in \mathbb{Z}, \operatorname{Im}(\overline{\tau}c) \in \mathbb{Z}\} \\ &= \{\alpha + \beta\tau : \operatorname{Im}(\alpha + \beta\tau) = \beta \cdot \operatorname{Im}(\tau) \in \mathbb{Z}, \\ &\qquad \operatorname{Im}(\alpha\overline{\tau} + \beta\tau\overline{\tau}) = -\alpha \cdot \operatorname{Im}(\tau) \in \mathbb{Z}\} \\ &= \frac{1}{\operatorname{Im}(\tau)}\Lambda \cong \Lambda.\end{aligned} \tag{6.79}$$

Therefore, the dual elliptic curve is isomorphic to the original.

Choose any semi-character χ with respect to H. We have the polarization

$$\mathcal{E} \xrightarrow{\Phi_{\mathcal{L}(H,\chi)}} \mathcal{E}^\vee \tag{6.80}$$

under which $y \mapsto e^{2\pi i E(y,-)}$. Let us calculate the Weil pairing:

$$\frac{1}{n}\Lambda \times \frac{1}{n}\Lambda \to \frac{1}{n}\Lambda \times \frac{1}{n}\hat{\Lambda} \to \mu_n. \tag{6.81}$$

By definition,

$$(x,y) \mapsto (x, \Phi_{\mathcal{L}(H,\chi)}(y)) \mapsto \Phi_{\mathcal{L}(H,\chi)}(y)(nx), \tag{6.82}$$

and in particular,

$$(1/n, \tau/n) \mapsto (1/n, e^{2\pi i E(\tau/n, -)}) \mapsto e^{2\pi i E(\tau/n, 1)} = e^{2\pi i/n}. \tag{6.83}$$

CHAPTER 2

Complex Abelian Varieties with Real Multiplication and Hilbert Modular Forms

In this chapter we treat the theory of abelian varieties with real multiplication over the complex numbers. We fix a totally real field L of degree g and consider g-dimensional abelian varieties over $\mathbb{C}$ endowed with an action of the ring of integers $\mathcal{O}_L$ of L. Certain algebraic preliminaries are given in Section 1; notably, the notion of $\mathcal{O}_L$-module with positivity that intervenes in the definition of the moduli problem for abelian varieties with real multiplication and is designed to rigidify that moduli problem so it becomes (coarsely) representable.

Abelian varieties with real multiplication and polarization module are parameterized by quotients of $\mathcal{H}^g$ by suitable discrete subgroups of $\mathrm{PGL}_2(L)$. They could be thought of as parameterizing complex structures on full lattices Λ, stable under $\mathcal{O}_L$, in the $2g$-dimensional real vector space $(L\otimes\mathbb{R})^2$. Such a lattice Λ is a projective $\mathcal{O}_L$-module of rank 2 and has a Steinitz class $\bigwedge^2_{\mathcal{O}_L}\Lambda$ in the class group $\mathrm{Cl}(L)$. This already suggests that there are at least as many components to the moduli space as the class number. In fact, the right number is the strict class number, obtained by taking into account polarization.

We take a somewhat pedestrian approach to the construction of the moduli spaces. Instead of using adelic methods, we give a very explicit description of the moduli space and of the correspondence between its complex points and complex abelian varieties with real multiplication. Experience shows this is very useful. All this is done in Section 2.

In Section 3 we define Hilbert modular forms over the complex numbers and prove some of their basic properties, including the Köcher principle. This principle states that for $g > 1$, a modular form which is holomorphic on $\mathcal{H}^g$ modulo a discrete group Γ extends to a holomorphic modular form on the Satake, or minimal, compactification of $\Gamma\backslash\mathcal{H}^g$. This is explained geometrically by the fact that the compactification $\Gamma\backslash\mathcal{H}^{g*}$ is a normal space and the complement of $\Gamma\backslash\mathcal{H}^g$ in $\Gamma\backslash\mathcal{H}^{g*}$ is zero-dimensional.

The next section provides some examples of Hilbert modular forms. We discuss Eisenstein series and the method of pulling back modular forms from the Siegel space $\mathcal{H}_g$ via a modular "embedding." In this chapter we approach Hilbert modular forms very concretely as holomorphic functions on $\mathcal{H}^g$ enjoying certain transformation properties, though we also indicate how to view them as sections of line bundles. In Chapter 5 we discuss Hilbert modular forms systematically as sections of line bundles and obtain a re-interpretation of complex Hilbert modular forms as well.

As an example of a modular embedding we take in Section 5 the case of a quadratic field L and consider the diagonal curve in $\mathrm{SL}_2(\mathcal{O}_L)\backslash\mathcal{H}^2$, arising from the diagonal map $\mathcal{H} \to \mathcal{H}^2$. In the last section we apply the construction of the diagonal curve and the construction of Eisenstein series for Hilbert modular varieties to obtain Siegel's formula for $\zeta_L(-1)$.

1. Algebraic preliminaries

Let L be a field extension of $\mathbb{Q}$.

DEFINITION 1.1. A field L is a *totally real field* of degree $[L:\mathbb{Q}] = g$, if every complex embedding $\sigma \in \mathrm{Emb}(L, \mathbb{C}) = \{\sigma_1, \dots, \sigma_g\}$ factors via $\mathbb{R}$:

$$L \underset{\sigma_i}{\hookrightarrow} \mathbb{R}, \quad \forall i \in \{1, \dots, g\}. \tag{1.1}$$

We will assume henceforth that L is totally real. Here are some important examples:

(1) $L = \mathbb{Q}$;
(2) $L = \mathbb{Q}(\sqrt{D})$, for $D > 1$ square-free integer;
(3) $L = \mathbb{Q}(\zeta_m)^+$, i.e., the subfield of $\mathbb{Q}(\zeta_m)$ fixed under complex conjugation, where ζ_m is a primitive m-th root of unity.

An element $s \in L$ is called *totally positive* if $\sigma_i(s) > 0$ for all i. We use the notation $s \gg 0$. Given any set $S \subset L$ we denote by S^+ the set of totally positive elements of S.

We let $\mathcal{O}_L$ denote the ring of integers of L. We let $\mathcal{D}_L = \mathcal{D}_{L/\mathbb{Q}}$ denote the different ideal of L over $\mathbb{Q}$. Thus,

$$\mathcal{D}_{L/\mathbb{Q}}^{-1} = \{\ell \in L : \mathrm{Tr}_{L/\mathbb{Q}}(\ell r) \in \mathbb{Z} \ \forall r \in \mathcal{O}_L\}. \tag{1.2}$$

We let $d_L = \mathrm{Norm}(\mathcal{D}_{L/\mathbb{Q}})$ denote the discriminant of L.

Let $\mathrm{Cl}(L)$ stand for the ideal class group of L and let h denote its order. Let $\mathrm{Cl}(L)^+$ stand for the strict (or narrow) ideal class group in L and denote its order by h^+. Thus, if $\mathfrak{A}$, $\mathfrak{B}$ are two fractional ideals of L then $\mathfrak{A} = \mathfrak{B}$ in $\mathrm{Cl}(L)$ (resp. $\mathrm{Cl}(L)^+$) if and only if there exists $\lambda \in L^\times$ (resp. $\lambda \in L^{\times +}$) such that $\mathfrak{A} = \lambda\mathfrak{B}$.

Note the exact sequence

$$1 \to L^\times / \mathcal{O}_L^\times L^{\times +} \to \mathrm{Cl}(L)^+ \to \mathrm{Cl}(L) \to 1. \tag{1.3}$$

There is also an isomorphism $L^\times / L^{\times +} \to \{\pm 1\}^g$ taking an element ℓ to the signs of its embeddings $\big[\mathrm{sign}\big(\sigma_1(\ell)\big), \dots, \mathrm{sign}\big(\sigma_g(\ell)\big)\big]$. Thus h^+/h divides 2^g.

EXERCISE 1.2. *Show that* $h^+/h = [(\mathcal{O}_L^\times)^+ : (\mathcal{O}_L^\times)^2]$.

There is an equivalent way of defining $\mathrm{Cl}(L)^+$. Recall that $\mathrm{Cl}(L)$ is the group of projective $\mathcal{O}_L$-modules of rank 1 up to isomorphism. Indeed, every such module is isomorphic to a fractional ideal and that ideal is determined up to multiplication by $\lambda \in L^\times$. We define an *$\mathcal{O}_L$-module with a notion of positivity* to be a projective $\mathcal{O}_L$-module M of rank 1 together with an orientation on the real one-dimensional vector space $M \otimes_{\sigma_i} \mathbb{R}$ for every i, i.e., a choice of connected component of $M \otimes_{\sigma_i} \mathbb{R} - \{0\}$ (the positive component).

We may then talk about the group of isomorphism classes of projective $\mathcal{O}_L$-modules of rank 1 with a notion of positivity that we call temporarily $\mathrm{Cl}(L)^{++}$. An isomorphism of $\mathcal{O}_L$-modules M_1, M_2 with notion of positivity is an isomorphism of modules $\phi\colon M_1 \to M_2$ such that the induced maps $\phi \otimes_{\sigma_i} 1$ respect the orientation.

One verifies that the natural map $\mathrm{Cl}(L)^{++} \longrightarrow \mathrm{Cl}(L)$ has fibers that are principal homogeneous spaces under $L^\times/\mathcal{O}_L^\times L^{\times +}$ and that there is a natural map

$$\mathrm{Cl}(L)^+ \rightarrow \mathrm{Cl}(L)^{++}, \tag{1.4}$$

obtained by taking on a fractional ideal $\mathfrak{A}$ the natural orientation induced from the embeddings $L \xrightarrow{\sigma_i} \mathbb{R}^g$, where $i = 1, \ldots, g$. It is easily checked that this map is injective and this identifies $\mathrm{Cl}(L)^+$ with $\mathrm{Cl}(L)^{++}$.

We also remark that two notions of positivity on M are equal if and only if they define the same positive cone. The positive cone consists of the elements of M that are non-negative in any oriented space $M \otimes_{\sigma_i} \mathbb{R}$.

EXERCISE 1.3. *Recall that the class group $\mathrm{Cl}(L)$ of L is isomorphic to the Galois group of the maximal unramified abelian extension of L. Here unramified also means unramified at infinity. That is, the maximal abelian extension of L that is unramified at all finite primes, which is also totally real (so no archimedean valuation extends to a complex one). Prove that $\mathrm{Cl}(L)^+$ has the interpretation of the Galois group of the maximal abelian extension of L that is unramified at every finite prime.*

2. Complex abelian varieties with real multiplication

Let L be a totally real field of degree g over $\mathbb{Q}$.

DEFINITION 2.1. A complex abelian variety with *real multiplication* (abbreviated RM) by $\mathcal{O}_L$ is a g-dimensional abelian variety over $\mathbb{C}$ together with a *given* embedding of rings

$$\iota \colon \mathcal{O}_L \hookrightarrow \mathrm{End}(A). \tag{2.1}$$

EXAMPLE 2.2. (1) Let $L = \mathbb{Q}$. Elliptic curves, and only elliptic curves, satisfy the conditions of the definition. Thus, the theory of abelian varieties with real multiplication encompasses the theory of elliptic curves.

(2) Consider the product $E \times E$ and the totally real field $L = \mathbb{Q}(\sqrt{D})$ with $D > 1$ square free. The action of $\sqrt{D}$ (respectively $(1 + \sqrt{D})/2$) is given by a matrix of the form $\left(\begin{smallmatrix} a & b \\ c & -a \end{smallmatrix}\right)$ (resp. $\left(\begin{smallmatrix} (1+a)/2 & b/2 \\ c/2 & (1-a)/2 \end{smallmatrix}\right)$, and a is odd, b, c are even), when $D \equiv 2, 3 \bmod 4$ (resp. $D \equiv 1 \bmod 4$). Note that $D = a^2 + bc$.

(3) For any elliptic curve E, say over $\mathbb{C}$, consider:

$$E \otimes_{\mathbb{Z}} \mathcal{O}_L \cong E^g \tag{2.2}$$

with the canonical right $\mathcal{O}_L$ action; the isomorphism being obtained by choosing a $\mathbb{Z}$-basis to $\mathcal{O}_L$. We get an abelian variety over $\mathbb{C}$ such that

$$(E \otimes_{\mathbb{Z}} \mathcal{O}_L)(\mathbb{C}) = E(\mathbb{C}) \otimes_{\mathbb{Z}} \mathcal{O}_L. \tag{2.3}$$

EXERCISE 2.3. *Let $g = 2$. By a choice of basis for $\mathcal{O}_L$ over $\mathbb{Z}$ we get $E \otimes \mathcal{O}_L \cong E^2$ and the action of $\sqrt{D}$ is given by a matrix as in the previous example. Which matrices are obtained this way?*

(4) If A has RM by $\mathcal{O}_L$, so does the dual abelian variety $A^\vee$. That is, given a map

$$\iota \colon \mathcal{O}_L \rightarrow \mathrm{End}(A), \tag{2.4}$$

we define

$$\iota^\vee \colon \mathcal{O}_L \to \operatorname{End}(A^\vee) \tag{2.5}$$

by

$$\iota^\vee(\ell) = \iota(\ell)^\vee, \quad \ell \in \mathcal{O}_L, \tag{2.6}$$

where $\iota(\ell)^\vee$ is the dual map to $\iota(\ell)$.

(5) We say that a complex abelian variety is *simple* if it has no proper sub-abelian variety. For example, an elliptic curve is simple. By replacing the notion of isomorphism by the notion of isogeny the simple abelian varieties become the basic building blocks of the category of abelian varieties. More precisely:

THEOREM 2.4 (Poincaré). *The category of abelian varieties considered up to isogeny is semi-simple with simple objects given by simple abelian varieties, i.e., every abelian variety is isogenous to a product*

$$A_1^{n_1} \times \cdots \times A_k^{n_k}, \quad A_i \textit{ simple}, \tag{2.7}$$

and the n_i, A_i, are uniquely determined by A up to isogeny.

For the proof over the complex numbers see [**159**, Corollary 2, p. 58], or [**90**, Theorem (3.7), p. 128]. We remark that the same definition can be made over an arbitrary base field k. The simple objects are now abelian varieties that are defined and simple over k. Note that such an abelian variety may become reducible over $\overline{k}$. Poincaré's theorem still holds for the category of abelian varieties over k. See [**107**, Theorem 1, p. 173].

We say that A is *iso-simple* if A is isogenous to B^r for some simple abelian variety B. We shall see in Corollary 2.7 that every abelian variety with RM is iso-simple.

Let $J_0(p) = \operatorname{Jac}(X_0(p)) = \operatorname{Jac}(\Gamma_0(p)\backslash\mathcal{H}^*)$. Then it is a fact that the simple factors (the A_i's) of $J_0(p)$ are abelian varieties with real multiplication (by some field L_i). More generally, the new part of the Jacobian $J_0(N)$ decomposes over $\mathbb{Q}$ (up to isogeny) into a product of abelian varieties with RM.

2.1. Complex and rational representations. The embedding

$$\iota \colon \mathcal{O}_L \hookrightarrow \operatorname{End}(A), \tag{2.8}$$

can be extended by tensoring with $\mathbb{Q}$ to:

$$\iota \colon L \hookrightarrow \operatorname{End}^0(A) = \operatorname{End}(A) \otimes \mathbb{Q}. \tag{2.9}$$

Suppose that $A = \mathbb{C}^g/\Lambda$. Any endomorphism r of A induces an endomorphism $\tilde{r}$ on the universal covering space $\mathbb{C}^g$ such that the following diagram commutes:

$$\begin{array}{ccc} \mathbb{C}^g & \xrightarrow{\tilde{r}} & \mathbb{C}^g \\ \downarrow & & \downarrow \\ A & \xrightarrow{r} & A. \end{array} \tag{2.10}$$

We define the complex representation to be the map:

$$\rho_{\mathbb{C}} \colon \operatorname{End}^0(A) \to \operatorname{GL}_g(\mathbb{C}), \quad r \mapsto \tilde{r}. \tag{2.11}$$

We have an integral representation:

$$\rho_{\mathbb{Q}} \colon \operatorname{End}(A) \to \operatorname{End}_{\mathbb{Z}}(\Lambda) \cong M_{2g}(\mathbb{Z}). \tag{2.12}$$

The last isomorphism requires a choice of a $\mathbb{Z}$-basis for Λ. It gives the rational representation as the map

$$\rho_{\mathbb{Q}}\colon \operatorname{End}^0(A) \to \operatorname{End}_{\mathbb{Q}}(\Lambda \otimes \mathbb{Q}) \cong M_{2g}(\mathbb{Q}). \tag{2.13}$$

PROPOSITION 2.5. *The rational representation $\rho_{\mathbb{Q}}$ is conjugate to $\rho_{\mathbb{C}} \oplus \overline{\rho_{\mathbb{C}}}$.*

PROOF. Let $N = \rho_{\mathbb{C}}(r)$ be the complex representation of an endomorphism $r \in \operatorname{End}^0(A)$: Put $\operatorname{Re}(N) = (N + \overline{N})/2$, $\operatorname{Im}(N) = (N - \overline{N})/(2i)$. We have the matrix

$$M = \begin{pmatrix} \operatorname{Re}(N) & -\operatorname{Im}(N) \\ \operatorname{Im}(N) & \operatorname{Re}(N) \end{pmatrix}. \tag{2.14}$$

The matrix M is describing the action of $N = \rho_{\mathbb{C}}(r)$ with respect to the basis

$$\begin{pmatrix} 1 \\ \vdots \\ 0 \end{pmatrix}, \dots, \begin{pmatrix} 0 \\ \vdots \\ 1 \end{pmatrix}, \begin{pmatrix} i \\ \vdots \\ 0 \end{pmatrix}, \dots, \begin{pmatrix} 0 \\ \vdots \\ i \end{pmatrix}. \tag{2.15}$$

Since $\Lambda \otimes \mathbb{R} = \mathbb{C}^g$, M is $\rho_{\mathbb{Q}}(r)$ up to conjugation. Thus, we only need to show that M is conjugate to $\left(\begin{smallmatrix} N & 0 \\ 0 & \overline{N} \end{smallmatrix}\right)$. But

$$\begin{pmatrix} I & iI \\ iI & I \end{pmatrix} \cdot \begin{pmatrix} \operatorname{Re} N & -\operatorname{Im}(N) \\ \operatorname{Im}(N) & \operatorname{Re}(N) \end{pmatrix} \cdot \frac{1}{2} \begin{pmatrix} I & -iI \\ -iI & I \end{pmatrix} = \begin{pmatrix} N & 0 \\ 0 & \overline{N} \end{pmatrix}. \tag{2.16}$$ □

COROLLARY 2.6. *The tangent space $\mathfrak{t}_{A,0}$ to A at the origin is a free $\mathcal{O}_L \otimes \mathbb{C}$-module of rank* 1.[1]

PROOF. Note that the tangent space $\mathfrak{t}_{A,0}$ to A at the origin is canonically $\mathbb{C}^g$. Thus it is left to check that each σ_i appears exactly (equivalently, at least) once in the action of L on $\mathfrak{t}_{A,0}$. This action is described by the complex representation of L as a subspace of $\operatorname{End}^0(A)$. Since L is totally real the complex conjugate of the complex representation of L is equivalent to itself, and hence, by Proposition 2.5, the rational representation is equal to twice the complex representation. However, in the rational representation (being rational) every character σ_i has to appear with the same multiplicity. Hence, every σ_i appears in the complex representation as well. □

COROLLARY 2.7. *Any complex abelian variety with RM is iso-simple.*[2]

PROOF. Let B^r be a maximal iso-simple factor of A. Then $L \hookrightarrow \operatorname{End}^0(B^r)$ and the tangent space to B^r at the origin is a free $\mathcal{O}_L \otimes \mathbb{C}$-module. Hence $g \mid \dim(B^r)$. □

2.2. Construction of families of abelian varieties with real multiplication. Let us fix some notation. *For the rest of the chapter, fix fractional ideals $\mathfrak{A}, \mathfrak{B}$ of $\mathcal{O}_L$.* For $\ell \in L$ and $t = (t_1, \dots, t_g) \in \mathbb{C}^g$ we put

$$\ell \cdot t = \ell \cdot (t_1, \dots, t_g) = (\sigma_1(\ell)t_1, \dots, \sigma_g(\ell)t_g). \tag{2.17}$$

Given $z = (z_1, \dots, z_g) \in \mathcal{H}^g$, put

$$\begin{aligned} \Lambda_z &= \mathfrak{A} \cdot (z_1, \dots, z_g) + \mathfrak{B} \cdot (1, \dots, 1) \\ &= \left\{ \big(\sigma_1(a)z_1 + \sigma_1(b), \dots, \sigma_g(a)z_g + \sigma_g(b)\big) : a \in \mathfrak{A}, b \in \mathfrak{B} \right\} \subseteq \mathbb{C}^g. \end{aligned} \tag{2.18}$$

---

[1]In the context of this corollary, see Chapter 3, §5.

[2]In fact, this holds for every abelian variety with real multiplication.

Note that Λ_z is a lattice: First,

$$\operatorname{rank}_{\mathbb{Z}}(\Lambda_z) \leq \operatorname{rank}_{\mathbb{Z}} \mathfrak{A} + \operatorname{rank}_{\mathbb{Z}} \mathfrak{B} \leq 2g. \tag{2.19}$$

On the other hand,

$$\Lambda_z \otimes_{\mathbb{Z}} \mathbb{R} \cong \prod_{i=1}^{g} (\mathbb{R}z_i + \mathbb{R}) \cong \mathbb{C}^g, \tag{2.20}$$

since $\operatorname{Im}(z_i) > 0$. Thus Λ_z must be of rank $2g$ and must contain an $\mathbb{R}$-basis to $\mathbb{C}^g$. That is, Λ_z is a lattice.

We let A_z be the complex torus $\mathbb{C}^g/\Lambda_z$. We now proceed to construct polarizations on A_z. Using the connection between Riemann forms and polarizations (see Chapter 1, §6.1), we turn our attention to constructing certain $\mathbb{R}$-bilinear antisymmetric forms.

Let $r \in L$ and define

$$\begin{gathered} E_r \colon \mathfrak{A} \oplus \mathfrak{B} \times \mathfrak{A} \oplus \mathfrak{B} \to \mathbb{Q}, \\ E_r\big((x_1, y_1), (x_2, y_2)\big) = \operatorname{Tr}_{L/\mathbb{Q}}\big(r(x_1 y_2 - x_2 y_1)\big). \end{gathered} \tag{2.21}$$

The proof of the following lemma is left to the reader.

LEMMA 2.8. (1) *E_r is an alternating bilinear form.*
(2) *The image of E_r is in $\mathbb{Z}$ if and only if $r \in (\mathcal{D}_{L/\mathbb{Q}}\mathfrak{A}\mathfrak{B})^{-1}$.*

Note that for all $\ell \in \mathcal{O}_L$, $\alpha \in \mathfrak{A} \oplus \mathfrak{B}$ and $\beta \in \mathfrak{A} \oplus \mathfrak{B}$,

$$E_r(\ell\alpha, \beta) = E_r(\alpha, \ell\beta). \tag{2.22}$$

Transport E_r to Λ_z and extend $\mathbb{R}$-linearly to $\mathbb{C}^g$. We get an antisymmetric $\mathbb{R}$-bilinear pairing

$$E_{r,z} \colon \mathbb{C}^g \times \mathbb{C}^g \to \mathbb{R}, \tag{2.23}$$

with the property

$$E_{r,z}(\ell \cdot t, t') = E_{r,z}(t, \ell \cdot t'). \tag{2.24}$$

Moreover

$$E_{r,z}(\Lambda_z, \Lambda_z) \subseteq \mathbb{Z} \Leftrightarrow r \in (\mathcal{D}_{L/\mathbb{Q}}\mathfrak{A}\mathfrak{B})^{-1}. \tag{2.25}$$

If $r \neq 0$, $E_{r,z}$ is also perfect, by property of the trace. It also follows from the following:

LEMMA 2.9. *Let*

$$H_{r,z}\big((x_1, \ldots, x_g), (y_1, \ldots, y_g)\big) = \sum_{i=1}^{g} \frac{x_i \overline{y_i} \sigma_i(r)}{\operatorname{Im}(z_i)}. \tag{2.26}$$

Then $H_{r,z}$ is an hermitian form and $\operatorname{Im} H_{r,z} = E_{r,z}$.

PROOF. This form is clearly hermitian. To prove the last statement, it is enough to show

$$\operatorname{Im} H_{r,z}\big((x_1, \ldots, x_g), (y_1, \ldots, y_g)\big) = E_{r,z}\big((x_1, \ldots, x_g), (y_1, \ldots, y_g)\big), \tag{2.27}$$

for $(x_1, \dots, x_g) \in \mathfrak{A} \cdot (z_1, \dots, z_g)$ and $(y_1, \dots, y_g) \in \mathfrak{B} \cdot (1, \dots, 1)$. (Because $E_{r,z}$ and $\operatorname{Im} H_{r,z}$ are antisymmetric bilinear forms, they are determined by their values on such couples). But

$$\begin{aligned}(2.28)\quad \operatorname{Im}\{H([\sigma_1(a)z_1, \dots, \sigma_g(a)z_g], [\sigma_1(b), \dots, \sigma_g(b)])\} &= \sum_{i=1}^{g} \operatorname{Im}\left\{\frac{\sigma_i(a)z_i\overline{\sigma_i(b)}\sigma_i(r)}{\operatorname{Im}(z_i)}\right\} \\ &= \sum_{i=1}^{g} \sigma_i(a)\sigma_i(b)\sigma_i(r) \\ &= \operatorname{Tr}_{L/\mathbb{Q}}(abr) \\ &= E_r\big((a,0),(0,b)\big) \\ &= E_{r,z}\big(a \cdot (z_1, \dots, z_g), b \cdot (1, \dots, 1)\big). \qquad \square\end{aligned}$$

We note that if $(x_2, y_2) \in \mathfrak{A} \oplus \mathfrak{B}$ the set

$$(2.29)\qquad \{(x_1, y_1) \in L \oplus L : E_r\big((x_1, y_1), (x_2, y_2)\big) \in \mathbb{Z},\ \forall (x_2, y_2) \in \mathfrak{A} \oplus \mathfrak{B}\}$$

is precisely

$$(2.30)\qquad (r\mathcal{D}_{L/\mathbb{Q}}\mathfrak{B})^{-1} \oplus (r\mathcal{D}_{L/\mathbb{Q}}\mathfrak{A})^{-1}.$$

Hence, in the case $E_{r,z}$ defines a polarization its degree is equal to

$$\begin{aligned}(2.31)\quad [(r\mathcal{D}_{L/\mathbb{Q}}\mathfrak{B})^{-1} \oplus (r\mathcal{D}_{L/\mathbb{Q}}\mathfrak{A})^{-1} : \mathfrak{A} \oplus \mathfrak{B}] &= [(r\mathcal{D}_{L/\mathbb{Q}}\mathfrak{B})^{-1} : \mathfrak{A}] \cdot [(r\mathcal{D}_{L/\mathbb{Q}}\mathfrak{A})^{-1} : \mathfrak{B}] \\ &= [(r\mathcal{D}_{L/\mathbb{Q}}\mathfrak{A}\mathfrak{B})^{-1} : \mathcal{O}_L]^2 \\ &= \operatorname{Norm}(r\mathcal{D}_{L/\mathbb{Q}}\mathfrak{A}\mathfrak{B})^2.\end{aligned}$$

Summing up, we get:

COROLLARY 2.10. *Let $z \in \mathcal{H}^g$ and let $r \in L$. The hermitian form $H_{r,z}$ is positive definite iff $r \gg 0$. Therefore, the real alternating form $E_{r,z}$ defines a polarization on A_z iff $r \in \big((\mathcal{D}_{L/\mathbb{Q}}\mathfrak{A}\mathfrak{B})^{-1}\big)^+$. If this holds, then A_z is an abelian variety with RM by $\mathcal{O}_L$, and the degree of the polarization given by $E_{r,z}$ is $\operatorname{Norm}(r\mathcal{D}_{L/\mathbb{Q}}\mathfrak{A}\mathfrak{B})^2$. In particular, there exists an r such that $E_{r,z}$ is principal iff $\mathfrak{A}\mathfrak{B} = \mathcal{D}_{L/\mathbb{Q}}^{-1}$ in $\operatorname{Cl}(L)^+$.*

DEFINITION 2.11. Let (A, ι) be an abelian variety with RM by $\mathcal{O}_L$. Let

$$\begin{aligned}(2.32)\qquad \mathcal{M}_A &:= \{\lambda\colon A \to A^\vee : \lambda = \lambda^\vee, \lambda \text{ is an } \mathcal{O}_L\text{-linear homomorphism}\} \\ &\subseteq \operatorname{NS}(A),\end{aligned}$$

$$(2.33)\qquad \mathcal{M}_A^+ := \{\lambda \in \mathcal{M}_A : \lambda \text{ is a polarization}\} \subseteq \operatorname{NS}(A)^+.$$

We call $\mathcal{M}_A$ the *polarization module* of A and $\mathcal{M}_A^+$ the positive cone of polarizations. These notions depend on the given embedding $\iota\colon \mathcal{O}_L \to \operatorname{End}(A)$.

EXERCISE 2.12. *Let $\lambda\colon A \to B$ be a homomorphism of two g-dimensional (complex) abelian varieties with real multiplication. Show that if λ is $\mathcal{O}_L$-linear and non-zero then λ is an isogeny.*

LEMMA 2.13. (1) *The module $\mathcal{M}_A$ is a projective $\mathcal{O}_L$-module of rank* 1. *In particular,*

$$\ell \in \mathcal{O}_L, \lambda \in \mathcal{M}_A \implies \lambda \circ \ell \in \mathcal{M}_A. \tag{2.34}$$

(2) *The set $\mathcal{M}_A^+$ is a positive cone. Moreover, $\mathcal{M}_A$ is generated by $\mathcal{M}_A^+$.*

PROOF. We first check that $\mathcal{M}_A$ is an $\mathcal{O}_L$-module. We write more pedantically $\lambda \circ \ell$ as $\lambda \circ \iota(\ell)$. Then $\left(\lambda \circ \iota(\ell)\right)^\vee = \iota(\ell)^\vee \circ \lambda^\vee = \iota^\vee(\ell) \circ \lambda = \lambda \circ \iota(\ell)$, using that $\lambda^\vee = \lambda$ and that λ is $\mathcal{O}_L$-linear. It is also clear that $\lambda \circ \iota(\ell)$ is $\mathcal{O}_L$-linear. Thus $\mathcal{M}_A$ is an $\mathcal{O}_L$-module, torsion free by Exercise 2.12 (hence projective).

Next, we remark that any abelian variety A over $\mathbb{C}$ with RM by $\mathcal{O}_L$ has an $\mathcal{O}_L$-linear polarization. This follows from considerations as in Theorem 2.17. See also [**135**, Proposition 1.17]. Thus $\mathcal{M}_A^+$ is not empty.

To see that $\mathcal{M}_A$ is of rank one, take some $\lambda \in \mathcal{M}_A^+$ and get an embedding

$$\mathcal{M}_A \hookrightarrow \mathrm{Cent}_{\mathrm{End}^0(A)}(L)^{\mathrm{sym}}, \quad \mu \mapsto \lambda^{-1}\mu, \tag{2.35}$$

where "sym" on the right-hand side signifies the elements fixed by the Rosati involution defined by λ. The classification of endomorphism algebras with positive involution shows that we must have the image of $\mathcal{M}_A \otimes \mathbb{Q}$ equal to L. See Chapter 1, §6.2.

The rest follows from the general principles given there. The Rosati involution induced by some element of $\mathcal{M}_A^+$ must induce the identity on L. Thus the embedding $\mathcal{M}_A \hookrightarrow L$ given by (2.35) identifies $\mathcal{M}_A$ with a fractional ideal $\mathfrak{a}$ of L, and identifies $\mathcal{M}_A^+$ with $\mathfrak{a}^+$. □

In particular, $(\mathcal{M}_A, \mathcal{M}_A^+)$ is naturally an $\mathcal{O}_L$-module with a notion of positivity (via the embedding in $\mathrm{End}^0(A)$).

In contrast to the case of elliptic curves, $E = \mathbb{C}/\Lambda$, where all lattices Λ are indistinguishable as $\mathbb{Z}$-modules, in the case of a totally real field L of degree $g > 1$ and abelian varieties $\mathbb{C}^g/\Lambda$ with RM the lattice Λ is a projective rank 2 module over $\mathcal{O}_L$ and as such has a canonically associated ideal class in $\mathrm{Cl}(L)$, called its Steinitz class; see [**43**, Chapter II.4]. If we write $\Lambda \cong \mathfrak{A} \oplus \mathfrak{B}$, and in fact every projective rank 2 module can be written this way, then the Steinitz class is $\mathfrak{A}\mathfrak{B}$. More canonically, it is $\bigwedge^2_{\mathcal{O}_L} \Lambda$. Two projective rank 2 $\mathcal{O}_L$-modules are isomorphic iff they have the same Steinitz class.

The Steinitz class gives a discrete invariant of abelian varieties with RM by $\mathcal{O}_L$ and already shows that the set of components of the moduli space of abelian varieties with RM maps into $\mathrm{Cl}(L)$. The consideration of polarization would show that there is a natural bijection between the components of the moduli space and $\mathrm{Cl}(L)^+$.

DEFINITION 2.14. The group $\mathrm{GL}(\mathfrak{A} \oplus \mathfrak{B})^+$ consists of the matrices

$$\left\{ \begin{pmatrix} a & b \\ c & d \end{pmatrix} : a, d \in \mathcal{O}_L, b \in \mathfrak{A}^{-1}\mathfrak{B}, c \in \mathfrak{A}\mathfrak{B}^{-1}, ad - bc \in (\mathcal{O}_L^\times)^+ \right\} \tag{2.36}$$

with matrix multiplication.

REMARK 2.15. Note that $\mathrm{GL}(\mathfrak{A} \oplus \mathfrak{B})^+$ is the set of matrices $\left(\begin{smallmatrix} a & b \\ c & d \end{smallmatrix}\right)$ in $\mathrm{GL}_2(L)$ with totally positive determinant, such that

$$(\mathfrak{A}, \mathfrak{B}) \begin{pmatrix} a & b \\ c & d \end{pmatrix} = (\mathfrak{A}, \mathfrak{B}). \tag{2.37}$$

The group $G = \mathrm{GL}(\mathfrak{A} \oplus \mathfrak{B})^+$ acts on the g-fold product of the upper half plane $\mathcal{H}^g$ from the left by

$$\begin{pmatrix} a & b \\ c & d \end{pmatrix} z = \begin{pmatrix} a & b \\ c & d \end{pmatrix} (z_1, \ldots, z_g) = \left(\ldots, \frac{\sigma_i(a) z_i + \sigma_i(b)}{\sigma_i(c) z_i + \sigma_i(d)}, \ldots \right). \tag{2.38}$$

We also define the special linear subgroup of $\mathrm{GL}(\mathfrak{A} \oplus \mathfrak{B})^+$:

DEFINITION 2.16. The group $\mathrm{SL}(\mathfrak{A} \oplus \mathfrak{B})$ is the subgroup of $\mathrm{GL}(\mathfrak{A} \oplus \mathfrak{B})^+$ composed of matrices with determinant 1.

THEOREM 2.17. (1) *The isomorphism classes of* $(A, \iota)/\mathbb{C}$ *such that there exists an isomorphism*

$$(\mathcal{M}_A, \mathcal{M}_A^+) \xrightarrow{\sim} (\mathfrak{C}, \mathfrak{C}^+), \quad \mathfrak{C} = (\mathcal{D}_L \mathfrak{A}\mathfrak{B})^{-1}, \tag{2.39}$$

are parameterized by $\mathrm{GL}(\mathfrak{A} \oplus \mathfrak{B})^+ \backslash \mathcal{H}^g$.

(2) *The isomorphism classes of* $(A, \iota)/\mathbb{C}$ *together with a* given *isomorphism*

$$m \colon (\mathcal{M}_A, \mathcal{M}_A^+) \xrightarrow{\sim} (\mathfrak{C}, \mathfrak{C}^+), \quad \mathfrak{C} = (\mathcal{D}_L \mathfrak{A}\mathfrak{B})^{-1} \tag{2.40}$$

are parameterized by $\mathrm{SL}(\mathfrak{A} \oplus \mathfrak{B}) \backslash \mathcal{H}^g$.

REMARK 2.18. There is of course a difference between the two statements. There is more elbow room in the first case since any $\epsilon \in (\mathcal{O}_L^\times)^+$ induces an isomorphism:

$$(\mathfrak{C}, \mathfrak{C}^+) \underset{\sim}{\xrightarrow{\times \epsilon}} (\mathfrak{C}, \mathfrak{C}^+). \tag{2.41}$$

The possible m's form in fact a principal homogeneous space under $(\mathcal{O}_L^\times)^+$. Fix (A, ι, m). For any isomorphism $\phi \colon (B, j) \xrightarrow{\sim} (A, \iota)$ the diagram:

$$\begin{array}{ccc} B & \xrightarrow{\phi^* f} & B^\vee \\ \phi \big\downarrow & & \big\uparrow \phi^\vee \\ A & \xrightarrow{f} & A^\vee \end{array} \tag{2.42}$$

identifies $(\mathcal{M}_A, \mathcal{M}_A^+)$ with $(\mathcal{M}_B, \mathcal{M}_B^+)$. Let $m \colon (\mathcal{M}_A, \mathcal{M}_A^+) \cong (\mathfrak{C}, \mathfrak{C}^+)$ be an isomorphism. Then the induced isomorphism $\phi^* m \colon (\mathcal{M}_B, \mathcal{M}_B^+) \cong (\mathfrak{C}, \mathfrak{C}^+)$ is the following: If $g \in \mathcal{M}_B$ then

$$\phi^* m(g) = m\big((\phi^{-1})^\vee g \phi^{-1}\big). \tag{2.43}$$

Now, take in particular $(B, j) = (A, \iota)$ and ϕ equal to multiplication by ϵ^{-1}, where $\epsilon \in \mathcal{O}_L^\times$. Then

$$\phi^* m(g) = m\big((\epsilon)^\vee g \epsilon\big) = m(g\epsilon^2). \tag{2.44}$$

Thus

$$(A, \iota, m) \cong (A, \iota, m\epsilon^2). \tag{2.45}$$

Therefore, for every $\mu \in (\mathcal{O}_L^\times)^2$,

$$(A, \iota, m) \cong (A, \iota, m\mu). \tag{2.46}$$

Generically, the only endomorphisms of (A, ι) are $\iota(\mathcal{O}_L)$, as one may verify directly by taking $(A, \iota) = A_z$, where $z = (z_1, \ldots, z_g)$ are independent variables over L. Thus, generically, the map

$$(A, \iota, m) \to (A, \iota, \exists m) \tag{2.47}$$

has degree $[\mathrm{PGL}(\mathfrak{A}\oplus\mathfrak{B})^+ : \mathrm{PSL}(\mathfrak{A}\oplus\mathfrak{B})] = [(\mathcal{O}_L^\times)^+ : (\mathcal{O}_L^\times)^2] = h^+/h$.

PROOF OF THEOREM 2.17. Let $\mu = \left(\begin{smallmatrix} a & b \\ c & d\end{smallmatrix}\right)$ be an element of $\mathrm{GL}(\mathfrak{A}\oplus\mathfrak{B})^+$, and let

$$z = (z_1, \dots, z_g) \in \mathcal{H}^g, \quad 1 = (1, \dots, 1). \tag{2.48}$$

We have

$$\Lambda_z = \mathfrak{A}\cdot z + \mathfrak{B}\cdot 1 \quad \Lambda_{\mu z} = \mathfrak{A}\cdot \mu z + \mathfrak{B}\cdot 1. \tag{2.49}$$

Consider the map

$$f\colon \mathbb{C}^g \to \mathbb{C}^g, \quad x \mapsto x\cdot M, \tag{2.50}$$

where M is the diagonal matrix

$$M = \operatorname{diag}\bigl(\sigma_1(c)z_1 + \sigma_1(d), \dots, \sigma_g(c)z_g + \sigma_g(d)\bigr). \tag{2.51}$$

Under this map

$$f(\Lambda_{\mu z}) = \Lambda_{\mu z} \operatorname{diag}(\dots, \sigma_i(c)z_i + \sigma_i(d), \dots) \tag{2.52}$$

is the lattice

$$\begin{aligned}\Bigl\{\Bigl(\dots, \sigma_i(\alpha)\Bigl(\frac{\sigma_i(a)z_i + \sigma_i(b)}{\sigma_i(c)z_i + \sigma_i(d)}\Bigr) + \sigma_i(\beta), \dots\Bigr) \\ \times \operatorname{diag}(\dots, \sigma_i(c)z_i + \sigma_i(d), \dots) : (\alpha,\beta)\in\mathfrak{A}\oplus\mathfrak{B}\Bigr\}.\end{aligned} \tag{2.53}$$

This lattice is equal to

$$\Bigl\{\Bigl(\dots, \sigma_i\Bigl((\alpha,\beta)\begin{pmatrix} a & b \\ c & d\end{pmatrix}\Bigr)\begin{pmatrix} z_i \\ 1\end{pmatrix}, \dots\Bigr) : (\alpha,\beta)\in\mathfrak{A}\oplus\mathfrak{B}\Bigr\}. \tag{2.54}$$

Because $(\mathfrak{A},\mathfrak{B})\mu = (\mathfrak{A},\mathfrak{B})$, for $\mu \in \mathrm{GL}(\mathfrak{A}\oplus\mathfrak{B})^+$, this set is precisely Λ_z. Thus f induces a map

$$f\colon A_{\mu z} \to A_z. \tag{2.55}$$

This map is an isomorphism of abelian varieties with $\mathcal{O}_L$-action, because L acts diagonally: for $\ell \in L$

$$\ell \mapsto \begin{pmatrix} \sigma_1(\ell) & & \\ & \ddots & \\ & & \sigma_g(\ell)\end{pmatrix}, \tag{2.56}$$

because the action of ℓ on any element in $\mathfrak{A}\cdot z + \mathfrak{B}\cdot 1$ is

$$\alpha\cdot z + \beta\cdot \mapsto \ell\alpha\cdot z + \ell\beta\cdot 1 = (\alpha\cdot z + \beta\cdot 1)\begin{pmatrix} \sigma_1(\ell) & & \\ & \ddots & \\ & & \sigma_g(\ell)\end{pmatrix} \tag{2.57}$$

and this holds also for the extended action of L on $\mathbb{R}\otimes_{\mathbb{Z}} (\mathfrak{A}\cdot z + \beta\cdot 1) = \mathbb{C}^g$.

Recall that by Lemma 2.9

$$\begin{aligned} &H_{r,z}\bigl((x_1,\dots,x_g),(y_1,\dots,y_g)\bigr) \\ &\qquad = (x_1,\dots,x_g)\operatorname{diag}\Bigl(\dots, \frac{\sigma_i(r)}{\operatorname{Im}(z_i)}, \dots\Bigr)\,{}^t(\overline{y_1},\dots,\overline{y_g}).\end{aligned} \tag{2.58}$$

Let us calculate $f^*H_{r,z}(x,y)$. We have

$$(2.59)\quad \begin{aligned} f^*H_{r,z}(x,y) &= (x_1,\dots,x_g)\,\mathrm{diag}(\dots,\sigma_i(c)z_i+\sigma_i(d),\dots)\,\mathrm{diag}\left(\dots,\frac{\sigma_i(r)}{\mathrm{Im}(z_i)},\dots\right) \\ &\quad\times \mathrm{diag}(\dots,\sigma_i(c)\overline{z_i}+\sigma_i(d),\dots)^t(\overline{y_1},\dots,\overline{y_g}) \\ &= (x_1,\dots,x_g)\,\mathrm{diag}(\dots,b_i,\dots)^t(\overline{y_1},\dots,\overline{y_g}), \end{aligned}$$

where

$$(2.60)\quad b_i = \sigma_i(r)\cdot\frac{|\sigma_i(c)z_i+\sigma_i(d)|^2}{\mathrm{Im}(z_i)} = \frac{\sigma_i(r)}{\mathrm{Im}(\mu z_i)}\sigma_i(ad-bc) = \frac{\sigma_i(r)}{\mathrm{Im}((\mu z)_i)}\sigma_i(ad-bc).$$

Hence

$$(2.61)\qquad f^*H_{r,z} = H_{(\det\mu)r,\mu z}.$$

Thus, if we want to get an isomorphism of (A_z,ι,m_z) to $(A_{\mu z},\iota,m_{\mu z})$, we need $\det\mu = 1$, while if we want only to preserve the polarization structure we need $\det\mu$ to be a totally positive unit of $\mathcal{O}_L$.

To prove the first statement of the theorem, we still need to prove the assertions A and B below.

A. *If $A_z \cong A_{z'}$ as abelian varieties with RM by $\mathcal{O}_L$, where $z, z' \in \mathcal{H}^g$, there exists a matrix μ in* $\mathrm{GL}(\mathfrak{A}\oplus\mathfrak{B})^+$ *such that $\mu z = z'$.*

Suppose that $A_z \cong A_{z'}$ as abelian varieties with RM by $\mathcal{O}_L$. Let $M\colon \mathbb{C}^g \to \mathbb{C}^g$ be the transformation inducing the isomorphism:

$$(2.62)\qquad A_{z'} = \mathbb{C}^g/\Lambda_{z'} \to \mathbb{C}^g/\Lambda_z = A_z.$$

The action of L on $\mathfrak{t}_{A_z,0} = \mathfrak{t}_{A_{z'},0} = \mathbb{C}^g$ is diagonal, and since M commutes with this action and the σ_i are independent, M must be a diagonal matrix $\mathrm{diag}(m_1,\dots,m_g)$. We can write $m_i = \sigma_i(c)z_i + \sigma_i(d)$, for some $c,d\in L$. Then M takes $\mathfrak{B}\cdot 1 \subseteq \Lambda_{z'}$ into Λ_z. That is, for every $\beta\in\mathfrak{B}$

$$(2.63)\qquad \big([\sigma_1(c)z_1+\sigma_1(d)]\sigma_1(\beta),\dots,[\sigma_g(c)z_g+\sigma_g(d)]\sigma_g(\beta)\big)\in\Lambda_z.$$

We must therefore have $c\in\mathfrak{B}^{-1}\mathfrak{A}$, $d\in\mathcal{O}_L$. On the other hand, any $\alpha\cdot z'$ (with $\alpha\in L$) is mapped into $\mathbb{Q}\otimes\Lambda_z$. That is, for suitable $a,b\in L$:

$$(2.64)\quad \begin{aligned} &\Big(\big(\sigma_1(c)z_1+\sigma_1(d)\big)\sigma_1(\alpha)\cdot z_1',\dots,\big(\sigma_g(c)z_g+\sigma_g(d)\big)\sigma_g(\alpha)\cdot z_g'\Big) \\ &\qquad = \big(\sigma_1(a)z_1+\sigma_1(b),\dots,\sigma_g(a)z_g+\sigma_g(b)\big). \end{aligned}$$

Take $\alpha = 1$, so $z_i' = \big(\sigma_i(a)z_1+\sigma_i(b)\big)/\big(\sigma_i(c)z_1+\sigma_i(d)\big)$, for a suitable matrix $\left(\begin{smallmatrix} a & b\\ c & d\end{smallmatrix}\right) = \mu\in\mathrm{GL}_2(L)$. Since μ preserves lattices, it is actually in $\mathrm{GL}(\mathfrak{A}\oplus\mathfrak{B})$, and $\mu\cdot z = z'$ gives $\sigma_i(\det\mu) > 0$ for all i. That is, $\mu\in\mathrm{GL}(\mathfrak{A}\oplus\mathfrak{B})^+$.

B. *Given (A,ι) such that $(\mathcal{M}_A,\mathcal{M}_A^+)\cong(\mathfrak{C},\mathfrak{C}^+)$, then $(A,\iota)\cong(A_z,\iota)$, for some $z\in\mathcal{H}^g$.*

Write $A = \mathbb{C}^g/\Lambda$. Since Λ is a projective rank 2 $\mathcal{O}_L$-module, we may write $\Lambda\cong\mathfrak{A}'\oplus\mathfrak{B}'$ for some fractional $\mathcal{O}_L$-ideals $\mathfrak{A}',\mathfrak{B}'$, and the isomorphism class of Λ is determined by its Steinitz class $\mathfrak{A}'\mathfrak{B}'$ [**43**, Theorem 13, p. 95]. See also p. 56. The polarization module of $\mathbb{C}^g/\Lambda$ is $(\bigwedge^2_{\mathcal{O}_L}\Lambda)^* = (\mathcal{D}_{L/\mathbb{Q}}\mathfrak{A}'\mathfrak{B}')^{-1}$. Hence, $\Lambda\cong\mathfrak{A}\oplus\mathfrak{B}$. We know that $\rho_{\mathbb{C}}\sim\mathrm{diag}(\sigma_1,\dots,\sigma_g)$.

We may therefore choose coordinates on $\mathbb{C}^g$ such that: $A = \mathbb{C}^g/\Lambda, \Lambda \cong \mathfrak{A} \oplus \mathfrak{B}$ and the action of L is given by the diagonal map. If necessary, we may still change coordinates by diagonal matrices, so that the map:

$$\phi\colon \mathfrak{A} \oplus \mathfrak{B} \to \Lambda, \tag{2.65}$$

extends to a linear map on:

$$\phi\colon L \oplus L \to \mathbb{Q} \cdot \Lambda, \tag{2.66}$$

satisfying $\phi(0,1) = (1,\ldots,1)$ and $\phi(1,0) = (z_1,\ldots,z_g)$. Every L-linear Riemann form is easily seen to be of the form:

$$H_{(r,z)}(x,y) = \sum \frac{x_i \overline{y_i} \sigma_i(r)}{\mathrm{Im}(z_i)}, \quad r \in (\mathcal{D}_{L/\mathbb{Q}}\mathfrak{A}\mathfrak{B})^{-1}. \tag{2.67}$$

It defines a polarization, i.e., is positive definite, if and only if for every i we have $\mathrm{sign}(\sigma_i(r)) = \mathrm{sign}(\mathrm{Im}(z_i))$. We conclude that if we choose $\mathfrak{A}, \mathfrak{B}$ such that $\Lambda \cong \mathfrak{A} \oplus \mathfrak{B}$ and $\mathrm{NS}(A)^+ \cong (\mathcal{D}_L\mathfrak{A}\mathfrak{B})^{-1+}$, then $\mathrm{Im}(z_i) > 0$ for all i, hence $A = A_z$ for some $z \in \mathcal{H}^g$. □

Fix representatives of $\mathrm{Cl}(L)^+$ of the form $\{(\mathfrak{A}, \mathfrak{A}^+)\}$.

COROLLARY 2.19. (1) *We have a natural bijection*

$$\{\text{Isomorphism classes of } (A,\iota)/\mathbb{C}\} \leftrightarrow \coprod_{(\mathfrak{A},\mathfrak{A}^+)} \mathrm{GL}(\mathcal{O}_L \oplus \mathfrak{A})^+ \backslash \mathcal{H}^g, \tag{2.68}$$

and (A,ι) is parameterized by $\mathrm{GL}(\mathcal{O}_L \oplus \mathfrak{A})^+$ iff there exists an isomorphism of $\mathcal{O}_L$-modules with a notion of positivity

$$(\mathcal{M}_A, \mathcal{M}_A^+) \cong ((\mathcal{D}_L\mathfrak{A})^{-1}, (\mathcal{D}_L\mathfrak{A})^{-1\,+}). \tag{2.69}$$

(2) *We have a natural bijection*

$$\{\text{Isomorphism classes of } (A,\iota,m)/\mathbb{C}\} \leftrightarrow \coprod_{(\mathfrak{A},\mathfrak{A}^+)} \mathrm{SL}(\mathcal{O}_L \oplus \mathfrak{A}) \backslash \mathcal{H}^g. \tag{2.70}$$

EXERCISE 2.20. *Let $\mathfrak{a}$ be an ideal of L. Consider the natural transformation of the functor of complex abelian varieties with RM defined by $(A,\iota) \mapsto (A \otimes_{\mathcal{O}_L} \mathfrak{a}, \iota \otimes 1)$. It induces an automorphism of $\coprod_{(\mathfrak{A},\mathfrak{A}^+)} \mathrm{GL}(\mathcal{O}_L \oplus \mathfrak{A})^+ \backslash \mathcal{H}^g$. Describe the action on the component set.*

REMARK 2.21. There is a way to compactify $\mathrm{GL}(\mathfrak{A} \oplus \mathfrak{B})^+ \backslash \mathcal{H}^g$ that follows the lines of the classical theory for elliptic curves: We add the finite set of points (called cusps) $\mathrm{GL}(\mathfrak{A} \oplus \mathfrak{B})^+ \backslash \mathbb{P}^1(L)$ to $\mathrm{GL}(\mathfrak{A} \oplus \mathfrak{B})^+ \backslash \mathcal{H}^g$. The action of $\mathrm{GL}(\mathfrak{A} \oplus \mathfrak{B})^+$ on $\mathbb{P}^1(L)$ is the usual one:

$$\begin{pmatrix} a & b \\ c & d \end{pmatrix} \begin{pmatrix} \alpha \\ \beta \end{pmatrix} = \begin{pmatrix} a\alpha + b\beta \\ c\alpha + d\beta \end{pmatrix}. \tag{2.71}$$

There is charm in the following

PROPOSITION 2.22. *For any two fractional ideals $\mathfrak{A}$, $\mathfrak{B}$ of L we have*

$$|\,\mathrm{GL}(\mathfrak{A} \oplus \mathfrak{B})^+ \backslash \mathbb{P}^1(L)| = h, \tag{2.72}$$

where h is the class number of L.

PROOF. Consider the map:

$$\mathrm{GL}(\mathfrak{A}\oplus\mathfrak{B})^+\backslash\mathbb{P}^1(L)\to \mathrm{Cl}(L), \tag{2.73}$$

given by

$$\begin{pmatrix}\alpha\\ \beta\end{pmatrix}\mapsto(\mathfrak{A},\mathfrak{B})\begin{pmatrix}\alpha\\ \beta\end{pmatrix}=\alpha\mathfrak{A}+\beta\mathfrak{B}. \tag{2.74}$$

Using $(\mathfrak{A},\mathfrak{B})\mu=(\mathfrak{A},\mathfrak{B})$ for $\mu\in\mathrm{GL}(\mathfrak{A}\oplus\mathfrak{B})^+$, we see that the map is well-defined. It is surjective because every fractional ideal $\mathfrak{C}$ of L is of the form $\mathfrak{C}=\alpha\mathfrak{A}+\beta\mathfrak{B}$, for some $\alpha,\beta\in L$. Indeed, every ideal class contains infinitely many prime ideals. Take α and β such that $\alpha\mathfrak{A}\mathfrak{C}^{-1}$ and $\beta\mathfrak{B}\mathfrak{C}^{-1}$ are distinct prime ideals. Or if one wishes to use less machinery: assume w.l.o.g. that $\mathfrak{C}=\mathcal{O}_L$ and that $\mathfrak{A}$ is integral. Consider $\mathcal{O}_L/\mathfrak{A}=\oplus(\mathcal{O}_{L\mathfrak{p}_i}/\mathfrak{p}_i{}^{e_i}\mathcal{O}_{L\mathfrak{p}_i})$, where $\mathfrak{A}=\prod\mathfrak{p}_i{}^{e_i}$ (Chinese Reminder Theorem). We may assume that $\mathfrak{B}$ is integral. Again by the CRT, we may choose $\beta\in L$ such that for every i we have $\beta\mathfrak{B}\mathcal{O}_{L\mathfrak{p}_i}=\mathcal{O}_{L\mathfrak{p}_i}$ and $\beta\mathfrak{B}\subset\mathcal{O}_L$.

We now show that this map is injective. Suppose that in the class group of L we have

$$\alpha\mathfrak{A}+\beta\mathfrak{B}=\alpha'\mathfrak{A}+\beta'\mathfrak{B}. \tag{2.75}$$

Multiplying ${}^t(\alpha,\beta)$ by a suitable $\lambda\in L^\times$, we may assume that (2.75) is an equality of fractional ideals

$$\alpha\mathfrak{A}+\beta\mathfrak{B}=\alpha'\mathfrak{A}+\beta'\mathfrak{B}. \tag{2.76}$$

Let $\mathfrak{C}^{-1}=\alpha\mathfrak{A}+\beta\mathfrak{B}$. Then

$$\mathcal{O}_L=\alpha\mathfrak{A}\mathfrak{C}+\beta\mathfrak{B}\mathfrak{C}=\alpha'\mathfrak{A}\mathfrak{C}+\beta'\mathfrak{B}\mathfrak{C}. \tag{2.77}$$

Note that $\alpha,\alpha'\in(\mathfrak{A}\mathfrak{C})^{-1}$ and $\beta,\beta'\in(\mathfrak{B}\mathfrak{C})^{-1}$. We may then find $A,A'\in\mathfrak{A}\mathfrak{C}$ and $B,B'\in\mathfrak{B}\mathfrak{C}$ such that

$$1=\alpha A+\beta B=\alpha'A'+\beta'B'. \tag{2.78}$$

Consider the matrices of determinant 1

$$M=\begin{pmatrix}\alpha & -B\\ \beta & A\end{pmatrix},\quad N=\begin{pmatrix}\alpha' & -B'\\ \beta' & A'\end{pmatrix}. \tag{2.79}$$

Note that $M^t(1,0)={}^t(\alpha,\beta)$ and $N^t(1,0)={}^t(\alpha',\beta')$. Thus, $NM^{-1t}(\alpha,\beta)={}^t(\alpha',\beta')$. Now,

$$NM^{-1}=\begin{pmatrix}\alpha'A+\beta B' & \alpha'B-\alpha B'\\ \beta'A-\beta A' & \beta'B+\alpha A'\end{pmatrix}. \tag{2.80}$$

This is a matrix in $\mathrm{GL}(\mathfrak{A}\oplus\mathfrak{B})^+$ (in fact, in $\mathrm{SL}(\mathfrak{A}\oplus\mathfrak{B})^+$), e.g., $\alpha'B-\alpha B'\in(\mathfrak{A}\mathfrak{C})^{-1}\mathfrak{B}\mathfrak{C}+(\mathfrak{A}\mathfrak{C})^{-1}\mathfrak{B}\mathfrak{C}=\mathfrak{A}^{-1}\mathfrak{B}$. □

To make $\mathrm{GL}(\mathfrak{A}\oplus\mathfrak{B})^+\backslash\mathbb{P}^1(L)\coprod\mathrm{GL}(\mathfrak{A}\oplus\mathfrak{B})^+\backslash\mathcal{H}^g$ into a topological space one defines a fundamental system of neighborhoods at the cusp $(i\infty,\dots,i\infty)$ by setting:

$$U_r:=\{z\in\mathcal{H}^g\mid \mathrm{Im}(z_i)>r\ \forall i\},\quad r\in\mathbb{R}. \tag{2.81}$$

See Figure 2.1. Acting on those neighborhoods by elements $\mu\in\mathrm{GL}_2(L)$, we get neighborhoods of all the cusps $(\alpha:\beta)\in\mathbb{P}^1(L)$, and this allows one to extend the topology to $\mathrm{GL}(\mathfrak{A}\oplus\mathfrak{B})^+\backslash\mathcal{H}^{g*}:=\mathrm{GL}(\mathfrak{A}\oplus\mathfrak{B})^+\backslash\mathbb{P}^1(L)\coprod\mathrm{GL}(\mathfrak{A}\oplus\mathfrak{B})^+\backslash\mathcal{H}^g$. One then shows that in fact this topological space has a structure of a normal compact complex variety. This compactification is called the Satake compactification of $\mathrm{GL}(\mathfrak{A}\oplus\mathfrak{B})^+\backslash\mathcal{H}^g$. See [**44**, Chapter II, p. 44].

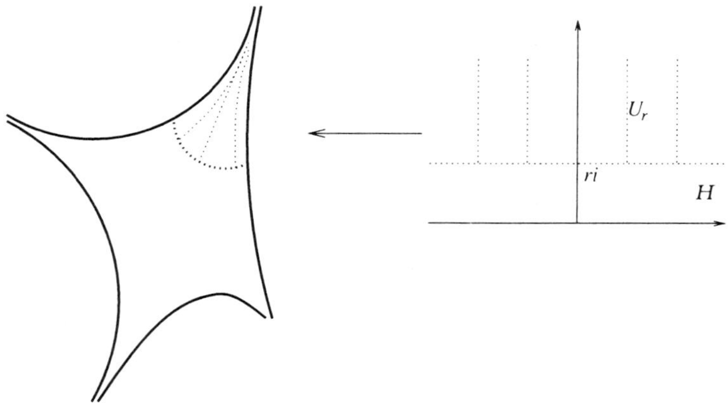

FIGURE 2.1. A neighborhood of the cusp at ∞.

From the point of view of moduli, one may ask whether the cusps have a moduli interpretation. As in the case of elliptic curves, the answer is yes. In the case of elliptic curves the moduli interpretation is given using generalized elliptic curves. The precise formulation requires some care, especially when introducing level structure. We refer to [**9, 26, 34**] and Appendix B.

Ignoring the case of level structure, we may very roughly say that the compactified moduli space classifies semi-abelian varieties with RM by $\mathcal{O}_L$. By a semi-abelian variety $\mathcal{G}$ over a base scheme S we understand a group scheme over S, $\pi\colon \mathcal{G} \to S$, such that π is smooth, of fixed relative dimension, and every geometric fiber of π is an extension of an abelian variety by a torus. We say that the semi-abelian scheme $\mathcal{G} \to S$ has real multiplication if there is given an embedding of rings $\mathcal{O}_L \hookrightarrow \mathrm{End}_S(\mathcal{G})$, such that the Lie algebra of $\mathcal{G} \to S$, i.e., the relative tangent space at the origin $\mathfrak{t}_{\mathcal{G}/S}$ is a locally free sheaf of $\mathcal{O}_L \otimes_{\mathbb{Z}} \mathcal{O}_S$-modules of rank 1.[3]

Now suppose that one of the geometric fibers of $\mathcal{G} \to S$, say $\mathcal{G}_s$, is not an abelian variety. Then we have an exact sequence

$$0 \to T \to \mathcal{G}_s \to A \to 0, \tag{2.82}$$

where T is a non-trivial torus and A is an abelian variety. The ring $\mathcal{O}_L$ acts non-trivially on T and therefore L acts non-trivially on the rational vector space $X(T) \otimes \mathbb{Q}$, where $X(T)$ are the characters of T. It follows that $\dim(X(T) \otimes \mathbb{Q}) = g$ and therefore $\mathcal{G}_s = T$. That is,

LEMMA 2.23. *A semi-abelian variety with RM has geometric fibers that are either tori or abelian varieties.*

EXERCISE 2.24. *Let A be a g-dimensional semi-abelian variety with a faithful action of R where R is either an order of a CM field of degree $2g$ over $\mathbb{Q}$, or an order in a division quaternion algebra over a totally real field L of degree g. Use the ideas above to show that A is an abelian variety. This result is usually used to show that an abelian variety (say over a dvr) with CM, or with quaternionic multiplication, has potentially good reduction.*

We further note that if the torus T is split then its RM structure is completely determined by the lattice of characters $X(T)$ as an $\mathcal{O}_L$-module. This lattice has

[3] One needs to modify the definition if d_L is not invertible on S. See Chapter 5.

rank g and is therefore a projective $\mathcal{O}_L$-module of rank 1. Thus, we expect that there would be h cusps to the compactified moduli space, corresponding to the split tori of dimension g with RM by $\mathcal{O}_L$. This is indeed the case.

Finally we note the following. The stabilizer of $(i\infty, \ldots, i\infty)$ in $\mathrm{SL}(\mathfrak{A} \oplus \mathfrak{B})$[4] is precisely the set of matrices $\left(\begin{smallmatrix} a & b \\ 0 & d \end{smallmatrix}\right)$ where a and d are in $\mathcal{O}_L$ and satisfy $ad = 1$, and b is any element in the ideal $\mathfrak{D} = \mathfrak{A}^{-1}\mathfrak{B}$. This group is isomorphic to

$$\mathcal{O}_L^\times \ltimes \mathfrak{D}. \tag{2.83}$$

Dividing first only by the action of $\mathfrak{D}$, we get local coordinates as follows: we choose a basis $\nu_1, \ldots, \nu_g$ to the dual of $\mathfrak{D}$ (with respect to the trace pairing). Then

$$x_1 = e^{2\pi i \cdot \mathrm{Tr}(\nu_1 \tau)}, \ldots, x_g = e^{2\pi i \cdot \mathrm{Tr}(\nu_g \tau)}, \tag{2.84}$$

($\tau = (\tau_1, \ldots, \tau_g)$ and $\mathrm{Tr}(\nu\tau) = \sigma_1(\nu)\tau_1 + \cdots + \sigma_g(\nu)\tau_g$) are local coordinates for $\mathfrak{D}\backslash\mathcal{H}^g$ around $(i\infty, \ldots, i\infty)$. Every function f invariant under the action of $\mathfrak{D}$, namely under translation by the elements of $\mathfrak{D}$ (as every modular form with respect to $\mathrm{GL}(\mathfrak{A} \oplus \mathfrak{B})^+$ is), has a Taylor expansion

$$f = \sum_{n \in \mathbb{Z}^g} a_n x^n \tag{2.85}$$

(using multi-index notation). But this is the same as writing

$$f = \sum_{\nu \in \mathfrak{D}^\vee} a_\nu e^{2\pi i \cdot \mathrm{Tr}(\nu \cdot \tau)}. \tag{2.86}$$

This is the q-expansion we shall discuss in more detail in Chapter 5, §2. One often puts $q^\nu := e^{2\pi i \cdot \mathrm{Tr}(\nu\tau)}$ and writes the q-expansion as $f(q) = \sum_{\nu \in \mathfrak{D}^\vee} a_\nu q^\nu$.

3. Hilbert modular forms

We keep the notation of the previous sections. Thus L is totally real of degree g, with different $\mathcal{D}_L$, discriminant d_L and ring of integers $\mathcal{O}_L$; $\mathrm{Emb}(L, \mathbb{C}) = \{\sigma_1, \ldots, \sigma_g\}$, $h^+ = |\mathrm{Cl}(L)^+|$, and $h = |\mathrm{Cl}(L)|$. In what follows, $\mathfrak{A}$ and $\mathfrak{B}$ are fractional ideals, and

$$\mathrm{GL}(\mathcal{O}_L \oplus \mathfrak{A})^+ = \left\{ \begin{pmatrix} a & b \\ c & d \end{pmatrix} : a, d \in \mathcal{O}_L, b \in \mathfrak{A}, c \in \mathfrak{A}^{-1}, ad - bc \in \mathcal{O}_L^{\times\,+} \right\}. \tag{3.1}$$

Given a matrix $\delta = \left(\begin{smallmatrix} \delta_1 & \delta_2 \\ \delta_3 & \delta_4 \end{smallmatrix}\right) \in \mathrm{GL}_2(\mathbb{R})^+$, put

$$j(\delta, z) = (\delta_3 z + \delta_4)(\det \delta)^{-1/2}. \tag{3.2}$$

For a vector $\boldsymbol{k} = (k_1, \ldots, k_g) \in \mathbb{Z}^g$, and a matrix $\mu \in \mathrm{GL}_2(L)^+$, put:

$$j_{\boldsymbol{k}}(\mu, z) = \prod_{i=1}^{g} j(\sigma_i(\mu), z_i)^{k_i}. \tag{3.3}$$

For $f\colon \mathcal{H}^g \to \mathbb{C}$, put

$$(f|_{\boldsymbol{k}}\mu)(z) = j_{\boldsymbol{k}}(\mu, z)^{-1} f(\mu z). \tag{3.4}$$

For a group $\Gamma \subseteq \mathrm{GL}(\mathcal{O}_L \oplus \mathfrak{A})^+$ of finite index we make the following

[4]The groups $\mathrm{SL}(\mathfrak{A} \oplus \mathfrak{B})$ and $\mathrm{GL}(\mathfrak{A} \oplus \mathfrak{B})^+$ have the same number of cusps and only minor modifications are needed to discuss the case of $\mathrm{GL}(\mathfrak{A} \oplus \mathfrak{B})^+$. The difference eventually amounts to some extra relations satisfied by the coefficients of the q-expansion.

DEFINITION 3.1. A *Hilbert modular form* of weight $\boldsymbol{k}$ and level Γ is a holomorphic function

$$f\colon \mathcal{H}^g \to \mathbb{C}, \tag{3.5}$$

such that

$$f|_{\boldsymbol{k}}\mu = f, \quad \forall \mu \in \Gamma. \tag{3.6}$$

Explicitly,

$$\begin{aligned} f\left(\frac{\sigma_1(a)z_1+\sigma_1(b)}{\sigma_1(c)z_1+\sigma_1(d)}, \dots, \frac{\sigma_g(a)z_g+\sigma_g(b)}{\sigma_g(c)z_g+\sigma_g(d)}\right) \\ = \left(\prod_{i=1}^{g}\big(\sigma_i(c)z_i+\sigma_i(d)\big)^{k_i}\det\big(\sigma_i(\mu)\big)^{-k_i/2}\right)f(z_1,\dots,z_g). \end{aligned} \tag{3.7}$$

We also require f to be holomorphic at infinity in the following sense:

Let $M = \{a : \left(\begin{smallmatrix}1 & a\\ 0 & 1\end{smallmatrix}\right) \in \Gamma\}$. (Note that $n\mathfrak{A}^{-1} \subseteq M$ for some integer $n > 0$, therefore M is a projective $\mathcal{O}_L$-module of rank 1). Then f is invariant under translation by elements of M,

$$\left(f|_{\boldsymbol{k}}\begin{pmatrix}1 & b\\ 0 & 1\end{pmatrix}\right)(z_1,\dots,z_g) = f\big(z_1+\sigma_1(b),\dots,z_g+\sigma_g(b)\big) = f(z_1,\dots,z_g), \tag{3.8}$$

and hence possesses a q-expansion:

$$f(z) = \sum_{\nu\in M^\vee} a_\nu e^{2\pi i\,\mathrm{Tr}(\nu\cdot z)}, \tag{3.9}$$

with $M^\vee = \{\ell \in L : \mathrm{Tr}_{L/\mathbb{Q}}(\ell m) \in \mathbb{Z}, \forall m \in M\}$ and using the notation $\mathrm{Tr}(\nu\cdot z) = \sigma_1(\nu)z_1 + \cdots + \sigma_g(\nu)z_g$. (When $\Gamma = \mathrm{GL}(\mathcal{O}_L \oplus \mathfrak{A}^{-1})^+$, we have $M \cong \mathfrak{A}^{-1}$ and $M^\vee \cong \mathfrak{A}\mathcal{D}_{L/\mathbb{Q}}^{-1}$; in the case $L = \mathbb{Q}$, $g = 1$, we retrieve:

$$f(\tau) = \sum_{n\in\mathbb{Z}} a_n e^{2\pi i n\tau}.) \tag{3.10}$$

We require that if f is a modular form of weight $\boldsymbol{k}$ and level Γ, then for every $\mu \in \mathrm{GL}_2(L)^+$, $g := f|_{\boldsymbol{k}}\mu$ has a q-expansion:

$$g(z) = \sum_{\nu\in M^\vee} a_\nu e^{2\pi i\,\mathrm{Tr}(\nu\cdot z)}, \tag{3.11}$$

with $a_\nu = 0$ unless $\nu = 0$ or $\nu \gg 0$. Here the module M is the one associated to the group $\mu^{-1}\Gamma\mu$.

EXERCISE 3.2. *Let Γ be a subgroup of finite index of $\mathrm{GL}(\mathcal{O}_L \oplus \mathfrak{A})^+$. Let f (resp. f') be a modular form of level Γ and weight $\boldsymbol{k}$ (resp. $\boldsymbol{k}'$). Show that if f and f' have the same q-expansion at $(i\infty,\dots,i\infty)$ then $f = f'$ and $\boldsymbol{k} = \boldsymbol{k}'$.*

Under the topology defined in Remark 2.21, the constraint on the q-expansion means that f possesses a holomorphic Taylor expansion at every cusp. The next theorem tells us that in fact for $g > 1$ the holomorphy requirement we imposed is automatic!

THEOREM 3.3 (Köcher's Principle). *Let $f(z) = \sum_{\nu\in M^\vee} a_\nu e^{2\pi i\,\mathrm{Tr}(\nu\cdot z)}$ be a modular form of weight $\boldsymbol{k}$ and level Γ. Assume $g > 1$, then*

$$a_\nu \neq 0 \implies \nu \gg 0 \text{ or } \nu = 0. \tag{3.12}$$

PROOF. First, a lemma:

LEMMA 3.4. *Put* $r_{\boldsymbol{k}}(\epsilon) = \prod_{i=1}^{g} \sigma_i(\epsilon)^{-k_i/2}$, *for* $\epsilon \in \mathcal{O}_L^{\times +}$. *If* $\left(\begin{smallmatrix} \epsilon & 0 \\ 0 & 1 \end{smallmatrix}\right) \in \Gamma$, *then*

$$a_\nu = r_{\boldsymbol{k}}(\epsilon) a_{\epsilon\nu}. \tag{3.13}$$

PROOF. Let $A(\epsilon) = \left(\begin{smallmatrix} \epsilon & 0 \\ 0 & 1 \end{smallmatrix}\right)$. On the one hand,

$$f(\epsilon z) = f(\sigma_1(\epsilon)z_1, \ldots, \sigma_g(\epsilon)z_g) = \sum_{\nu \in M^\vee} a_\nu e^{2\pi i \operatorname{Tr}(\epsilon\nu \cdot z)}. \tag{3.14}$$

On the other hand,

$$f(\epsilon z) = j_{\boldsymbol{k}}(A(\epsilon), z) f(z) = r_{\boldsymbol{k}}(\epsilon) \sum_{\nu \in M^\vee} a_\nu e^{2\pi i \operatorname{Tr}(\nu \cdot z)}. \tag{3.15}$$

Equating coefficients of $e^{2\pi i \operatorname{Tr}(\epsilon\nu \cdot z)}$ we are done. □

Assume now that $a_{\nu_0} \neq 0$ for some $\nu_0 \neq 0$ such that ν_0 is not totally positive and, without loss of generality, $\sigma_1(\nu_0) < 0$. Find $\epsilon \gg 0$ such that $A(\epsilon) \in \Gamma$ and

$$\sigma_1(\epsilon) > 1, \quad \sigma_i(\epsilon) < 1 \quad \text{for all } 2 \leq i \leq g. \tag{3.16}$$

The existence of such ϵ follows from the fact that Γ is of finite index in $\mathrm{GL}(\mathcal{O} \oplus \mathfrak{A})^+$ and the lattice structure of the units modulo roots of unity.

Consider the terms of $\sum_{\nu \in M^\vee} a_\nu e^{2\pi i \operatorname{Tr}(\nu \cdot z)}$, parameterized by $\nu_0 \epsilon^m$, $m = 1, 2, \ldots$. Put $z = i$. Then

$$a_{\nu_0 \epsilon^m} e^{-2\pi \operatorname{Tr}(\nu_0 \epsilon^m)} = a_{\nu_0} r_{\boldsymbol{k}}(\epsilon)^{-m} e^{-2\pi \operatorname{Tr}(\nu_0 \epsilon^m)}. \tag{3.17}$$

But when $m \to \infty$, $e^{-2\pi \operatorname{Tr}(\nu_0 \epsilon^m)} \sim e^{-2\pi \sigma_1(\nu_0) \cdot \sigma_1(\epsilon)^m}$, and the exponential growth gives $r_{\boldsymbol{k}}(\epsilon)^{-m} e^{-2\pi \operatorname{Tr}(\nu_0 \epsilon^m)} \to \infty$. So the general term does not approach zero and thus the sum is divergent; contradiction. □

There is a geometric explanation to the Köcher principle: If Γ is a torsion free subgroup of finite index of $\mathrm{SL}(\mathfrak{A} \oplus \mathfrak{B})$, a modular form f of weight $\boldsymbol{k}$ is a section a of line bundle on $\Gamma \backslash \mathcal{H}^g$. In fact, $j_{\boldsymbol{k}}(\mu, z)$ is the factor of automorphy for this line bundle. This is very similar to the case of elliptic curves discussed in Chapter 1, §4. One can interpret this line bundle in terms of moduli. Recall that for every $z \in \mathcal{H}^g$ we constructed an abelian variety

$$A_z = \mathbb{C}^g / (\mathcal{O}_L \cdot z + \mathfrak{A} \cdot 1) \tag{3.18}$$

with $\mathcal{O}_L$-action and polarization module. The trivial line bundle

$$\mathbb{C}^g \times \mathcal{H}^g \to \mathcal{H}^g \tag{3.19}$$

may be interpreted as either the tangent space $\mathfrak{t}_{A_z,0}$ or the cotangent space $\mathfrak{t}^*_{A_z,0}$ to A_z at the origin. The usual decomposition $\mathbb{C}^g = \oplus_{i=1}^g \mathbb{C}$ is in fact a decomposition of $\mathfrak{t}^*_{A_z,0}$ into 1-dimensional vector spaces $\oplus_{i=1}^g L_{i,z}$ such that $\mathcal{O}_L$ acts on $L_{i,z}$ via σ_i.

Let $\boldsymbol{k} = (0, \ldots, 1, \ldots, 0)$ (1 in the i-th place) then $j_{\boldsymbol{k}}(\mu, z)$ is the automorphy factor for the line bundle $L_{i,z}$. That is, the relative cotangent space to the "universal" polarized abelian scheme with $(\mathcal{D}_L \mathfrak{A})^{-1}$-polarization $\mathcal{X} \to \Gamma \backslash \mathcal{H}^g$ denoted $\mathfrak{t}^*_{\mathcal{X}, \Gamma \backslash \mathcal{H}^g}$ decomposes into line bundles: $\mathfrak{t}^*_{\mathcal{X}, \Gamma \backslash \mathcal{H}^g} = \bigoplus_{i=1}^g L_i$ such that $\mathcal{O}_L$ acts on L_i via σ_i.

The transformation law (3.6) shows that a modular form f of weight $\boldsymbol{k}$ is a global section of the line bundle determined by the factor of automorphy $j_{\boldsymbol{k}}(\mu, z)$. That is, $f \in H^0(\Gamma \backslash \mathcal{H}^g, \bigotimes_{i=1}^g L_i^{\otimes k_i})$.

As discussed above, one may add finitely many cusps to $\Gamma\backslash\mathcal{H}^g$ and get a compact normal complex manifold $\Gamma\backslash\mathcal{H}^{g*}$ on which the divisor of f is well-defined. If f is *not* holomorphic then normality implies that its divisor of poles is of codimension 1 (and thus "visible" on $\Gamma\backslash\mathcal{H}^g$). Thus, f holomorphic on $\Gamma\backslash\mathcal{H}^g$ implies that f is holomorphic on $\Gamma\backslash\mathcal{H}^{g*}$ if $g > 1$.

DEFINITION 3.5. A Hilbert modular form f with respect to Γ is a *cusp form* if the constant a_0 in the Fourier expansion of $f|_{\boldsymbol{k}}\mu$ is zero for all $\mu \in \mathrm{GL}_2(L)^+$.

Put

(3.20) $\mathcal{M}(\mathbb{C}, \boldsymbol{k}, \Gamma) =$ complex vector space of modular forms of weight $\boldsymbol{k}$, level Γ

and

(3.21) $\mathcal{S}(\mathbb{C}, \boldsymbol{k}, \Gamma) =$ complex vector space of cusp forms of weight $\boldsymbol{k}$, level Γ.

Let $\mathcal{M}(\mathbb{C}, \Gamma) = \bigoplus_{\boldsymbol{k}\in\mathbb{Z}^g} \mathcal{M}(\mathbb{C}, \boldsymbol{k}, \Gamma)$ and $\mathcal{S}(\mathbb{C}, \Gamma) = \bigoplus_{\boldsymbol{k}\in\mathbb{Z}^g} \mathcal{S}(\mathbb{C}, \boldsymbol{k}, \Gamma)$. The ring $\mathcal{M}(\mathbb{C}, \Gamma)$ is a $\mathbb{Z}^g$-graded ring and $\mathcal{S}(\mathbb{C}, \Gamma)$ is a graded ideal of it.

PROPOSITION 3.6. (1) *If some $k_i \leq 0$ and $\boldsymbol{k} \neq (0, \ldots, 0)$ then we have $\mathcal{M}(\mathbb{C}, \boldsymbol{k}, \Gamma) = \{0\}$.*
(2) $\mathcal{M}(\mathbb{C}, 0, \Gamma) = \mathbb{C}$, $\mathcal{S}(\mathbb{C}, 0, \Gamma) = \{0\}$.
(3) *If $\mathcal{S}(\mathbb{C}, \boldsymbol{k}, \Gamma) \neq \mathcal{M}(\mathbb{C}, \boldsymbol{k}, \Gamma)$, then $k_1 = \cdots = k_g$.*

PROOF (cf. [**44**]). The case $g = 1$ is classical and we shall assume it known, though the argument below could be used to prove this case as well.

We note that we may replace Γ by a smaller group and hence we may assume that Γ has no elliptic elements. We shall use the fact, mentioned above, that for $g > 1$ the non-singular complex variety $\Gamma\backslash\mathcal{H}^g$ has a compactification $\Gamma\backslash\mathcal{H}^{g*}$ and it is a normal variety to which every modular form extends and thus the divisor of every modular form on $\Gamma\backslash\mathcal{H}^{g*}$ is already visible in $\Gamma\backslash\mathcal{H}^g$.

Part 2 of the proposition follows easily, because modular forms of weight 0 are identified with holomorphic functions on a normal compact connected variety and are therefore constant.

Let $f \in \mathcal{M}(\mathbb{C}, \boldsymbol{k}, \Gamma)$, $f \neq 0$. A *twisted diagonal curve* mod Γ' is a curve of the form $\{\mu(z, \ldots, z) : z \in \mathcal{H}\}$ mod Γ' (for μ a fixed element of $\mathrm{GL}_2(L)^+$). Note that letting μ range over $\mathrm{GL}_2(L)^+$, $\{\mu(z, \ldots, z) : z \in \mathcal{H}\}$ is dense in $\mathcal{H}^g$ (Note that $\mathrm{GL}_2(L)^+$ is dense in $\prod_{i=1}^g \mathrm{GL}_2(\mathbb{R})^+$). Thus $\exists \mu \in \mathrm{GL}_2(L)^+$ such that

(3.22) $$f|_{\boldsymbol{k}}\mu \text{ is not zero on } \{(z, \ldots, z) \in \mathcal{H}\} \bmod \Gamma'.$$

Take $\Gamma' \subseteq \Gamma$ with Γ' commensurable with Γ, such that $f|_{\boldsymbol{k}}\mu$ is modular with respect to Γ'. Consider the diagonal map

(3.23) $$\mathcal{H}/\Gamma'' \xrightarrow{\Phi} \mathcal{H}^g/\Gamma',$$

well-defined for some congruence subgroup $\Gamma'' \subset \mathrm{SL}_2(\mathbb{Z})$. Then $\Phi^*(f|_{\boldsymbol{k}}\mu)$ is a non-zero modular form of weight $\sum_{i=1}^g k_i$ with respect to Γ''. Hence

(3.24) $$\sum_{i=1}^g k_i \geq 0.$$

Consider first the case $\sum_{i=1}^g k_i > 0$ and some k_i, say k_1, is 0. Let $f \in \mathcal{M}(\mathbb{C}, \boldsymbol{k}, \Gamma)$. We shall ultimately show that $f = 0$. We have $a_0 = r_{\boldsymbol{k}}(\epsilon)a_0$, for

$\left(\begin{smallmatrix}\epsilon & 0\\ 0 & 1\end{smallmatrix}\right) \in \Gamma$. Since $\boldsymbol{k} \neq \boldsymbol{0}$:

$$r_{\boldsymbol{k}}(\epsilon) = \prod_{i=2}^{g} \big(\sigma_i(\epsilon)\big)^{k_i/2} < 1, \tag{3.25}$$

for a suitable ϵ, and that implies that $a_0 = 0$. The same argument holds for $(f|_{\boldsymbol{k}}\mu)$ for any $\mu \in \mathrm{GL}_2(L)^+$; that is, under $\mathrm{GL}_2(L)^+$, all the cusps are equivalent, so we just need to consider the q-expansion at one cusp. Thus, f is a cusp form.

Before proving that $f = 0$ we note that the same argument proves Part 3: If $\boldsymbol{k}$ is not parallel, then we can always find ϵ such that:

$$r_{\boldsymbol{k}}(\epsilon) = \prod_{i=1}^{g} \big(\sigma_i(\epsilon)\big)^{k_i/2} \neq 1, \tag{3.26}$$

and $\left(\begin{smallmatrix}\epsilon & 0\\ 0 & 1\end{smallmatrix}\right) \in \Gamma$. Hence the constant coefficient of the expansion is zero, and we have a cusp form. That is, if $f \in \mathcal{M}(\mathbb{C}, \boldsymbol{k}, \Gamma)\backslash\mathcal{S}(\mathbb{C}, \boldsymbol{k}, \Gamma)$ then $k_1 = \cdots = k_g$.

Let us finish the proof of Part 1. Recall the formula:

$$\mathrm{Im}(\delta\tau) = \frac{\det(\delta)\,\mathrm{Im}(\tau)}{|\delta_3\tau + \delta_4|^2}, \qquad \delta = \begin{pmatrix}\delta_1 & \delta_2\\ \delta_3 & \delta_4\end{pmatrix} \in \mathrm{GL}_2(\mathbb{R})^+. \tag{3.27}$$

Put

$$g(z) = \left(\prod_{i=1}^{g} (\mathrm{Im}\, z_i)^{k_i/2}\right) f(z). \tag{3.28}$$

Then,

- $g(z)$ is *holomorphic* in z_1, ($k_1 = 0!$).
- $|g|$ is Γ-invariant:

$$\begin{aligned}|g(\mu z)| &= \left(\prod_{i=1}^{g} \big(\mathrm{Im}(\sigma_i(\mu)z_i)\big)^{k_i/2}\right)|f(\mu z)| \\ &= \prod_{i=1}^{g}\left(\frac{\det\big(\sigma_i(\mu)\big)\cdot \mathrm{Im}(z_i)}{|\sigma_i(c)z_i + \sigma_i(d)|^2}\right)^{k_i/2} \prod_{i=1}^{g} |\sigma_i(c)z_i + \sigma_i(d)|^{k_i} \\ &\qquad\qquad \times \big(\det\sigma_i(\mu)\big)^{-k_i/2}|f(z)| \\ &= |g(z)|.\end{aligned} \tag{3.29}$$

- $|g| = 0$ at the cusps ($a_0 = 0$).

The two last statements imply that $|g|$ is bounded on $\mathcal{H}^g$ (because $\Gamma\backslash\mathcal{H}^{g*}$ is compact), hence, by Liouville's theorem, g is constant as a function of z_1. Fix some $(z_1, \ldots, z_g)$. The assumption yields:

$$\forall \mu \in \Gamma \quad |g(z_1, \sigma_2(\mu)z_2, \ldots, \sigma_g(\mu)z_g)| = |g(z)|. \tag{3.30}$$

However, $\{(z_1, \sigma_2(\mu)z_2, \ldots, \sigma_g(\mu)z_g)\}_{\mu\in\Gamma}$ is *dense* in $\{z_1\}\times\mathcal{H}^{g-1}$, hence $|g|$ constant implies $|g| = 0$, so $f = 0$.

Consider now the case $\sum_{i=1}^{g} k_i \geq 0$ and some k_i is negative; say $k_1 < 0$. There exists a non-zero modular form G_4 of weight $\mathbf{4}$ on Γ (see §4.1). Consider $\tilde{f} = f^4 G_4^{-k_1}$. It is a cusp form of weight $\tilde{\boldsymbol{k}} = (\widetilde{k_1}, \ldots, \widetilde{k_g})$, and $\sum \widetilde{k_i} > 0$, $\widetilde{k_1} = 0$. Therefore $\tilde{f} = 0$, hence $f = 0$. ∎

EXERCISE 3.7. *Let $g > 1$. Prove the following facts (used in the proof above).*

(1) *The group* $\mathrm{GL}_2(L)$ *is dense in* $\prod_{i=1}^{g}\mathrm{GL}_2(\mathbb{R})$ *under the usual map* $\mu \mapsto (\sigma_1(\mu), \ldots, \sigma_g(\mu))$.
(2) *Let* Γ *be a group of finite index in* $\mathrm{GL}_2(\mathcal{O}_L)$. *Prove that* Γ *is dense in* $\prod_{i=2}^{g}\mathrm{GL}_2(\mathbb{R})$ *under the map* $\mu \mapsto (\sigma_2(\mu), \ldots, \sigma_g(\mu))$.

DEFINITION 3.8. Let $\mathfrak{A}_1, \ldots, \mathfrak{A}_{h^+}$ be a complete set of representatives to $\mathrm{Cl}(L)^+$. A Hilbert modular form of level $\Gamma_0(\mathfrak{B})$ and weight $\boldsymbol{k}$ is an h^+ tuple

$$(f_1, \ldots, f_{h^+}), \tag{3.31}$$

where each f_i is a modular form of weight $\boldsymbol{k}$ with respect to the subgroup $\Gamma(\mathfrak{A}_i, \mathfrak{B})$ of $\mathrm{GL}(\mathcal{O}_L \oplus \mathfrak{A}_i^{-1})^+$, where

$$\Gamma(\mathfrak{A}_i, \mathfrak{B}) = \left\{ \begin{pmatrix} a & b \\ c & d \end{pmatrix} : a, d \in \mathcal{O}_L, b \in \mathfrak{A}_i^{-1}, c \in \mathfrak{A}_i\mathfrak{B}, ad - bc \in \mathcal{O}_L^{\times +} \right\}. \tag{3.32}$$

4. Construction of Hilbert modular forms

4.1. Eisenstein series. Fix a fractional ideal $\mathfrak{A}$ of L.
The complex manifold

$$\mathrm{GL}(\mathcal{O}_L \oplus \mathfrak{A}^{-1})^+ \backslash \mathcal{H}^g \tag{4.1}$$

parameterizes complex abelian varieties with RM by $\mathcal{O}_L$ and polarization module $(\mathfrak{A}\mathcal{D}_L^{-1}, (\mathfrak{A}\mathcal{D}_L^{-1})^+)$. Take $\mathfrak{B}$ a fractional ideal of L with class B in $\mathrm{Cl}(L)$. Put $\boldsymbol{k} = (k, \ldots, k)$, $k \geq 2$, $k \in 2\mathbb{Z}$.

DEFINITION 4.1. The Eisenstein series of weight $\boldsymbol{k}$ and class B is:

$$G_{\boldsymbol{k},B}(z) := \mathbb{N}(\mathfrak{B})^k \sum_{(\alpha,\beta)\in\mathfrak{B}\mathfrak{A}\oplus\mathfrak{B}}{}^{\prime\prime} \mathbb{N}(\alpha z + \beta)^{-\boldsymbol{k}}, \tag{4.2}$$

where $\mathbb{N}(\mathfrak{B})$ is the norm to $\mathbb{Q}$ of the ideal $\mathfrak{B}$ and where the symbol $''$ indicates the following: We say that (α, β) and (μ, δ) are *associated* if there exists an $\epsilon \in \mathcal{O}_L^\times$ such that:

$$(\alpha, \beta) = (\epsilon\mu, \epsilon\delta). \tag{4.3}$$

This is an equivalence relation, and the summation is over representatives of equivalence classes different from $\{(0,0)\}$. We use the notation $\mathbb{N}(v) = \mathbb{N}(v_1, \ldots, v_g) = v_1 \cdots v_g$.

Let us show that the Eisenstein series of weight $\boldsymbol{k}$ and class B is indeed well-defined: If $(\alpha, \beta) \sim (\mu, \delta)$, say $(\alpha, \beta) = (\epsilon\mu, \epsilon\delta)$, the norms satisfy

$$\mathbb{N}(\alpha z + \beta)^{-\boldsymbol{k}} = \mathbb{N}(\epsilon)^{-\boldsymbol{k}}\mathbb{N}(\mu z + \delta)^{-\boldsymbol{k}} = \mathbb{N}(\mu z + \delta)^{-\boldsymbol{k}}, \tag{4.4}$$

because $\boldsymbol{k}$ is parallel and even implies $\mathbb{N}(\epsilon)^{-\boldsymbol{k}} = 1$.

REMARK 4.2. The sum converges for $k > 2$, and also for $k = 2$, provided $g > 1$, and defines a holomorphic function on $\mathcal{H}^g$. Note that $G_{\boldsymbol{k},B}$ depends only on the ideal class of B.

An Eisenstein series is a modular form with respect to $\mathrm{GL}(\mathcal{O}_L \oplus \mathfrak{A}^{-1})^+$. To see that note first that the group $\mathrm{GL}(\mathcal{O}_L \oplus \mathfrak{A}^{-1})^+$ is contained in the automorphism group of $\mathfrak{B}\mathfrak{A} \oplus \mathfrak{B}$:

$$(\mathfrak{B}\mathfrak{A} \oplus \mathfrak{B})\mu = \mathfrak{B}\mathfrak{A} \oplus \mathfrak{B}, \quad \forall \mu \in \mathrm{GL}(\mathcal{O}_L \oplus \mathfrak{A}^{-1})^+. \tag{4.5}$$

This follows from the definition of $\mathrm{GL}(\mathcal{O}_L \oplus \mathfrak{A}^{-1})^+$ as invertible elements with totally positive determinant of the ring

$$\begin{pmatrix} \mathcal{O}_L & \mathfrak{A}^{-1} \\ \mathfrak{A} & \mathcal{O}_L \end{pmatrix}. \tag{4.6}$$

The action of $\mathrm{GL}(\mathcal{O}_L \oplus \mathfrak{A}^{-1})^+$ respects the equivalence relation we have defined on pairs $(\alpha, \beta) \in \mathfrak{B}\mathfrak{A} \oplus \mathfrak{B}$. Furthermore, let $\mu = \left(\begin{smallmatrix} a & b \\ c & d \end{smallmatrix}\right)$. Then

$$\begin{aligned} \mathbb{N}(\alpha\mu z + \beta)^{-k} &= \left[\mathbb{N}\big(\alpha(az+b) + \beta(cz+d)\big) \cdot \mathbb{N}(cz+d)^{-1}\right]^{-k} \\ &= \mathbb{N}\left((\alpha,\beta)\begin{pmatrix} a & b \\ c & d \end{pmatrix}\begin{pmatrix} z \\ 1 \end{pmatrix}\right)^{-k} \mathbb{N}(cz+d)^k. \end{aligned} \tag{4.7}$$

Thus

$$G_{k,B}(\mu z) = \mathbb{N}\big((cz+d)\big)^k G_{k,B}(z) = j_{\boldsymbol{k}}(\mu, z) G_{k,B}(z), \tag{4.8}$$

where, as usual,

$$j_{\boldsymbol{k}} = \mathbb{N}(cz+d)^k \mathbb{N}(\det \mu)^{-k/2}, \tag{4.9}$$

which is equal to $\mathbb{N}(cz+d)^k$ if $\boldsymbol{k}$ is parallel and even.

DEFINITION 4.3. Let A be an ideal class, $\mathfrak{b}$ an integral ideal. Put

$$\zeta_A(k) = \sum_{\substack{\mathfrak{c}\in A \\ \mathfrak{c}\subseteq \mathcal{O}_L}} \mathbb{N}(\mathfrak{c})^{-k}; \quad \sigma_{k-1,A}(\mathfrak{b}) = \sum_{\substack{\mathfrak{c}\in A \\ \mathfrak{b}\subseteq\mathfrak{c}\subseteq \mathcal{O}_L}} \mathbb{N}(\mathfrak{c})^{k-1}. \tag{4.10}$$

THEOREM 4.4 (see [**44**, Chapter I, §6]). (1) *The modular forms $G_{k,B}$ on $\mathrm{GL}(\mathcal{O}_L \oplus \mathfrak{A}^{-1})^+ \backslash \mathcal{H}^g$ are linearly independent over $\mathbb{C}$, and span a vector space of dimension $h = |\operatorname{Cl}(L)|$.*

(2) *The Fourier expansion of $G_{k,B}$ with respect to the cusp $(i\infty, \ldots, i\infty)$ is*

$$G_{k,B} = c\left\{\frac{\zeta_{\mathcal{D}_L B}(1-k)}{2^g} + \sum_{\substack{\nu\in\mathfrak{A}\mathcal{D}_L^{-1} \\ \nu\gg 0}} \sigma_{k-1,B\mathcal{D}_L}\big((\nu)\mathfrak{A}^{-1}\mathcal{D}_L\big) e^{2\pi i \operatorname{Tr}(\nu z)}\right\}, \tag{4.11}$$

with $c = \dfrac{(2\pi i)^{kg}}{(k-1)!^g} d_L^{1/2-k}$.

REMARK 4.5. The number $2^{-g}\zeta_{\mathcal{D}_L B}(1-k)$ is a rational number by a theorem of Siegel, and $\sigma_{k-1,B\mathcal{D}_L}\big((\nu)\mathfrak{A}^{-1}\mathcal{D}_L\big)$ are integers by definition.

In what follows, we give a heuristic explanation for the rationality of $\zeta_{\mathcal{D}_L B}(1-k)$:

There exists an algebraic variety V defined over $\mathbb{Q}$, whose underlying analytic variety is isomorphic analytically to $\mathrm{GL}(\mathcal{O}_L \oplus \mathfrak{A}^{-1})^+ \backslash \mathcal{H}^g$. The variety V is a coarse moduli space for abelian varieties with RM and polarization module $\mathfrak{A}\mathcal{D}_L^{-1}$. If $\mathcal{X} \to V$ is the "universal object" then the space of modular forms of parallel weight $\boldsymbol{k}$ is $\Gamma(V, \det \mathfrak{t}_{\mathcal{X},V}^{*\otimes k})$ and therefore the space of modular forms of parallel weight $\boldsymbol{k}$ has $\mathbb{Q}$-rational structure. This vector space is contained in $\operatorname{Mor}(V, \det \mathfrak{t}_{\mathcal{X},V}^{*\otimes k})$, a space with a natural $\operatorname{Gal}(\mathbb{C}/\mathbb{Q})$-action since both $\mathfrak{A}$ and $\det \mathfrak{t}_{\mathcal{X},V}^{*\otimes k}$ are defined over $\mathbb{Q}$.

If f is a modular form, so is f^σ for $\sigma \in \operatorname{Gal}(\mathbb{C}/\mathbb{Q})$. The q-expansion principle says that the q-expansion of f is in some sense algebraic. It is really the development of f into a power series with respect to local parameters in the local ring at infinity that are defined over $\mathbb{Q}$. That is, if $f = \sum_{\nu \in \mathfrak{A}\mathcal{O}_L^{-1}} a_\nu e^{2\pi i \operatorname{Tr} \nu z}$, one finds that $f^\sigma = \sum_{\nu \in \mathfrak{A}\mathcal{O}_L^{-1}} a_\nu^\sigma e^{2\pi i \operatorname{Tr} \nu z}$. Thus, the constant

$$\left(\frac{\zeta_{B,\mathcal{D}_L}(1-k)}{2^g}\right) - \left(\frac{\zeta_{B,\mathcal{D}_L}(1-k)}{2^g}\right)^\sigma = \left(\frac{1}{c}G_{k,B}\right) - \left(\frac{1}{c}G_{k,B}\right)^\sigma \tag{4.12}$$

is a modular form of weight $k > 0$, which is possible only if this constant is zero.

EXERCISE 4.6. *Prove that a non-zero modular form of weight $\boldsymbol{k} \neq (0, \ldots, 0)$ has infinitely many terms in its q-expansion.*

We also define the following Eisenstein series:

$$E_k^{L,*} = \frac{1}{c} \sum_{B \in \mathrm{Cl}(L)} G_{k,B}, \tag{4.13}$$

and

$$E_k^L = 2^g \zeta_L(1-k)^{-1} E_k^{L,*}. \tag{4.14}$$

Thus, the q-expansion of $E_k^{L,*}$ at the cusp $(i\infty, \ldots, i\infty)$ is

$$\frac{\zeta_L(1-k)}{2^g} + \sum_{\substack{\nu \in \mathfrak{A}\mathcal{D}_L^{-1} \\ \nu \gg 0}} \sigma_{k-1}\big((\nu)\mathfrak{A}^{-1}\mathcal{D}_L\big) e^{2\pi i \operatorname{Tr}(\nu z)}, \tag{4.15}$$

where

$$\sigma_{k-1}(\mathfrak{b}) = \sum_{\mathfrak{b} \subseteq \mathfrak{c} \subseteq \mathcal{O}_L} \mathbb{N}(\mathfrak{c})^{k-1}. \tag{4.16}$$

4.2. Other methods of constructing modular forms. We indicate below a method of constructing modular forms. In essence, all modular forms come one way or another from theta series or from Eisenstein series. The method that we now demonstrate by a particular example is to pull back modular forms from other modular varieties.

Let $\mathfrak{A}$ be a fractional ideal of L, taken with its natural notion of positivity, and consider the (coarsely) representable functor Φ of abelian varieties with RM and polarization module $\mathfrak{A}$. That is, the functor assigning to a scheme S over $\mathbb{C}$ the isomorphism classes of triples over S

$$(A, \iota, m\colon (\mathcal{M}_A, \mathcal{M}_A^+) \cong (\mathfrak{A}, \mathfrak{A}^+)). \tag{4.17}$$

Choose an element $\lambda \in \mathfrak{A}^+$, consider it as an element of $\mathcal{M}_A^+$, and let $\boldsymbol{d} = (d_1 | \ldots | d_g)$ be the elementary divisors of $\mathrm{Ker}(\lambda)$ (Thus, the kernel of the polarization is isomorphic to $(\bigoplus_{i=1}^g \mathbb{Z}/d_i\mathbb{Z})^2$. Note that the d_i's may well be equal to 1). There is a natural transformation of functors from the functor Φ to the (coarsely) representable functor Ψ of complex abelian varieties of dimension g with a polarization of elementary divisors $\boldsymbol{d} = (d_1, \ldots, d_g)$. If we let $\mathcal{M}^{\mathfrak{A}}(\mathbb{C})$ denote the moduli scheme over $\mathbb{C}$ that (coarsely) represents Φ, and we let $\mathcal{S}_{\boldsymbol{d}}(\mathbb{C})$ denote the moduli scheme that (coarsely) represents Ψ,[5] we get a morphism of schemes

$$\varphi\colon \mathcal{M}^{\mathfrak{A}}(\mathbb{C}) \to \mathcal{S}_{\boldsymbol{d}}(\mathbb{C}). \tag{4.18}$$

[5]Schemes like $\mathcal{M}^{\mathfrak{A}}$ usually go under the name of *Hilbert modular schemes* (not to be confused with Hilbert schemes!), while the schemes $\mathcal{S}_{\boldsymbol{d}}$ usually go under the name of Siegel modular schemes.

(Such morphisms and their generalizations are usually called modular embeddings, though they are rarely *embeddings*. Another example of such a map is the diagonal curve discussed in §5). One may write (4.18) explicitly. In particular, $\mathcal{S}_{\boldsymbol{d}}(\mathbb{C}) = \Gamma\backslash\mathcal{H}_g$ for a suitable discrete subgroup of $\mathrm{Sp}(2g,\mathbb{Q})$. See [**155**, pp. 640–641]. It can be formulated as coming from a suitable embedding of $\mathrm{GL}_2(L)$ in $\mathrm{Sp}(2g,\mathbb{Q})$, and the interpretation of the moduli spaces as double cosets spaces of the adelic points of these algebraic groups.

Every modular form of weight k on $\mathcal{S}_{\boldsymbol{d}}(\mathbb{C})$ (that means: $f\colon \mathcal{H}_g \to \mathbb{C}$ such that

$$f(\gamma\tau) = f\big((A\tau+B)(C\tau+D)^{-1}\big) = \det(C\tau+D)^k f(\tau), \quad \gamma = \begin{pmatrix} A & B \\ C & D \end{pmatrix} \tag{4.19}$$

in a suitable subgroup of $\mathrm{Sp}(2g,\mathbb{Z})$) pulls back to a modular form on $\mathcal{M}^{\mathfrak{A}}(\mathbb{C})$ of parallel weight $(k,\dots,k)$. While this may be checked by a laborious calculation, using the explicit form of the map $\varphi\colon \mathcal{M}^{\mathfrak{A}}(\mathbb{C}) \to \mathcal{S}_{\boldsymbol{d}}(\mathbb{C})$, it may also be justified easily using Katz's view of modular forms. We discussed that point of view in the case of elliptic curves in Chapter 1, §4. The case of abelian varieties with RM is similar and is discussed in Chapter 5, §1.

Based on that we argue as follows. Let f be a modular form on $\mathcal{S}_{\boldsymbol{d}}(\mathbb{C})$. Let $\lambda \in \mathfrak{A}^+$ as above. Given

$$(A, \iota\colon \mathcal{O}_L \hookrightarrow \mathrm{End}_S(A), m\colon (\mathcal{M}_A, \mathcal{M}_A^+) \to (\mathfrak{A}, \mathfrak{A}^+), \omega)/S, \tag{4.20}$$

where ω is an $\mathcal{O}_L \otimes \mathcal{O}_S$ basis to $\mathbf{t}^*_{A/S}$, we get the data (A, λ, ω'), with ω' a basis for $\det \mathbf{t}^*_{A/S}$ and λ considered as an element of $\mathcal{M}_A^+$. Here, we first decompose $\mathcal{O}_L \otimes \mathcal{O}_S$ as $\bigoplus_{\sigma\in\mathrm{Emb}(L,\mathbb{C})} \mathcal{O}_S$ ($\mathcal{O}_S$ is a sheaf of $\mathbb{C}$-algebras), and get from ω an $\mathcal{O}_S$ basis $\{\omega_1,\dots,\omega_g\}$ to $\mathbf{t}^*_{A/S}$. We then let $\omega' = \omega_1 \wedge \cdots \wedge \omega_g$. The transformation of functors $(A,\iota,m,\omega) \mapsto (A,\lambda,\omega')$ is φ. Thus, $\varphi^* f$ is given by

$$(A,\iota,m,\omega)/S \overset{\varphi^* f}{\longmapsto} f(A,\lambda,\omega'). \tag{4.21}$$

It is immediate that this is a modular form!

Take for example the case when $\mathfrak{A} = \mathcal{O}_L$ and λ is the principal polarization corresponding to 1. Then we consider a map

$$\mathrm{GL}(\mathcal{O}_L \oplus \mathcal{D}_{L/\mathbb{Q}}^{-1})^+\backslash\mathcal{H}^g \to \mathrm{Sp}(2g,\mathbb{Z})\backslash\mathcal{H}_g. \tag{4.22}$$

On $\mathcal{H}_g$ one can define Riemann's theta functions with characteristic $\left[\begin{smallmatrix}\epsilon\\ \epsilon'\end{smallmatrix}\right] \in \mathbb{Q}^{2g}$ by

$$\Theta\begin{bmatrix}\epsilon\\ \epsilon'\end{bmatrix}(\tau) = \sum_{N\in\mathbb{Z}^g} \exp\Big(2\pi i\Big\{\frac{1}{2}\,{}^t(N+\epsilon)\tau(N+\epsilon) + {}^t(N+\epsilon)\epsilon'\Big\}\Big). \tag{4.23}$$

They are holomorphic functions on $\mathcal{H}_g$. One can show, using the transformation formula for theta functions ([**90**, Chapter 8.6]) that if $n\epsilon, n\epsilon' \in \mathbb{Z}^g$ then $\Theta^{8n}\left[\begin{smallmatrix}\epsilon\\ \epsilon'\end{smallmatrix}\right]$ is a modular form of weight $4n$ on the principal congruence subgroup of level $2n$; if n is even $\Theta^{2n}\left[\begin{smallmatrix}\epsilon\\ \epsilon'\end{smallmatrix}\right]$ is a modular form of weight n on the principal congruence subgroup of level n; $\Theta^2\left[\begin{smallmatrix}\epsilon\\ \epsilon'\end{smallmatrix}\right]$ is a modular form of weight 1 on the principal congruence subgroup of level $4n^2$.

By pulling back these modular forms we get Hilbert modular forms, of some level. We note that the construction is arithmetic. The morphism

$$f\colon \mathcal{M}_L^{\mathcal{O}} \to \mathcal{S}_{(1,\dots,1)} \tag{4.24}$$

exists over any scheme. The theta functions often have the g.c.d. of their q-expansion coefficients equal to one, and in any event the g.c.d. divides $2n$ if $n\epsilon, n\epsilon' \in \mathbb{Z}^g$.

We remark that Shimura used vector valued modular forms on Siegel spaces in [**155**, pp. 640–642] to obtain Hilbert modular forms of weight $(1, \ldots, 1, 2, 1, \ldots, 1)$.

EXERCISE★ 4.7. *For which weights are there non-zero modular forms? (Same question in positive characteristic).*

5. More on the diagonal curve

Consider the case $[L : \mathbb{Q}] = 2$. To ease notation we assume that $L \subset \mathbb{C}$ and we write $\mathrm{Emb}(L, \mathbb{C}) = \{1, \sigma\}$. We have a commutative diagram:

$$\begin{array}{ccc} \mathcal{H} & \xrightarrow{\Phi} & \mathcal{H}^2 \\ \downarrow & & \downarrow \\ \mathrm{SL}_2(\mathbb{Z})\backslash\mathcal{H} & \xrightarrow{\Phi} & \mathrm{SL}_2(\mathcal{O}_L)\backslash\mathcal{H}^2, \end{array} \tag{5.1}$$

where Φ is the diagonal map: $\Phi(z) = (z, z)$, and $\mathrm{SL}_2(\mathcal{O}_L) = \mathrm{SL}(\mathcal{O}_L \oplus \mathcal{O}_L)$.

PROPOSITION 5.1. *The map Φ is generically injective.*

PROOF. Let $\lambda \in \mathrm{SL}_2(\mathcal{O}_L)$ be a matrix such that for all $z \in \mathcal{H}$ there exists $\tau \in \mathcal{H}$ such that $\lambda(z, z) = (\tau, \tau)$. Then $\forall z$ we have $\sigma(\lambda)^{-1}\lambda z = z$. So $\sigma(\lambda)^{-1}\lambda = \left(\begin{smallmatrix} a & b \\ c & d \end{smallmatrix}\right)$ and $cz^2 + dz = az + b$ for every $z \in \mathcal{H}$. Hence $b = c = 0$ and $a = d$, with $ad = 1$. That is, $\sigma(\lambda)^{-1}\lambda = \pm 1$. We have to consider two cases:

(1) $\sigma(\lambda) = \lambda \implies \lambda \in \mathrm{SL}_2(\mathbb{Z})$,
(2) $\sigma(\lambda) = -\lambda \implies \lambda = \sqrt{d}\left(\begin{smallmatrix} a & b \\ c & d \end{smallmatrix}\right)$, $a, b, c, d \in \mathbb{Z} \implies \det \lambda \neq 1$,

whence a contradiction. Thus, the stabilizer of the diagonal in $\mathrm{SL}_2(\mathcal{O}_L)$ is $\mathrm{SL}_2(\mathbb{Z})$ and the proof is complete. □

EXERCISE 5.2. *Find the fibers of the map $\Phi\colon \mathrm{SL}_2(\mathbb{Z})\backslash\mathcal{H} \to \mathrm{SL}_2(\mathcal{O}_L)\backslash\mathcal{H}^2$.*

5.1. Modular interpretation. Let $\tau \in \mathcal{H}$ and let Λ be the lattice $\mathbb{Z} + \mathbb{Z}\tau$ in $\mathbb{C}$; let $\tilde{\Lambda}$ be the lattice $\mathcal{O}_L \cdot (\tau, \tau) + \mathcal{O}_L \cdot (1, 1)$ in $\mathbb{C}^2$. Consider the map

$$\mathbb{C} \to \mathbb{C} \otimes \mathcal{O}_L, \quad z \mapsto z \otimes 1. \tag{5.2}$$

It induces a map

$$\mathbb{C}/\Lambda \xrightarrow{\Phi} (\mathbb{C} \otimes \mathcal{O}_L)/(\Lambda \otimes \mathcal{O}_L). \tag{5.3}$$

Under the isomorphism $\mathbb{C} \otimes \mathcal{O}_L \cong \mathbb{C}^2$, given by $z \otimes \ell \mapsto (\ell z, \sigma(\ell) z)$, the lattice $\Lambda \otimes \mathcal{O}_L$ is mapped to the lattice $\tilde{\Lambda}$. Let $E = \mathbb{C}/\Lambda$ and $A = \mathbb{C}^2/\tilde{\Lambda}$. We note that the map $\Phi\colon \mathrm{SL}_2(\mathbb{Z})\backslash\mathcal{H} \to \mathrm{SL}_2(\mathcal{O}_L)\backslash\mathcal{H}^2$ is none other then the map

$$E \to E \otimes_{\mathbb{Z}} \mathcal{O}_L, \tag{5.4}$$

given by $r \mapsto r \otimes 1$.

Assume that $\mathcal{O}_L = \mathbb{Z}[\sqrt{d}]$, with basis $1, \sqrt{d}$. Then

$$\tilde{\Lambda} = \left\langle \begin{pmatrix} \tau \\ \tau \end{pmatrix}, \begin{pmatrix} \sqrt{d}\tau \\ -\sqrt{d}\tau \end{pmatrix}, \begin{pmatrix} 1 \\ 1 \end{pmatrix}, \begin{pmatrix} \sqrt{d} \\ -\sqrt{d} \end{pmatrix} \right\rangle, \tag{5.5}$$

and $\sqrt{d}$ acts on $\mathbb{C}^2$ by $\left(\begin{smallmatrix}\sqrt{d} & 0\\ 0 & -\sqrt{d}\end{smallmatrix}\right)$. The rational representation of $\sqrt{d}$ with respect to the basis given in (5.5) is:

$$\sqrt{d} \mapsto \begin{pmatrix} 0 & d & 0 & 0\\ 1 & 0 & 0 & 0\\ 0 & 0 & 0 & d\\ 0 & 0 & 1 & 0 \end{pmatrix}.$$

We may change the coordinate on $\mathbb{C}^2$ by:

$$\begin{pmatrix} 1 & \sqrt{d}\\ 1 & -\sqrt{d}\end{pmatrix}^{-1} = -\frac{1}{2\sqrt{d}}\begin{pmatrix} -\sqrt{d} & -\sqrt{d}\\ -1 & 1\end{pmatrix}. \tag{5.6}$$

Under that change of coordinates $\tilde{\Lambda}$ is mapped to

$$\tilde{\tilde{\Lambda}} = \left\langle \begin{pmatrix}\tau\\ 0\end{pmatrix}, \begin{pmatrix}0\\ \tau\end{pmatrix}, \begin{pmatrix}1\\ 0\end{pmatrix}, \begin{pmatrix}0\\ 1\end{pmatrix} \right\rangle, \tag{5.7}$$

(This lattice is obviously defining $E \times E$) and $\sqrt{d}$ acts now on $\mathbb{C}^2$ by

$$\begin{pmatrix} 1 & \sqrt{d}\\ 1 & -\sqrt{d}\end{pmatrix}^{-1} \begin{pmatrix} \sqrt{d} & 0\\ 0 & -\sqrt{d}\end{pmatrix} \begin{pmatrix} 1 & \sqrt{d}\\ 1 & -\sqrt{d}\end{pmatrix} = \begin{pmatrix} 0 & d\\ 1 & 0\end{pmatrix} \tag{5.8}$$

(with the same rational representation).

What are the $\mathcal{O}_L$-linear symmetric homomorphisms $\mu\colon E \times E \to (E \times E)^\vee$?

Let $\lambda\colon E \times E \to (E \times E)^\vee$ be the canonical product polarization. That is, if we let $p_i\colon E \times E \to E$ be the projection on the i-th component then $\lambda = \phi_{\mathcal{L}}$, where $\mathcal{L} = p_1^*\mathcal{O}_E([0]) \otimes p_2^*\mathcal{O}_E([0])$.

REMARK 5.3. The map μ is $\mathcal{O}_L$-linear iff $\mu m = m^\vee \mu$ for all $m \in \mathcal{O}_L$, iff $(\lambda^{-1}\mu)m = (\lambda^{-1}m^\vee\lambda)(\lambda^{-1}\mu)$.

Recall that $\mu \mapsto \lambda^{-1}\mu$ gives an embedding $\mathrm{NS}^0 \hookrightarrow \mathrm{End}^0(A)^{\mathrm{sym}}$, and that $m \mapsto \lambda^{-1}m^\vee\lambda$ is the Rosati involution defined by λ. In our case, identifying $\mathrm{End}(E^2)$ with $M_2(\mathrm{End}(E))$, where $\mathrm{End}(E) = \mathbb{Z}$ or an order in an imaginary quadratic field, one finds that the Rosati involution is given by

$$\begin{pmatrix} a & b\\ c & d\end{pmatrix} \mapsto \begin{pmatrix} a^\vee & c^\vee\\ b^\vee & d^\vee\end{pmatrix}, \tag{5.9}$$

where $^\vee$ is complex conjugation. Thus, the symmetric elements are the elements of the form

$$\begin{pmatrix} \alpha & \beta\\ \beta^\vee & \delta\end{pmatrix}, \tag{5.10}$$

with $\alpha, \delta \in \mathbb{Z}$ and $\beta \in \mathrm{End}(E)$.

Write $\lambda^{-1}\mu = \left(\begin{smallmatrix}\alpha & \beta\\ \beta^\vee & \delta\end{smallmatrix}\right)$, take $m = \left(\begin{smallmatrix}0 & d\\ 1 & 0\end{smallmatrix}\right)$. The map μ is $\mathcal{O}_L$-linear iff

$$\begin{pmatrix} \alpha & \beta\\ \beta^\vee & \delta\end{pmatrix}\begin{pmatrix} 0 & d\\ 1 & 0\end{pmatrix} = \begin{pmatrix} 0 & 1\\ d & 0\end{pmatrix}\begin{pmatrix} \alpha & \beta\\ \beta^\vee & \delta\end{pmatrix}. \tag{5.11}$$

That is, if and only if

$$\begin{pmatrix} \beta & d\alpha\\ \delta & d\beta^\vee\end{pmatrix} = \begin{pmatrix} \beta^\vee & \delta\\ d\alpha & d\beta\end{pmatrix}. \tag{5.12}$$

So $\beta = \beta^\vee$, $\delta = d\alpha$, and this implies:

$$\mathcal{M}_A = \left\{ \begin{pmatrix} \alpha & \beta \\ \beta & d\alpha \end{pmatrix} : \alpha, \beta \in \mathbb{Z} \right\}, \tag{5.13}$$

$$\mathcal{M}_A^+ = \left\{ \begin{pmatrix} \alpha & \beta \\ \beta & d\alpha \end{pmatrix} \in \mathcal{M}_A : \alpha > 0, d\alpha^2 - \beta^2 > 0 \right\}. \tag{5.14}$$

But $\mathcal{D}_L^{-1} = \langle \frac{1}{2}, \frac{1}{2\sqrt{d}} \rangle$ and $\mathcal{M}_A \cong \mathcal{D}_L^{-1}$ as $\mathcal{O}_L$-modules via

$$\begin{pmatrix} \alpha & \beta \\ \beta & d\alpha \end{pmatrix} \mapsto \alpha \cdot \frac{1}{2} + \beta \cdot \frac{1}{2\sqrt{d}}, \tag{5.15}$$

because

$$\begin{aligned} \begin{pmatrix} \alpha & \beta \\ \beta & d\alpha \end{pmatrix} \begin{pmatrix} 0 & d \\ 1 & 0 \end{pmatrix} &= \begin{pmatrix} \beta & d\alpha \\ d\alpha & d\beta \end{pmatrix} \\ &\mapsto \beta \cdot \frac{1}{2} + d\alpha \cdot \frac{1}{2\sqrt{d}} = \left(\alpha \cdot \frac{1}{2} + \beta \cdot \frac{1}{2\sqrt{d}} \right) \sqrt{d}. \end{aligned} \tag{5.16}$$

Under this isomorphism,

$$\begin{aligned} \mathcal{M}_A^+ \mapsto & \left\{ \alpha \cdot \frac{1}{2} + \beta \frac{1}{2\sqrt{d}} : \alpha > 0, d\alpha^2 > \beta^2 \right\} \\ &= \left\{ \alpha \cdot \frac{1}{2} + \beta \frac{1}{2\sqrt{d}} : \alpha + \frac{\beta}{\sqrt{d}} > 0, \alpha - \frac{\beta}{\sqrt{d}} > 0 \right\} \\ &= (\mathcal{D}_L^{-1})^+. \end{aligned} \tag{5.17}$$

Note that this fits with $\mathrm{SL}_2(\mathcal{O}_L)\backslash\mathcal{H}^2 = \mathrm{SL}(\mathcal{O}_L \oplus \mathcal{O}_L)\backslash\mathcal{H}^2$ parameterizing abelian varieties, constructed from lattices $\mathcal{O}_L \cdot \tau + \mathcal{O}_L$, and having polarization module $(\mathcal{D}_L^{-1}, (\mathcal{D}_L^{-1})^+)$.

EXERCISE 5.4. *It seems as if the identity map $E \otimes \mathcal{O}_L \to E \otimes \mathcal{O}_L$ is an $\mathcal{O}_L$-linear polarization, where we identify $(E \otimes \mathcal{O}_L)^\vee$ with $E \otimes \mathcal{O}_L$. (This identification comes from $E^\vee = E$ and is canonical). Yet it is not! Explain.*

6. Siegel's formula

Consider the case $\mathfrak{A} = \mathcal{O}_L$. Let

$$\Phi\colon \mathrm{SL}_2(\mathbb{Z})\backslash\mathcal{H} \to \mathrm{SL}_2(\mathcal{O}_L)\backslash\mathcal{H} \tag{6.1}$$

be the diagonal curve. Let $F_k = \Phi^* E_k^{L,*}$. It is a modular form on $\mathrm{SL}_2(\mathbb{Z})$ of weight gk with q-expansion:

$$F_k = \frac{\zeta_L(1-k)}{2^g} + \sum_{n=1}^{\infty} a_k(n) q^n, \tag{6.2}$$

where

$$a_k(n) = \sum_{\substack{\nu \in (\mathcal{D}_L^{-1})^+ \\ \mathrm{Tr}(\nu) = n}} \sigma_{k-1}\big((\nu)\mathcal{D}_L\big) = \sum_{\substack{\nu \in (\mathcal{D}_L^{-1})^+ \\ \mathrm{Tr}(\nu) = n}} \sum_{(\nu)\mathcal{D}_L \subseteq \mathfrak{A} \subseteq \mathcal{O}_L} \mathrm{N}(\mathfrak{A})^{k-1}. \tag{6.3}$$

Lets apply this when $g = k = 2$. Since the space of modular forms on $\mathrm{SL}_2(\mathbb{Z})$ of weight 4 is one-dimensional, F_2 is a multiple of

$$G_{4,\mathrm{SL}_2(\mathbb{Z})} = \frac{\zeta_{\mathbb{Q}}(-3)}{2} + \sum_{n=1}^{\infty} c_3(n) q^n = \frac{1}{240} + q + \cdots. \tag{6.4}$$

Hence $\frac{1}{4}\zeta_L(-1)/(1/240) = a_2(1)$, and

$$\zeta_L(-1) = \frac{1}{60} \sum_{\substack{\nu \in (\mathcal{D}_L^{-1})^+ \\ \mathrm{Tr}(\nu)=1}} \sum_{(\nu)\mathcal{D}_L \subseteq \mathfrak{A} \subseteq \mathcal{O}_L} \mathbb{N}(\mathfrak{A}). \tag{6.5}$$

LEMMA 6.1. *We have the following identity:*

$$a_2(1) = \sum_{\substack{a\in\mathbb{Z} \\ |a|<\sqrt{d_L} \\ a \equiv d_L \bmod 2}} \sigma_1\left(\frac{d_L - a^2}{4}\right). \tag{6.6}$$

PROOF. We have

$$\mathcal{D}_L^{-1} = \left\{ \frac{1}{\sqrt{d_L}} \left(\frac{a + b\sqrt{d_L}}{2} \right) : a, b \in \mathbb{Z}, a, b \equiv d_L \bmod 2 \right\} = \frac{1}{\sqrt{d_L}} \mathcal{O}_L. \tag{6.7}$$

EXERCISE 6.2. *Write the totally positive elements ν of trace 1 in $\mathcal{D}_L^{-1}$ explicitly. Find the ideals $\nu\mathcal{D}_L$ and obtain the formula*

$$a_2(1) = \sum_{\substack{|a|<\sqrt{d_L} \\ a \equiv d_L \bmod 2}} \sum_{([a+\sqrt{d_L}]/2) \subseteq \mathfrak{A} \subseteq \mathcal{O}_L} \mathbb{N}(\mathfrak{A}). \tag{6.8}$$

We rewrite (6.8) as

$$a_2(1) = \sum_{\substack{|a|<\sqrt{d_L} \\ a \equiv d_L \bmod 2}} \sum_{d|(d_L - a^2)/4} d \cdot S_{a,d_L}(d), \tag{6.9}$$

where

$$S_{a,d_L}(d) = \sum_{\substack{([a+\sqrt{d_L}]/2) \subseteq \mathfrak{A} \subseteq \mathcal{O}_L, \\ \mathbb{N}\mathfrak{A} = d}} 1. \tag{6.10}$$

We claim that $S_{a,d_L}(d) = 1$. Write

$$\left(\frac{a + \sqrt{d_L}}{2}\right) = \prod \mathcal{S}_i^{a_i} \overline{\mathcal{S}}_i^{c_i} \prod \mathcal{R}_i^{\beta_i} \prod \mathcal{T}_i^{\mu_i}, \tag{6.11}$$

where the first product runs over split primes, the second over ramified primes and the last one over inert primes.

Recall that the primes dividing $\sqrt{d_L}$ are precisely the ramified primes. If $\mathrm{val}_{\mathcal{T}_i}([a+\sqrt{d_L}]/2) > 0$ then $\mathrm{val}_{\mathcal{T}_i}([a-\sqrt{d_L}]/2) > 0$, which implies that $\mathrm{val}_{\mathcal{T}_i}(\sqrt{d_L})$, being equal to $\mathrm{val}_{\mathcal{T}_i}([a+\sqrt{d_L}]/2 + [a-\sqrt{d_L}]/2)$ is positive. Contradiction. If both $\mathrm{val}_{\mathcal{S}_i}([a-\sqrt{d_L}]/2) = \mathrm{val}_{\overline{\mathcal{S}_i}}([a+\sqrt{d_L}]/2)$, and $\mathrm{val}_{\mathcal{S}_i}([a+\sqrt{d_L}]/2)$ are positive, then $\mathrm{val}_{\mathcal{S}_i}(\sqrt{d_L}) > 0$. Again, a contradiction. We conclude that we may write:

$$\left(\frac{a + \sqrt{d_L}}{2}\right) = \prod \mathcal{S}_i^{a_i} \prod \mathcal{R}_i^{\beta_i}, \quad \text{for } i \neq j, \mathcal{S}_i \neq S_j, \overline{S_j}. \tag{6.12}$$

Therefore

$$\frac{d_L - a^2}{4} = \prod s_i^{a_i} \prod r_i^{\beta_i}, \quad (s_i) = \mathcal{S}_i\overline{\mathcal{S}_i}, (r_i) = \mathcal{R}_i^2. \tag{6.13}$$

The inclusion $([a+\sqrt{d_L}]/2) \subseteq \mathfrak{A}$, $\mathbb{N}\mathfrak{A} = d$ implies,

$$d = \prod s_i^{a_i'} \prod r_i^{\beta_i'}, \quad a_i' \leq a_i, \beta_i' \leq \beta_i. \tag{6.14}$$

And therefore

$$\mathfrak{A} = \prod \mathcal{S}_i^{a_i'} \prod \mathcal{R}_i^{\beta_i'}. \tag{6.15}$$

That is, $\mathfrak{A}$ is *uniquely* determined by d and the condition $([a + \sqrt{d_L}]/2) \subseteq \mathfrak{A}$, $\mathbb{N}\mathfrak{A} = d$, hence the claim. ☐

COROLLARY 6.3 (Siegel's formula).

$$\zeta_L(-1) = \frac{1}{60} \sum_{\substack{a \in \mathbb{Z} \\ a \equiv d_L \bmod 2 \\ |a| < \sqrt{d_L}}} \sigma_1\left(\frac{d_L - a^2}{4}\right). \tag{6.16}$$

EXERCISE 6.4 ([**13**]). *Prove the identity*

$$\zeta_L(-3) = \frac{1}{120} \sum_{\substack{a \in \mathbb{Z} \\ a \equiv d_L \bmod 2 \\ |a| < \sqrt{d_L}}} \sigma_3\left(\frac{d_L - a^2}{4}\right). \tag{6.17}$$

CHAPTER 3

Abelian Varieties with Real Multiplication over General Fields

The purpose of this chapter is to introduce some basic techniques to study abelian varieties over arbitrary fields and to apply them to obtain fundamental facts regarding certain moduli spaces. Over the complex numbers we have seen that the theory of abelian varieties is to a large extent the theory of lattices and Riemann forms on them. Over a general field we have no such simple device to study abelian varieties and more sophisticated tools must be introduced.

The first section contains a rigidity lemma implying the uniqueness and commutativity of the group law on an abelian variety. It also treats the dual abelian variety, the dual exact sequence and polarizations. Our discussion is restricted to the bare minimum we consider necessary to read the book and the reader would certainly profit from consulting [**107**] and other sources for a careful and complete exposition of this material.

In Section 2 we discuss finite Heisenberg groups that allow one to study line bundles over abelian varieties, and replace, in this setting, the Appell–Humbert Theorem. We also indicate how one obtains coordinates (that turn out to be Riemann theta functions) for moduli spaces of abelian varieties using the theory of almost canonical bases. This explains why modular forms tend to arise from theta functions.

The third section is devoted to stating and illustrating the Honda–Tate Theorem. This theorem is a powerful tool in the study abelian varieties, and in particular ordinary abelian varieties, over finite fields. Since these are dense in many moduli spaces of abelian varieties, including those discussed in this book (Siegel and Hilbert moduli spaces), this theorem has an important role in the theory. Later on (Section 6) we describe a refinement due to Deligne of this theory and explain, following Ribet, how it may be used to prove irreducibility of moduli spaces.

The deformation theory of ordinary abelian varieties is studied most effectively using Serre–Tate coordinates. This is the subject of Section 4. Following Chai–Norman we use this theory to determine the local structure of some moduli spaces at ordinary points.

Section 5 is devoted to discussing abelian varieties with real multiplication over a general field. In particular, we provide information on the endomorphism rings and define and compare two kinds of "regularity conditions" that appear in the definition of moduli problems for abelian varieties with real multiplication.

The last section deals with moduli spaces of abelian varieties with real multiplication using some tools introduced in the previous sections. We conclude by proving the irreducibility of moduli spaces of abelian varieties with μ_{p^n}-level. This fact plays a very important role in the theory of congruences for modular forms and in the construction of p-adic L-functions.

1. Abelian varieties over a general field

In this section, we will develop some general features of abelian varieties defined over any field k. Recall (Chapter 1, Definition 6.1) that an *abelian variety* A over k is a connected, reduced projective algebraic group.

LEMMA 1.1 (Rigidity Lemma). *Let $f\colon V \times W \to U$ be a morphism of algebraic varieties over k. Assume*

(1) *V is projective.*
(2) *There exist $v_0 \in V$, $w_0 \in W$ such that*

$$f(\{v_0\} \times W) = f(V \times \{w_0\}) = \{u_0\}. \tag{1.1}$$

Then $f(V \times W) = \{u_0\}$.

PROOF. Let $U_0 \subseteq U$ be an open affine set such that $u_0 \in U_0$, let

$$\pi_W : V \times W \to W \tag{1.2}$$

be the projection and let

$$Z = \pi_W\big(f^{-1}(U \backslash U_0)\big). \tag{1.3}$$

The set $U\backslash U_0$ is closed. Since V is projective, hence proper over k, π_W is a closed map, and thus Z is closed. The set Z is also characterized by the following property:

$$w \notin Z \iff f(V \times \{w\}) \subseteq U_0. \tag{1.4}$$

Hence $W\backslash Z$ is a non-empty open, hence dense, set in W. For $w \notin Z$, $f(V \times \{w\})$ is a point, being the image of a projective variety in an affine subset. Combining this with

$$f(\{v_0\} \times W) = \{u_0\}, \tag{1.5}$$

we get,

$$f\big(V \times (W\backslash Z)\big) = \{u_0\}. \tag{1.6}$$

Since $V \times (W\backslash Z)$ is dense in $V \times W$, we conclude: $f(V \times W) = \{u_0\}$. □

COROLLARY 1.2. *Every morphism $f\colon A \to B$ of abelian varieties over k is of the form $T_b \circ h$ where $h\colon A \to B$ is a homomorphism of abelian varieties, $b \in B$, and T_b is the translation-by-b-map: $T_b(b') = b' + b$.*

PROOF. Replacing f by $T_b \circ f$, with $b = -f(0)$, we may assume $f(0) = 0$ and we need to prove that f is a homomorphism. We apply the Rigidity Lemma to the morphism

$$F\colon A \times A \to B, \tag{1.7}$$

given by

$$F = f \circ m_A - m_B \circ (f, f), \tag{1.8}$$

(i.e., on points, $F(x,y) = f(x+y) - f(y) - f(x)$). Note that although we are using notation suggesting commutativity, this is not used in the proof. Thus, one could write $F(x,y) = f(xy)f(y)^{-1}f(x)^{-1}$.

The assumptions of the Rigidity Lemma hold: $F(A \times \{0\}) = 0_B = F(\{0\} \times A)$. Hence F is the constant map with image 0_B, i.e., f is a homomorphism. □

COROLLARY 1.3. *The group law on an abelian variety is commutative.*

PROOF. Apply Corollary 1.2 to the inverse morphism:

$$\mathrm{inv}\colon A \to A. \tag{1.9}$$

It gives that inv is a homomorphism, hence the group law is commutative. □

COROLLARY 1.4. *There is a unique group law on A for which 0_A is the identity.*

PROOF. Apply the Corollary 1.2 to the identity map! □

One can prove that the multiplication by n map, denoted $[n]$ (or simply n), is an isogeny of degree n^{2g}. That is, it is a surjective homomorphism and $A[n]$, the Kernel of $[n]$, defined by the Cartesian diagram:

$$\begin{array}{ccc} A[n] & \xrightarrow[\text{immersion}]{\text{closed}} & A \\ \downarrow & & \downarrow{\scriptstyle [n]} \\ \mathrm{Spec}(k) & \longrightarrow & A, \end{array} \tag{1.10}$$

is an affine group scheme of order n^{2g} (where $g = \dim A$), i.e., $A[n] = \mathrm{Spec}(R)$ where R is a Hopf algebra of dimension n^{2g} over k. See [**100**, Theorem 8.2]. If $\mathrm{char}(k) = 0$, or $(\mathrm{char}(k), n) = 1$, then $A[n]$ is étale, i.e., $A[n] \otimes_k k^s \cong (\mathbb{Z}/n\mathbb{Z})^{2g}$. If $\mathrm{char}(k) = p$ and $n = p$, then

$$|A[p](\bar{k})| \leq p^g. \tag{1.11}$$

By a theorem of Oort [**126**, (15.11)] every finite commutative group scheme over a perfect field embeds as a subgroup of some abelian variety. This was strengthened by Raynaud. See [**4**, Theorem 3.1.1].

EXERCISE 1.5. *Write the automorphism group of the underlying variety of an abelian variety A/k as a semidirect product of translations and group automorphisms. Find the connected component of this algebraic group. Deduce that the automorphism group need not be of finite type.*

1.1. The dual abelian variety. Let A be an abelian variety over a field k. Let $\mathrm{Pic}(A)$ denote the Picard functor of A. For a locally Noetherian scheme S over k, $\mathrm{Pic}(A)(S)$ is the group of isomorphism classes of invertible sheaves $\mathcal{L}$ over $A \times_{\mathrm{Spec}(k)} S$ together with a "trivialization along the base," i.e., together with an isomorphism of $e_S^*\mathcal{L}$ with $\mathcal{O}_S$, where $e_S\colon S \to A \times_{\mathrm{Spec}(k)} S$ is the zero section. Equivalently, $\mathrm{Pic}(A)(S)$ is the group of invertible sheaves $\mathcal{L}$ on $A \times_{\mathrm{Spec}(k)} S$ modulo the subgroup of the sheaves of the form $p_2^*\mathcal{M}$, where $p_2\colon A \times_{\mathrm{Spec}(k)} S \to S$ is the projection.

The Picard functor is representable by a group scheme over k that we denote by $\mathrm{Pic}(A)$ as well. See [**113**, Chapter 0, §5] for discussion and references.

A line bundle $\mathcal{L} \to A$ gives a morphism

$$\Phi_{\mathcal{L}}\colon A \to \mathrm{Pic}(A), \tag{1.12}$$

whose value on points is

$$\Phi_{\mathcal{L}}(x) = T_x^*\mathcal{L} \otimes \mathcal{L}^{-1}. \tag{1.13}$$

This is in fact a homomorphism of groups:

THEOREM 1.6 (Theorem of the Square). *We have:*

$$T_{x+y}^*\mathcal{L}\otimes\mathcal{L}^{-1}\cong T_x^*\mathcal{L}\otimes\mathcal{L}^{-1}\otimes T_y^*\mathcal{L}\otimes\mathcal{L}^{-1}. \tag{1.14}$$

PROOF. See [**107**, §6, Corollary 4]. □

Let $\mathrm{Pic}^0(A)$ denote the identity component of $\mathrm{Pic}(A)$. It is a scheme that represents the subfunctor of those invertible sheaves $\mathcal{L}$ such that $\Phi_{\mathcal{L}}\equiv 0$. Note that for all $\mathcal{L}$, $\mathrm{Im}(\Phi_{\mathcal{L}})\subseteq \mathrm{Pic}^0(A)$:

$$T_y^*(T_x^*\mathcal{L}\otimes\mathcal{L}^{-1})\otimes(T_x^*\mathcal{L}\otimes\mathcal{L}^{-1})^{-1}=T_{x+y}^*\mathcal{L}\otimes T_y^*\mathcal{L}^{-1}\otimes T_x^*\mathcal{L}^{-1}\otimes\mathcal{L}=0. \tag{1.15}$$

Note that $\mathrm{Pic}^0(A)$ is a subgroup scheme of $\mathrm{Pic}(A)$. This follows from the formula:

$$\Phi_{\mathcal{L}\otimes\mathcal{M}}=\Phi_{\mathcal{L}}+\Phi_{\mathcal{M}}. \tag{1.16}$$

One defines the Néron–Severi group, $\mathrm{NS}(A)$, by

$$\mathrm{NS}(A)=\mathrm{Pic}(A)/\mathrm{Pic}^0(A). \tag{1.17}$$

Then:

$$0\to\mathrm{Pic}^0(A)\to\mathrm{Pic}(A)\to\mathrm{NS}(A)\to 0. \tag{1.18}$$

EXERCISE 1.7. *Let $k=\mathbb{C}$. Use the Appell–Humbert Theorem to show that the definition of* $\mathrm{Pic}^0(A)$ *given here agrees with the previous definition in Chapter* 1, *Definition* 6.28.

One denotes $\mathrm{Pic}^0(A)$ also by $A^\vee$, or A^t, and calls it the *dual abelian variety*. The name is justified by the following theorem and further properties we list below. The proofs of these properties will take us too far astray, and we content ourselves with referring to [**107**], especially Chapters 8 and 13.

THEOREM 1.8. *There exists an abelian variety $A^\vee/k$ representing* $\mathrm{Pic}^0(A)$. *Moreover, on $A\times A^\vee$ there is a line bundle $\mathcal{P}$, unique up to isomorphism, called the* Poincaré bundle, *with the following properties*:

(1) *For every $\alpha\in A^\vee(\bar{k})$ the line bundle $\mathcal{P}|_{A\times\{\alpha\}}$ represents that element of* $\mathrm{Pic}^0(A)(\bar{k})$ *given by α.*
(2) *$\mathcal{P}|_{\{0\}\times A^\vee}$ is trivial.*

FACT 1.9. (1) If $\mathcal{L}$ is ample then $\Phi_{\mathcal{L}}\colon A\to A^\vee$ is an isogeny, i.e., a surjective homomorphism with finite kernel (but the converse is not necessarily true).

(2) $A\cong(A^\vee)^\vee$ canonically. The isomorphism is given in fact by $t\mapsto\mathcal{P}|_{t\times A^\vee}$. Moreover, the Poincaré bundle on $A\times A^\vee$, when A is interpreted as $A^{\vee\vee}$, is the same bundle.

(3) Let f_1, f_2 be homomorphisms and $f_1^\vee$, $f_2^\vee\colon B^\vee\to A^\vee$ the corresponding homomorphisms of the dual abelian varieties ($f_i^\vee(\mathcal{L})$ is just the pull-back $f_i^*(\mathcal{L})$). Then it is a theorem that $(f_1+f_2)^\vee=f_1^\vee+f_2^\vee$.

EXERCISE 1.10. *Use the fact that $A^\vee$ is an abelian variety to deduce the Theorem of the Square from Corollary* 1.2.

THEOREM 1.11. *Let*

$$0\to H\to A\xrightarrow{f}B\to 0 \tag{1.19}$$

be an exact sequence with A, B abelian varieties and let H be a finite group scheme, all defined over k. Then we have an exact sequence

$$0 \to H^\vee \to B^\vee \xrightarrow{f^\vee} A^\vee \to 0, \tag{1.20}$$

where $f^\vee(\mathcal{L}) = f^*(\mathcal{L})$, and $H^\vee$ is the k-group scheme dual to H, i.e., representing the functor on k-schemes $S \mapsto \mathrm{Hom}_{S-Gps}(H \times_{\mathrm{Spec}(k)} S, \mathbb{G}_{m/S})$.

PROOF. See [**107**, §15, Theorem 1]. □

REMARK 1.12. Though we use $\vee$ to denote both the dual of a finite group scheme and the dual of an abelian variety, the reader should note that these are two totally different notions of duality.

REMARK 1.13. The duality theory of abelian varieties and the special role of the Poincaré bundle can be developed much further. In particular, there is a Fourier transform. See [**31, 103, 114**].

REMARK 1.14. Let $f\colon A \to A^\vee$ be an isogeny. The relation $f \circ f^\vee = [\deg(f)]$ holds for elliptic curves, but does not hold in general if $\dim(A) > 1$.

DEFINITION 1.15. We define a *polarization* of A to be a homomorphism,

$$f\colon A \to A^\vee, \tag{1.21}$$

such that $f = \Phi_{\mathcal{L}}$ for some *ample* line bundle $\mathcal{L}$ over $A/\bar{k}$.

FACT 1.16. If f is a polarization, then f is an isogeny and $f = f^\vee$ (under $(A^\vee)^\vee \cong A$). Hence, the kernel of a polarization is a self-dual group scheme.

The polarization is called *principal* if f is an isomorphism. Since A is a variety, $A \subseteq \mathbb{P}^N$, for some N, by definition. Thus A possesses an ample line bundle. In particular, A always carries some polarization and is isogenous to its dual $A^\vee$.

For every finite commutative group scheme H over k, there exists a canonical perfect pairing

$$H \times H^\vee \to \mathbb{G}_m. \tag{1.22}$$

Now,

$$0 \to A[n] \to A \xrightarrow{n} A \to 0, \tag{1.23}$$

gives:

$$0 \to (A[n])^\vee \to A^\vee \xrightarrow{n^\vee} A^\vee \to 0. \tag{1.24}$$

On the other hand, note that $1^\vee = 1$, and since $(f+g)^\vee = f^\vee + g^\vee$, it follows by induction that $n^\vee = n$. Thus we conclude that

$$A^\vee[n] = A[n]^\vee, \tag{1.25}$$

and we therefore obtain a perfect pairing:

$$A[n] \times A^\vee[n] \to \mathbb{G}_m. \tag{1.26}$$

If $\lambda\colon A \to A^\vee$ is a polarization, we get a bilinear, antisymmetric, $\mathrm{Gal}(\bar{k}/k)$-invariant pairing:

$$\langle\ ,\ \rangle_\lambda\colon A[n] \times A[n] \to \mathbb{G}_m, \tag{1.27}$$

by letting

$$\langle x, y\rangle_\lambda = \langle x, \lambda(y)\rangle. \tag{1.28}$$

This pairing is called the *Weil pairing*. It is perfect iff $(\deg\lambda, n) = 1$.

We mentioned in §1 that if $\mathrm{char}(k) = p > 0$ then

$$|A[p](\bar{k})| \leq p^g. \tag{1.29}$$

One may prove this by showing that if A is isogenous to B, then

$$|A[p](\bar{k})| = |B[p](\bar{k})|. \tag{1.30}$$

Take $B = A^\vee$, and use the pairing:

$$A[p] \times A^\vee[p] \to \mathbb{G}_m. \tag{1.31}$$

If we put $H = A[p](\bar{k})$, $K = A^\vee[p](\bar{k})$. The map

$$H \times K \to \mathbb{G}_m(\bar{k}) \tag{1.32}$$

factors through μ_p and we get:

$$H \times K \to \mu_p(\bar{k}) = 1. \tag{1.33}$$

Hence $K \subseteq H^\perp$. The perfectness implies $|K| \leq p^{2g}/|H|$. On the other hand, $|K| = |H|$, thus $|H| \leq p^g$.

EXERCISE 1.17. *Let A/k be an abelian variety and let $T_pA(\bar{k}) = \varprojlim A[p^n](\bar{k})$. Prove that it is a flat $\mathbb{Z}_p$ module and if $f\colon A \to B$ is an isogeny then f induces an injection with finite cokernel $f\colon T_pA(\bar{k}) \to T_pB(\bar{k})$. Deduce Equation* (1.30).

1.2. Examples. At this point we would like to give some examples of abelian varieties so that they become less mysterious objects.

The simplest abelian varieties to describe are those of dimension one. That is, the elliptic curves. Over any field k an elliptic curve could be described as a plane cubic given by a Weierstrass equation $y^2 + a_1xy + a_3y = x^3 + a_2x^4 + a_4x + a_6$ with $a_i \in k$ and zero being "the point at ∞." The group law is given by the familiar "cord-secant" procedure.

A product of elliptic curves is an abelian variety. However, for any integer g there exist abelian varieties of dimension g over any algebraically closed field that are simple: their have no non-trivial abelian subvarieties (or quotients).

If k is an algebraically closed field and C/k is a non-singular curve of genus g then the Jacobian of C may be described as divisors of degree zero on C modulo principal divisors. Choose a base point $b \in C$ and let $\phi_b\colon C \to \mathrm{Jac}(C)$ be the map $\phi_b(c) = [c] - [b]$. The map ϕ_b extends to a surjective morphism

$$\phi_b\colon \underbrace{C \times \cdots \times C}_{g} \to \mathrm{Jac}(C), \quad \phi_b(c_1, \ldots, c_g) = [c_1] + \cdots + [c_g] - g[b]. \tag{1.34}$$

Moreover, the image of

$$\begin{gathered} \phi_b\colon \underbrace{C \times \cdots \times C}_{g-1} \to \mathrm{Jac}(C), \\ \phi_b(c_1, \ldots, c_{g-1}) = [c_1] + \cdots + [c_{g-1}] - (g-1)[b], \end{gathered} \tag{1.35}$$

is a divisor on $\mathrm{Jac}(C)$ called the theta divisor Θ. The line bundle corresponding to Θ gives a principal polarization λ_Θ on $\mathrm{Jac}(C)$ and the pair $(\mathrm{Jac}(C), \lambda_\Theta)$ is simple, though $\mathrm{Jac}(C)$ need not be simple in general. Torelli's theorem asserts that the pair $(\mathrm{Jac}(C), \lambda_\Theta)$ determines C up to isomorphism. One can prove that every abelian variety is a quotient of a Jacobian, usually of much higher dimension. For an excellent exposition of the theory of Jacobians see [**101**], and for the statements made above see especially §§4, 6, 10, 12 of loc. cit.

Consider the functor of isomorphism classes of, say, principally polarized g-dimensional abelian varieties. There exists a coarse moduli scheme for this functor. Let us denote it by $\mathcal{A}_g$. The complex points of $\mathcal{A}_g$ correspond to isomorphism classes of complex principally polarized abelian varieties of dimension g. In fact, as complex manifolds $\mathcal{A}_g(\mathbb{C}) \cong \mathrm{Sp}_{2g}(\mathbb{Z})\backslash\mathcal{H}_g$.

Let $n \geq 3$ be an integer. If one throws in a full symplectic level n structure[1] as well, then one obtains a fine moduli scheme $\mathcal{A}_{g,n} \to \mathrm{Spec}(\mathbb{Z}[\zeta_n, n^{-1}])$ whose complex points are $\Gamma(n)\backslash\mathcal{H}_g$, where $\Gamma(n)$ is the subgroup of matrices in $\mathrm{Sp}_{2g}(\mathbb{Z})$ congruent to 1 modulo n. Moreover, the morphism $\mathcal{A}_{g,n} \to \mathrm{Spec}(\mathbb{Z}[\zeta_n, n^{-1}])$ is *smooth*. This has the following corollary: "Every principally polarized abelian variety in characteristic p can be lifted to characteristic zero." Over the complex numbers principally polarized abelian varieties are more concrete. As we have seen, they correspond to lattices Λ in $\mathbb{C}^g$ that afford a positive definite Riemann pairing H with $\det(\mathrm{Im}\, H) = 1$.

Later on we shall see other methods to understand abelian varieties in positive characteristic. We shall describe in detail their deformation theory, and the Honda–Tate Theorem will provide a tool for understanding isogeny classes of abelian varieties over a finite field.

1.3. Abelian schemes. The definition of abelian scheme is one that aims to capture the notion of a "continuously varying family of abelian varieties." The notion of continuous is captured by a smoothness condition. The notion of a group scheme tells us that each fiber is a group variety, while being fiber-wise abelian is guaranteed by asking the family to be proper over the base. The reason for that is that one can prove [**100**, §7] that a complete group variety is projective (thus agreeing with Definition 6.1 in Chapter 1). One remarks that abelian schemes need not be projective. See [**137**, §XII].

Basic examples to keep in mind: (i) An elliptic curve $y^2 + a_1xy + a_3y = x^3 + a_2x^2 + a_4x + a_6$ with $a_i \in \mathbb{Z}$ and discriminant Δ could be viewed as an abelian scheme over $\mathbb{Z}[\Delta^{-1}]$. The fiber over a prime p of that abelian scheme is given by the curve $y^2 + a_1xy + a_3y = x^3 + a_2x^2 + a_4x + a_6$ where the coefficients are read modulo p; (ii) An elliptic curve $y^2 + a_1(t)xy + a_3(t)y = x^3 + a_2(t)x^2 + a_4(t)x + a_6(t)$ with $a_i(t) \in \mathbb{C}(t)$ and discriminant $\Delta(t)$ could be viewed as an abelian scheme over a suitable open subset of $\mathbb{P}^1(\mathbb{C})$ with variable t. The fiber over a complex number z of that abelian scheme is given by the curve $y^2 + a_1(z)xy + a_3(z)y = x^3 + a_2(z)x^2 + a_4(z)x + a_6(z)$ where the coefficients are complex numbers.

DEFINITION 1.18. Let S be a scheme. A group scheme $\pi \colon \mathcal{A} \to S$ is called an abelian scheme if π is smooth and proper, and the geometric fibers of π are connected.

There is a Rigidity Lemma for abelian schemes [**113**, Prop. 6.1] that allows one to prove—using the same ideas as in §1—that abelian schemes are commutative group schemes and have a unique group law once the zero section is fixed [**113**, Chapter 6, §1].

[1]By a full symplectic level n structure on a principally polarized abelian variety $(A, \lambda)/k$ we mean an isomorphism $(\mathbb{Z}/n\mathbb{Z})^{2g} \cong A[n]$ taking the standard symplectic pairing on $(\mathbb{Z}/n\mathbb{Z})^{2g}$ to the Weil Pairing induced by λ on $A[n]$.

Given an abelian scheme, one can construct the dual "abelian scheme." To be precise, one first constructs the functor $\mathrm{Pic}(\mathcal{A}/S)$ of line bundles rigidified along the base, much as explained in §1.1. See [**113**, Chapter 6, §1]. One then considers the subfunctor $\mathrm{Pic}^0(\mathcal{A}/S)$ classifying rigidified line bundles algebraically equivalent to zero on each geometric fiber. In the utmost generality $\mathrm{Pic}^0(\mathcal{A}/S)$ is not known to be an abelian scheme but merely an abelian "algebraic space." In any event, for the purpose of this book these issues never play a role and the reader may well always assume that $\mathrm{Pic}^0(\mathcal{A}/S)$, denoted by $\mathcal{A}^\vee$ or $\mathcal{A}^t$, is representable by an abelian scheme. We refer the reader to [**39**, Chapter 1, §1] for discussion and references.

DEFINITION 1.19. Let $\pi\colon \mathcal{A} \to S$ be an abelian scheme. A polarization of $\mathcal{A}$ is a homomorphism

$$\lambda\colon \mathcal{A} \to \mathcal{A}^\vee, \tag{1.36}$$

such that for every geometric point s the induced homomorphism $\lambda_s\colon \mathcal{A}_s \to \mathcal{A}_s^\vee$ is of the form $\phi_{\mathcal{L}}$ for some ample line bundle $\mathcal{L} \to \mathcal{A}_s$.[2]

DEFINITION 1.20. An isogeny between two abelian schemes $\mathcal{A}$, $\mathcal{B}$ over S is a homomorphism of group schemes,

$$f\colon \mathcal{A} \to \mathcal{B}, \tag{1.37}$$

such that f is a finite, flat, surjective map.

In this case it is known that the kernel of f is a finite flat group scheme over S. Conversely, given a finite flat group scheme $\mathcal{K} \subset \mathcal{A}$ over S, the quotient $\mathcal{A}/\mathcal{K}$ is an abelian scheme and the canonical morphism $\mathcal{A} \to \mathcal{A}/\mathcal{K}$ is an isogeny.

2. Finite Heisenberg groups

Over the complex numbers, the Appell–Humbert Theorem allows one complete understanding of line bundles on abelian varieties and thus of polarizations and their behaviour under isogenies. The Appell–Humbert Theorem rests on the fact that we have a surjective analytic map $\mathbb{C}^g \to A$ where $\mathbb{C}^g$ is a contractible space that is independent of A. In characteristic p, there is no similar construction. Thus, one needs other methods to manage line bundles and polarizations. This is provided by the finite Heisenberg group defined by Mumford. For an extensive treatment, see [**90, 105**]. See also [**112**].

Let A be an abelian variety over an algebraically closed field k, and $\mathcal{L}$ an *ample* line bundle on A. We assume, for simplicity, that either $\mathrm{char}(k) = 0$, or $\big(\deg \mathcal{L}, \mathrm{char}(k)\big) = 1$.

DEFINITION 2.1. Let $G(\mathcal{L})$ be the set of automorphisms $\phi\colon \mathcal{L} \to \mathcal{L}$ such that ϕ covers a translation map on the base and induces a linear map on fibers. That is, there exists $a \in A$ such that the following diagram commutes and ϕ is linear on fibers:

$$\begin{array}{ccc} \mathcal{L} & \xrightarrow{\phi} & \mathcal{L} \\ \downarrow & & \downarrow \\ A & \xrightarrow{T_a} & A. \end{array} \tag{2.1}$$

[2]This is equivalent to asking that locally étale on S the homomorphism λ is of the form $\phi_{\mathcal{L}}$ for some ample line bundle $\mathcal{L}$.

The group law is composition: $\phi_2 \circ \phi_1$ is given by

$$\begin{array}{ccccc} \mathcal{L} & \xrightarrow{\phi_1} & \mathcal{L} & \xrightarrow{\phi_2} & \mathcal{L} \\ \downarrow & & \downarrow & & \downarrow \\ A & \xrightarrow{T_{a_1}} & A & \xrightarrow{T_{a_2}} & A. \end{array} \tag{2.2}$$

One calls $G(\mathcal{L})$ the *Heisenberg group*, or *Theta group*, associated to $(A, \mathcal{L})$.

An equivalent way of defining $G(\mathcal{L})$ is as the set of pairs (ψ, a), where $a \in A$ and $\psi\colon \mathcal{L} \to T_a^*\mathcal{L}$ is an isomorphism of line bundles on A. Indeed, let $\rho\colon \mathcal{L} \to A$ be the projection. Then $T_a^*\mathcal{L} = \{(x, \ell) : \rho(\ell) = T_a(x) = x + a\}$. The map $\psi\colon \mathcal{L} \to T_a^*\mathcal{L}$ has the form $(\rho(\ell), \ell) \mapsto \big(\rho(\ell), \phi(\ell)\big)$, for some ϕ, where $\rho\big(\phi(\ell)\big) = \rho(\ell) + a$. Thus,

$$\begin{array}{ccc} \mathcal{L} & \xrightarrow{\phi} & \mathcal{L} \\ {\scriptstyle\rho}\downarrow & & \downarrow{\scriptstyle\rho} \\ A & \xrightarrow{T_a} & A. \end{array} \tag{2.3}$$

Conversely, given ϕ such that (2.3) holds, define $\psi(\rho(\ell), \ell) = (\rho(\ell), \phi(\ell))$. We note that in this language we have $(\phi_2, a_2) \circ (\phi_1, a_1) = (T_{a_1}^*\phi_2 \circ \phi_1, a_1 + a_2)$.

EXERCISE 2.2. *A less pedestrian approach to establish the equivalence of the two definitions of $G(\mathcal{L})$ is to relate Diagram (2.3) with the diagram*

$$\begin{array}{ccc} T_a^*\mathcal{L} & \longrightarrow & \mathcal{L} \\ \downarrow & & \downarrow{\scriptstyle\rho} \\ A & \xrightarrow{T_a} & A \end{array} \tag{2.4}$$

in which $T_a^*\mathcal{L}$ is the fiber product.

The group $G(\mathcal{L})$ sits in an exact sequence:

$$0 \to \mathbb{G}_m \to G(\mathcal{L}) \xrightarrow{\pi} K(\mathcal{L}) \to 0. \tag{2.5}$$

The group $K(\mathcal{L})$ is defined as $\{a : T_a^*\mathcal{L} \cong \mathcal{L}\} = \operatorname{Ker}(\Phi_{\mathcal{L}})$. Fixing a, ϕ is unique up to a scalar. To see that, note that if ϕ' is another such map, then

$$\phi \circ {\phi'}^{-1} : \mathcal{L} \to \mathcal{L} \tag{2.6}$$

is an isomorphism of line bundles, and thus for every $a \in A$ the morphism $f : \mathcal{L}_a \to \mathcal{L}_a$ is multiplication by a constant $f(a)$ and $f(a) \neq 0, \infty$. We get therefore a morphism $f\colon A \to \mathbb{G}_m$. But A is projective and $\mathbb{G}_m$ is affine, hence f is constant.

We have the following:

FACT 2.3. One has $\deg(\mathcal{L}) = \dim_k \Gamma(A, \mathcal{L})$, and $|K(\mathcal{L})| = \big(\deg(\mathcal{L})\big)^2$.

One finds immediately that $\mathbb{G}_m \subset Z\big(G(\mathcal{L})\big)$, the center of the Heisenberg group, and we may define an alternating bilinear pairing, the *Mumford pairing*

$$e^{\mathcal{L}}\colon K(\mathcal{L}) \times K(\mathcal{L}) \to \mathbb{G}_m, \tag{2.7}$$

by

$$e^{\mathcal{L}}(x, y) = [\tilde{x}, \tilde{y}], \tag{2.8}$$

where $\tilde{x}, \tilde{y}$ are lifts of x and y, respectively, to $G(\mathcal{L})$, and $[a, b] = aba^{-1}b^{-1}$.

DEFINITION 2.4. A subgroup $H \subseteq K(\mathcal{L})$ is *isotropic* if $e^{\mathcal{L}}(H, H) = 1$.

DEFINITION 2.5. A subgroup $\widetilde{H}$ of $G(\mathcal{L})$ is called a *level subgroup* if $\pi(\widetilde{H}) \cong \widetilde{H}$. If $H = \pi(\widetilde{H})$, we say that $\widetilde{H}$ *is above* H.

EXERCISE 2.6. *There exists a level subgroup $\widetilde{H}$ above H iff H is isotropic.*

Let $f\colon A \to B$ be an isogeny of abelian varieties, $\mathcal{M}$ an ample line bundle on B. Let $\mathcal{L} = f^*\mathcal{M}$. Then we have a canonical level subgroup $H_{\mathcal{M}}$ above $H = \operatorname{Ker} f$. It is constructed as follows: Given $a \in \operatorname{Ker} f$, we have

$$T_a^*\mathcal{L} \cong T_a^* f^*\mathcal{M} \cong (f \circ T_a)^*\mathcal{M} \cong f^*\mathcal{M} \cong \mathcal{L} \tag{2.9}$$

(all isomorphisms being *canonical*).

This yields a canonical isomorphism $\phi_a\colon \mathcal{L} \cong T_a^*\mathcal{L}$, for every $a \in H$ and $H_{\mathcal{M}} = \{\phi_a : a \in H\}$ is a level subgroup of $G(\mathcal{L})$ above H. Conversely, given a level subgroup $\widetilde{H} \subset G(\mathcal{L})$, let $H = \pi(\widetilde{H})$ and consider:

$$\begin{array}{ccc} \mathcal{L} & \longrightarrow & \mathcal{L}/\widetilde{H} \\ \downarrow & & \downarrow \\ A & \xrightarrow{p_H} & A/H. \end{array} \tag{2.10}$$

This is an inverse construction:

$$H_{\mathcal{L}/\widetilde{H}} = \widetilde{H}. \tag{2.11}$$

The construction is in fact a manifestation of the theory of descent. Using it one shows that there is bijection between the level subgroups of $G(\mathcal{L})$ over H and the isomorphism classes of line bundles $\mathcal{M}$ on A/H such that $p_H^*(\mathcal{M}) \cong \mathcal{L}$. This is the true content of the two exact sequences:

$$0 \to H \to A \xrightarrow{p_H} A/H \to 0, \tag{2.12}$$

and

$$0 \to H^\vee \to \operatorname{Pic}(A/H) \xrightarrow{p_H^*} \operatorname{Pic}(A). \tag{2.13}$$

(The last arrow is also surjective on Pic^0). Note that any two level subgroups H_1, H_2 over H differ by a character χ of H: Let $x \in H$ and x', x'' be its lifts to H_1 and H_2 respectively. Let $\chi(x) = (x')^{-1}(x'')$. Then χ is a character. Conversely, if $\chi \in H^\vee$ and H_1 is a level subgroup over H define H_2 by $\{x \cdot \chi(\pi(x)) : x \in H_1\}$.

LEMMA 2.7. *Let $f\colon A \to B$ be an isogeny with kernel H, $\mathcal{M}$ an ample line bundle on B, and $\mathcal{L} = f^*\mathcal{M}$. Then*

$$K(\mathcal{M}) \cong H^\perp/H. \tag{2.14}$$

PROOF. Let $H_{\mathcal{M}}$ be the level subgroup corresponding to $\mathcal{M}$ defined above. To prove the claim, consider the set of all possible liftings of isomorphisms

$$\begin{array}{ccc} \mathcal{M} & \xrightarrow{\phi} & \mathcal{M} \\ \downarrow & & \downarrow \\ B & \xrightarrow{T_b} & B \end{array} \tag{2.15}$$

to elements of $G(\mathcal{L})$. It is easy to see that for $\phi \in G(\mathcal{M})$ the lifting $p_H^*\phi$ is an element of $G(\mathcal{L})$ commuting with $H_{\mathcal{M}}$, and $p_H^*\phi$ is well-defined up to a choice of an element of $\mathrm{Ker}(p_H) = H$. Indeed, given $a \in A$ such that $f(a) = b$ there is a unique morphism ϕ' completing the following diagram:

$$\begin{array}{ccccccc} p_H^*\mathcal{M} & \xrightarrow{\phi'} & p_H^*\mathcal{M} & & \mathcal{M} & \xrightarrow{\phi} & \mathcal{M} \\ \downarrow & & \downarrow & \xrightarrow{f} & \downarrow & & \downarrow \\ A & \xrightarrow{T_a} & A & & B & \xrightarrow{T_b} & B. \end{array} \tag{2.16}$$

Conversely, every automorphism in $\pi^{-1}(H^\perp) = \mathrm{Cent}_{G(\mathcal{L})}(H_{\mathcal{M}})$ will descend to an automorphism of $\mathcal{M}$. Thus, in fact $G(\mathcal{M}) \cong \mathrm{Cent}_{G(\mathcal{L})}(H_{\mathcal{M}})/H_{\mathcal{M}}$ and hence $K(\mathcal{M}) \cong H^\perp/H$. □

FACT 2.8.

$$\deg \mathcal{M} = \deg \mathcal{L}/|\mathrm{Ker}\, f|. \tag{2.17}$$

This also follows from descent theory: $\Gamma(B, \mathcal{M}) \cong \Gamma(A, f^*\mathcal{M})^{H_{\mathcal{M}}}$.

COROLLARY 2.9. (1) *We have the following equalities*:

$$(\deg \mathcal{M})^2 = |K(\mathcal{M})| = |H^\perp|/|H| = (\deg \mathcal{L}/|H|)^2 = |K(\mathcal{L})|/|H|^2. \tag{2.18}$$

(2) *We have* $|K(\mathcal{L})| = |H| \cdot |H^\perp|$. *Hence,* $e^{\mathcal{L}}$ *is a perfect pairing and* $\mathbb{G}_m = Z\big(G(\mathcal{L})\big)$.

COROLLARY 2.10. *Every abelian variety A/k is isogenous to a principally polarized abelian variety.*

PROOF. Let $\mathcal{L}$ be an ample line bundle on A. Let $H \subset K(\mathcal{L})$ be a maximal isotropic subgroup.[3] Let $\widetilde{H}$ be a level subgroup above H. Let $\mathcal{M}$ be the line bundle $\mathcal{L}/\widetilde{H}$ on A/H. Clearly (A, ϕ_L) is isogenous to $(A/H, \phi_{\mathcal{M}})$. Moreover, by Corollary 2.9, $\deg(\mathcal{M}) = |H^\perp/H| = 1$. □

EXERCISE 2.11. *Prove that there exists a principally polarized complex abelian surface A such that $\mathrm{NS}(A) \cong \mathbb{Z}$. Let H be a subgroup of order p of A, where p is any prime. Prove that A/H has no principal polarization.*

There is a connection between the Mumford pairing $e^{\mathcal{L}}$ and the Weil pairing: Assume that $\mathcal{L}$ is an ample line bundle on A and $\mathcal{M} = \mathcal{L}^{\otimes n}$. Then,

$$\begin{aligned} x \in K(\mathcal{M}) &\iff T_x^*\mathcal{M} \cong \mathcal{M} \\ &\iff (T_x^*\mathcal{L})^n \cong \mathcal{L}^n \\ &\iff (T_x^*\mathcal{L} \otimes \mathcal{L}^{-1})^n \cong 0 \\ &\iff n \cdot \phi_{\mathcal{L}}(x) = 0 \\ &\iff x \in [n]^{-1}K(\mathcal{L}). \end{aligned} \tag{2.19}$$

Therefore, if $\mathcal{L}$ is of degree 1, then $K(\mathcal{L}) = \{0\}$ and $K(M) = A[n]$. One can prove that Mumford's pairing $e^{\mathcal{M}}$ and Weil's pairing $\langle\ ,\ \rangle_{\phi_L}$ are equal:

$$e^{\mathcal{M}}(x, y) = \langle x, y\rangle_{\phi_{\mathcal{L}}} = \langle x, \phi_{\mathcal{L}}(y)\rangle, \quad x, y \in A[n]. \tag{2.20}$$

[3]We only discussed these notions in the case where the degree of $\mathcal{L}$ is prime to the characteristic of k. However, the theory may be extended to cover the general case.

If one accepts (2.20) then one obtains that the $e^{\mathcal{M}}$ pairing is Galois equivariant and depends only on the class of $\mathcal{M}$ mod Pic^0, and, on the other hand, that the Weil pairing is an alternating pairing. The proof of (2.20) is not hard but requires going in detail into the identification of $H^\vee$ as the kernel of $\mathrm{Pic}(A/H) \to \mathrm{Pic}(A)$.

The finite Heisenberg groups are instrumental in the construction of moduli spaces for abelian varieties and in studying isogenies between them. Over the complex numbers this translates to identities between Riemann's theta functions. An exhaustive treatment may be found in Mumford's trilogy, [**109, 110, 112**]. Our discussion below is intended mainly to whet the reader's appetite.

Suppose that $\mathcal{L}$ is a very ample line bundle on A. Then a choice of basis $s_0, \dots, s_r$ of $\Gamma(A, \mathcal{L})$ gives a projective embedding

$$\begin{aligned} A &\to \mathbb{P}^r, \\ a &\mapsto \big(s_0(a) : \cdots : s_r(a)\big). \end{aligned} \tag{2.21}$$

For a general variety A there is no canonical way to choose the basis. In the case of abelian varieties, one can use the fact that $\Gamma(A, \mathcal{L})$ is an irreducible representation of $G(\mathcal{L})$ to try and narrow down the choices of bases, making such a choice "almost canonical."

DEFINITION 2.12. A representation ρ of $G(\mathcal{L})$ into GL_n is called of *weight* 1 if for every $x \in \mathbb{G}_m$, $\rho(x)$ acts as multiplication by x.

THEOREM 2.13 (Stone–von Neumann). *Under the assumptions above, there exists a unique irreducible representation of weight* 1.[4]

PROOF. Let V be such a representation, and choose a maximal isotropic subgroup H of $K(\mathcal{L})$. Let $\widetilde{H}$ be a level subgroup above H. We may decompose V according to characters of $\widetilde{H}$

$$V = \bigoplus_{\chi \in \widetilde{H}^*} V^\chi. \tag{2.22}$$

We denote the summand corresponding to the trivial character by $V^{\mathbf{1}}$. Let

$$\kappa \colon G(\mathcal{L}) \to \widetilde{H}^*, \tag{2.23}$$

be defined by

$$\kappa(g) = [\cdot, g]. \tag{2.24}$$

Let $v \in V^\chi$ and $g \in G(\mathcal{L})$. Then for $h \in \widetilde{H}$,

$$\begin{aligned} h(gv) = (hg)v &= ([h,g]gh)v \\ &= [h,g]\chi(h)gv \\ &= (\kappa(g)\chi)(h) \cdot gv. \end{aligned} \tag{2.25}$$

Hence,

$$gV^\chi = V^{\kappa(g)\cdot\chi}. \tag{2.26}$$

Therefore, if one $V^\chi \neq 0$, then every $V^\chi \neq 0$ and they are all isomorphic as vector spaces. Let $v \in V^{\mathbf{1}}, v \neq 0$. Then (2.25) implies that $\mathrm{Span}\{gv : g \in G(\mathcal{L})\}$ is a sub

[4] This theorem was greatly generalized by Weil [**171, 172**], who used it to derive and generalize Siegel's theorem on quadratic forms (a theorem now known as the Siegel–Weil Theorem).

representation W of $G(\mathcal{L})$ whose intersection with $V^{\mathbf{1}}$ is $k \cdot v$. Hence, $W = V$ and $V^{\mathbf{1}}$ is one-dimensional.

Choose a section $[g_\psi : \psi \in \widetilde{H}^*]$ to κ; and choose a basis v to the one-dimensional vector space $V^{\mathbf{1}}$. Then

$$V = \bigoplus_{\psi \in \widetilde{H}^*} g_\psi k \cdot v. \tag{2.27}$$

Every element in $G(\mathcal{L})$ can be written uniquely as xyz, with $x \in k^\times$, $y \in \{g_\psi\}$ and $z \in \widetilde{H}$. ($\mathrm{Ker}(\kappa) = k^\times \cdot \widetilde{H}$, because H was chosen to be maximal isotropic). We see that the representation is completely determined by (2.27). The only thing that matters is that $zg_\psi = rg_{\psi'}z$ for some ψ'. We remark that in fact (2.27) shows that $V \cong \mathrm{Ind}^G_{k^\times \widetilde{H}}\, k$ (where k is the trivial one-dimensional representation). □

REMARK 2.14 (Almost canonical bases). We now apply the representation theory of $G(\mathcal{L})$ to geometry. First note that $\Gamma(A, \mathcal{L})$ is a representation of $G(\mathcal{L})$, the action being described by the following diagram:

$$\begin{array}{ccc} \mathcal{L} & \xrightarrow{\ \phi\ } & \mathcal{L} \\ {\scriptstyle s}\uparrow & & \uparrow{\scriptstyle \phi\cdot s} \\ A & \xleftarrow{\ T_a^{-1}\ } & A. \end{array} \tag{2.28}$$

That is,

$$\phi \cdot s = \phi \circ s \circ T_a^{-1}. \tag{2.29}$$

The center of $G(\mathcal{L})$ acts by scalars and $\Gamma(A, \mathcal{L})$ has the dimension of the unique irreducible representation of weight 1. Therefore $\Gamma(A, \mathcal{L})$ *is* that unique irreducible representation.

Let $\mathcal{L}$ be a symmetric ample line bundle. That is, $[-1]^*\mathcal{L} \cong \mathcal{L}$. This induces a canonical automorphism ι of $G(\mathcal{L})$, where $\iota(\phi) = [-1]^*\phi$.

Let $d_1, \dots, d_g$ with $d_1 \mid d_2 \mid \cdots \mid d_g$ be the elementary divisors of $K(\mathcal{L})$, i.e., $K(\mathcal{L}) \cong \bigoplus_{i=1}^g (\mathbb{Z}/d_i\mathbb{Z})^2$. One can show that $G(\mathcal{L}) \cong G_{(d_1,\dots,d_g)}$, where

$$G_{(d_1,\dots,d_g)} = k^\times \times \left(\bigoplus_{i=1}^g \mathbb{Z}/d_i\mathbb{Z}\right) \times \left(\bigoplus_{i=1}^g \mathbb{Z}/d_i\mathbb{Z}\right)^*, \tag{2.30}$$

and

$$(\alpha, x, \phi)(\beta, y, \psi) = (\alpha\beta\psi(x), x + y, \phi\psi). \tag{2.31}$$

An isomorphism $\theta\colon G(\mathcal{L}) \to G_{(d_1,\dots,d_g)}$ restricting to the identity on $k^\times$, and satisfying $\theta \circ \iota = [-1] \circ \theta$ (where $[-1]$ on the right-hand side is the automorphism of $G_{(d_1,\dots,d_g)}$ given by $(\alpha, x, \phi) \mapsto (\alpha, -x, \phi^{-1})$) is called a *theta level structure*. If $\mathcal{M}$ is an ample symmetric line bundle of degree one and $\mathcal{L} = \mathcal{M}^{\otimes n}$, a theta level structure on $G(\mathcal{L})$ is intermediate between full symplectic level n structure and full symplectic level $2n$ structure (see [**105**]).

One may consider the moduli of triples $(A, \mathcal{L}, \theta)$. Here A is abelian variety (or, more generally, an abelian scheme), $\mathcal{L}$ is an ample symmetric line bundle of fixed elementary divisors $d_1, \dots, d_g$, and $\theta\colon G(\mathcal{L}) \to G_{(d_1,\dots,d_g)}$ is a theta level structure. Because of the relation to the usual level structures, proving the representability of such a functor is virtually equivalent to proving the representability of the usual moduli functors.

Given θ, there is an evident choice of a level subgroup $\widetilde{H}$, the one corresponding to $\{(1,x,0)\}$, and there is an evident choice of a section to κ, the one corresponding to $\{(1,0,\psi)\}$. Then choosing a non-zero element (unique up to scalar) Θ of $\Gamma(A,\mathcal{L})^{\widetilde{H}}$, we get a basis $\{g_\psi\Theta\}$ to $\Gamma(A,\mathcal{L})$, where the g_ψ are representatives of the cosets of $k^\times\widetilde{H}$ in $G(\mathcal{L})$ corresponding to $\{(1,0,\psi)\}$. The bases obtained by this method are "almost canonical"; they depend on the finitely many choices for θ.

Define a map

$$A \to \mathbb{P}\bigl(\Gamma(A,\mathcal{L})\bigr), \tag{2.32}$$

by

$$a \mapsto \bigl((g_\psi\Theta)(a)\bigr)_{\psi\in H^*}. \tag{2.33}$$

According to Mumford, when $\mathcal{L}$ is ample enough (e.g., an 8-th power of an ample symmetric line bundle of degree 1), this embeds our moduli space as a quasi-projective variety. The point corresponding to $(A,\mathcal{M},\theta)$ is $(g_\psi\Theta)(0_A)$, where 0_A is the identity. More than that, this construction actually gives the universal family over the moduli space.

Over $\mathbb{C}$, for the line bundles considered in the discussion of the Appell–Humbert Theorem, those $g_\psi\Theta$ are, up to a common exponent factor, Riemann's theta functions $\Theta\left[\begin{smallmatrix}\epsilon\\ \epsilon'\end{smallmatrix}\right]$, for suitable characteristics $\epsilon,\epsilon'\in\mathbb{Q}^g$, where

$$\Theta\begin{bmatrix}\epsilon\\ \epsilon'\end{bmatrix}(\tau) = \sum_{N\in\mathbb{Z}^g} \exp(2\pi i\{\tfrac{1}{2}{}^t(N+\epsilon)\tau(N+\epsilon)+{}^t(N+\epsilon)\epsilon'\}). \tag{2.34}$$

3. Honda–Tate Theorem

We now study abelian varieties over finite fields. Recall, once again, that over $\mathbb{C}$ we may regard an abelian variety as $\mathbb{C}^g/\Lambda$, where we may also interpret $\mathbb{C}^g$ as $H_1(A,\mathbb{R})$ and Λ as $H_1(A,\mathbb{Z})$. Moreover, the action of "real Frobenius"—complex conjugation—is that which gives the complex structure and thus determines A up to isomorphism. Given an abelian variety A over a field k, and a prime ℓ not equal to the characteristic of k, one may consider its ℓ-adic Tate module $T_\ell(A)$ as a partial replacement for the lattice Λ. There are certain analogies: (i) Over $\mathbb{C}$ we have $T_\ell(A)\cong H_1(A,\mathbb{Z})\otimes\mathbb{Z}_\ell = H_1(A,\mathbb{Z}_\ell)$ while in general $T_\ell(A)$ is $H_1(A_{\text{ét}},\mathbb{Z}_\ell)$—the $\mathbb{Z}_\ell$-dual to the first étale cohomology $H^1(A_{\text{ét}},\mathbb{Z}_\ell)$; (ii) Over $\mathbb{C}$ we have $T_\ell(A) = \varprojlim \ell^{-n}\Lambda/\Lambda$.

The main point of this approach is that if A is defined over a finite field $k=\mathbb{F}_q$, then the q-th Frobenius map $x\mapsto x^q$ induces a homomorphism $A\to A$ and therefore an endomorphism π of $T_\ell(A)$. If A is a k-simple abelian variety the "lattice" $T_\ell(A)$ together with the endomorphism π coming from Frobenius determine A up to k-isogeny. Honda–Tate theory makes the situation even more effective by giving a description of all the possible endomorphisms π.

Let k be a field, A/k an abelian variety of dimension g. Let ℓ be a prime such that $\bigl(\ell,\mathrm{char}(k)\bigr)=1$ if $\mathrm{char}(k)\neq 0$.

Define the ℓ-adic *Tate module*

$$T_\ell(A) = \varprojlim A[\ell^n](\bar{k}). \tag{3.1}$$

An element of $T_\ell(A)$ is thus a sequence $(x_n)_{n\geq 0}$, where $x_0=0$, and $\ell x_{n+1}=x_n$. It has a natural $\mathrm{Gal}(\bar{k}/k)$ action.

EXERCISE 3.1. *Let Γ be an abelian group, then we may define $T_\ell(\Gamma)$ in the same fashion: $T_\ell(\Gamma) = \varprojlim \Gamma[\ell^n]$. It may again be described as sequences $(x_n)_{n\geq 0}$, where $x_n \in \Gamma$, $x_0 = 0$, and $\ell x_{n+1} = x_n$. The addition is by component-wise addition and the zero element is the sequence consisting of $x_n = 0$ for all n.*

Show that $T_\ell(\Gamma)$ is a torsion-free $\mathbb{Z}_\ell$ module and that $\mathrm{Aut}(\Gamma)$ acts as automorphisms on $T_\ell(\Gamma)$. (N.B.: When going back to the case of abelian varieties note that $\Gamma = A(\bar{k})$ and not A. In particular, $\mathrm{Gal}(\bar{k}/k)$ acts on $A(\bar{k})$, and it does not act as automorphisms of the variety A! The representation $\mathrm{Gal}(\bar{k}/k) \to \mathrm{Aut}\big(T_\ell(A)\big)$ is called the ℓ-adic representation).

An important example is taking Γ to be the group of roots of unity of ℓ-power order in $\bar{k}$. We shall denote $T_\ell(\Gamma) = \varprojlim \mu_{\ell^n}(\bar{k})$ by $\mathbb{Z}_\ell(1)$ in this case.

Let G be a group. If Γ and Δ are G-modules and

$$\Gamma \times \Gamma \to \Delta \tag{3.2}$$

is a G-equivariant alternating pairing, we have an induced G-equivariant pairing

$$T_\ell(\Gamma) \times T_\ell(\Gamma) \to T_\ell(\Delta), \tag{3.3}$$

and hence a G-equivariant homomorphism

$$\textstyle\bigwedge^2_{\mathbb{Z}_\ell} T_\ell(\Gamma) \to T_\ell(\Delta). \tag{3.4}$$

That is, the following diagram commutes:

$$\begin{array}{ccc} \bigwedge^2 T_\ell(\Gamma) & \longrightarrow & T_\ell(\Delta) \\ {\scriptstyle \wedge^2 g}\downarrow & & \downarrow{\scriptstyle g} \\ \bigwedge^2 T_\ell(\Gamma) & \longrightarrow & T_\ell(\Delta). \end{array} \tag{3.5}$$

Use this to show that for an elliptic curve E over a totally real number field k, taking

$$G = \mathrm{Gal}(\bar{k}/k), \quad \Gamma = E(\bar{k}), \quad \Delta = \varprojlim_n \mu_{\ell^n} \tag{3.6}$$

and the Weil pairing, the ℓ-adic representation is odd. *That is, complex conjugation has determinant -1.*

We come back to the case of abelian varieties. If A and B are two abelian varieties over k, then we have a natural map commuting with the Galois group action,

$$\mathrm{Hom}_k(A, B) \otimes_{\mathbb{Z}} \mathbb{Z}_\ell \to \mathrm{Hom}_{G(\bar{k}/k)}\big(T_\ell(A), T_\ell(B)\big). \tag{3.7}$$

We remark that the module $T_\ell(A)$ is a free $\mathbb{Z}_\ell$-module of rank $2g$. The map in (3.7) is *always* injective with torsion free cokernel. When k is a finite field (Tate [**160**]), or a function field over a finite field (Zarhin [**176, 177**]) or a number field (Faltings [**40**]) or a finitely generated field over $\mathbb{Q}$ (Faltings [**40**]) then it is an isomorphism.

Let A/k be an abelian variety, $k = \mathbb{F}_{p^r}$. We refer the reader to Appendix A for definition and discussion of the Frobenius and Verschiebung morphisms and for properties of group schemes. We define $\Phi = \mathrm{Fr}_A^r = F_{A^{(p^{r-1})}} \circ \cdots \circ \mathrm{Fr}_{A^{(p)}} \circ \mathrm{Fr}_A$. Then

$$\Phi\colon A \to A^{(p^r)} = A \tag{3.8}$$

is a homomorphism of degree $(p^g)^r = p^{gr}$ and $\ker(\Phi)$ is a finite connected group scheme of A of order p^{gr}. Using the Verschiebung morphism Ver we see that

$$\mathrm{Ker}(\Phi) \subset A[p^r]. \tag{3.9}$$

Moreover, since $\mathrm{Ker}(\Phi)$ is a connected group scheme,

$$\mathrm{Ker}(\Phi)(\bar{k}) = \{0\}. \tag{3.10}$$

This may also be seen directly: $x \in A(\bar{k})$ implies that there exists an ℓ such that $\Phi^\ell(x) = x$. Incidentally, since $A[p]$ is a group scheme of order p^{2g}, this shows again that $|A[p](\bar{k})| \leq p^g$.

The action of the endomorphism Φ gives an endomorphism:

$$\Phi \in \mathrm{End}_k\big(T_\ell(A)\big). \tag{3.11}$$

Let Δ be the characteristic polynomial of Φ. One can prove that Δ has coefficients in $\mathbb{Z}$, is independent of ℓ and has degree $2g$.

THEOREM 3.2 (Honda–Tate). (1) *Let A be a $\mathbb{F}_q$-simple abelian variety. Then $E = \mathrm{End}_{\mathbb{F}_q}(A) \otimes_{\mathbb{Z}} \mathbb{Q}$ is a division algebra with center $\mathbb{Q}[\pi]$, where $\pi = \mathrm{Fr}_q = (\mathrm{Fr}_p)^r$, $q = p^r$. The number π is an algebraic integer such that for every embedding $\mathbb{Q}[\pi] \hookrightarrow \mathbb{C}$, we have $|\pi| = \sqrt{q}$. (We call such an algebraic integer π a q-Weil number.)*

(2) *For any prime ν of $\mathbb{Q}[\pi]$,*

$$\mathrm{inv}_\nu(E) = \begin{cases} \frac{1}{2}, & \nu \text{ real}, \\ \frac{\mathrm{ord}_\nu(\pi)}{\mathrm{ord}_\nu(q)}[\mathbb{Q}[\pi]_\nu{:}\mathbb{Q}_p], & \nu \mid p, \\ 0, & \text{otherwise}. \end{cases} \tag{3.12}$$

Moreover, $\dim A = \frac{1}{2}[\mathbb{Q}[\pi] : \mathbb{Q}] \cdot \sqrt{[E : \mathbb{Q}[\pi]]}$ and $e = \sqrt{[E : \mathbb{Q}[\pi]]}$ is the lowest common denominator of the $\mathrm{inv}_v(E)$'s. The characteristic polynomial of π is $m(T)^e$, where $m(T)$ is the minimal polynomial of π over $\mathbb{Q}$.

(3) *Two simple abelian varieties A and A' over $\mathbb{F}_q$ are isogenous over $\mathbb{F}_q$ iff there exists an isomorphism $\phi\colon \mathbb{Q}[\pi_A] \to \mathbb{Q}[\pi_{A'}]$ such that $\phi(\pi_A) = \pi_{A'}$.*

(4) *Every q-Weil number comes from a $\mathbb{F}_q$-simple abelian variety.*

REMARK 3.3. (1) In Part 1 of the theorem, the abelian variety need not be $\overline{\mathbb{F}_q}$-simple. For example, put $\pi = p$. Hence $q = p^2$ and

$$\mathrm{inv}_\nu(E) = \begin{cases} \frac{1}{2}, & \nu \text{ real}, \\ \frac{1}{2}, & \nu = p, \\ 0, & \text{otherwise}. \end{cases} \tag{3.13}$$

We infer that π corresponds to an elliptic curve X over $\mathbb{F}_{p^2}$, and $E = \mathrm{End}_{\mathbb{F}_{p^2}}(X) \cong B_{p,\infty}$, the "unique" quaternion algebra over $\mathbb{Q}$ ramified at p and ∞. This is a supersingular elliptic curve.

Now do the calculation with $\pi = \sqrt{p}$, so $q = p$ and $\mathbb{Q}[\pi] = \mathbb{Q}[\sqrt{p}]$.

$$\mathrm{inv}_\nu(E) = \begin{cases} \frac{1}{2}, & \nu \text{ real}, \\ 0, & \text{otherwise}. \end{cases} \tag{3.14}$$

The quaternion algebra E over $\mathbb{Q}[\sqrt{p}]$ is split at all finite places and ramified at the two real places. Then, the corresponding abelian variety A is 2-dimensional, and A is a simple abelian surface over $\mathbb{F}_p$. But A is not simple over $\mathbb{F}_{p^2}$, because over

$\mathbb{F}_{p^2}$ we get the previous computation leading to the elliptic curve X. In fact, A is isogenous to a product $X_1 \times X_2$, where the X_i are elliptic curves. Moreover, since $\pi^2 = p$ induces the Frobenius morphism of each $X_i/\mathbb{F}_{p^2}$, we see that X_1 and X_2 are both $\mathbb{F}_{p^2}$-isogenous to X above.

(2) There is a bijection between $\mathbb{F}_q$-isogeny classes of simple abelian varieties over $\mathbb{F}_q$ and minimal polynomials of q-Weil numbers.

(3) Take a supersingular elliptic curve over $\mathbb{F}_q$. Enlarging q, we have $E = B_{p,\infty}$. Then $\pi \in \mathbb{Q}$ and replacing q by q^2, we get $\pi = q$. Conclusion: Every two supersingular elliptic curves are isogenous over an algebraically closed field. An isogeny class of elliptic curves is supersingular if and only if the corresponding Weil number π satisfies $\pi^k \in \mathbb{Z}$ for some k.

(4) Examples of q-Weil numbers giving elliptic curves: Let π be a solution to $X^2 - \beta X + q = 0$ for $|\beta| < 2\sqrt{q}, \beta \in \mathbb{Z}$. Then $\pi = (\beta \pm \sqrt{\beta^2 - 4q})/2$ is totally imaginary, $|\pi| = \sqrt{q}$, and $q = \pi\overline{\pi}$. If π is not associated to $\overline{\pi}$, then p is split in $\mathbb{Q}[\pi]$.

See [**98, 141, 162, 169, 174**] for more on this fascinating subject.

EXERCISE 3.4. *Find the Weil numbers over $\mathbb{F}_p$ that correspond to supersingular elliptic curves.*

EXERCISE★ 3.5. *Estimate the number of isogeny classes of elliptic curves over $\mathbb{F}_q$.*

EXERCISE 3.6. *Fix q, $(q,2) = 1$, and an elliptic curve $E\colon y^2 = x^3 + ax + b$ over $\mathbb{F}_q$. For every $d \in \mathbb{F}_q$ let E_d be the quadratic twist $dy^2 = x^3 + ax + b$ of E. Among all quadratic twists, how many isomorphism classes are there? How many isogeny classes?*

4. Ordinary abelian varieties and Serre–Tate coordinates

4.1. Ordinary abelian varieties. Let k be an algebraically closed field of characteristic $p > 0$. An abelian variety A over k is *ordinary* if

$$|A[p](k)| = p^g. \tag{4.1}$$

As explained in §1, this is the maximal power of p possible. The property of being ordinary is stable under isogeny. If $A/\mathbb{F}_q$ is an abelian variety and Δ is the characteristic polynomial of Fr_q, then we will see later that A is ordinary iff the Newton polygon of Δ with respect to the p-adic valuation has two slopes, 0 and 1. N.B.: Being ordinary is an assertion about the Newton polygon of Δ, while being in the isogeny class is an assertion about Δ itself, and hence stronger.

EXERCISE 4.1. *Use Honda–Tate, and in particular Remark 3.3(4), to show that there are infinitely many isogeny classes of elliptic curves over $\overline{\mathbb{F}_p}$. Note that over $\mathbb{C}$ the assertion that there are infinitely many isogeny classes of elliptic curves follows from a cardinality argument. What about over $\mathbb{Q}$?*

Ordinary abelian varieties are of paramount importance because they are almost always dense subsets of moduli spaces ([**170**]), ordinary is an open property, and ordinary abelian varieties can be studied most effectively using Serre–Tate theory.

Let $\mathcal{C}_k$ be the category with:

Objects: Consisting of triples $(R, \mathfrak{m}_R, \phi)$, where R is a local artinian ring, $\mathfrak{m}_R$ is its maximal ideal, and $\phi\colon R/\mathfrak{m}_R \xrightarrow{\sim} k$.

Morphisms: $\mathrm{Mor}(R_1, R_2)$ are local homomorphisms $\psi\colon R_1 \to R_2$ inducing the identity on k. Recall that a local homomorphism of local rings is a homomorphism satisfying $\mathfrak{m}_{R_1} = \phi^{-1}(\mathfrak{m}_{R_2})$, i.e., taking the unique closed point of $\mathrm{Spec}(R_2)$ to the unique closed point of $\mathrm{Spec}(R_1)$.

Consider the functor $\mathcal{F}$ of local deformations of A/k:

$$\mathcal{F}\colon \mathcal{C}_k \to \mathbf{Sets}, \tag{4.2}$$

given by

$$\mathcal{F}(R) = \{\mathcal{A}/R \text{ an abelian scheme with an identification } \Phi\colon \mathcal{A}\otimes_R k \xrightarrow{\sim} A, \text{ up to isomorphisms inducing the identity on } A\}. \tag{4.3}$$

By fundamental results of Grothendieck, Mumford and Schlessinger (see [**127**] for discussion and references), there exists a complete Noetherian local ring $\mathcal{O} = \widehat{\mathfrak{M}}_{A/k}$[5] with residue field k such that the functor of points h of $\mathrm{Spec}(\mathcal{O})$ pro-represents $\mathcal{F}$. That is

$$h(R) = \mathrm{Mor}_{\mathrm{Spec}(k)}(\mathrm{Spec}\, R, \mathrm{Spec}\, \mathcal{O}) = \mathrm{Hom}(\mathcal{O}, R) = \mathcal{F}(R), \quad \forall R \in \mathcal{C}_k, \tag{4.4}$$

where the ring homomorphisms are local homomorphisms. We shall discuss these notions in more detail in Chapter 6. In particular, we shall state precisely a theorem of Serre and Tate that asserts that the deformations of an abelian variety A over an algebraically closed field k of characteristic p are the same as the deformations of its p-divisible group $\varinjlim A[p^n]$. For ordinary abelian varieties over k the p-divisible group is composed of the local part, which is the completion of a (split) torus at its origin, and the étale part which is a constant group scheme isomorphic to $(\mathbb{Q}_p/\mathbb{Z}_p)^g$. Rigidity for tori implies that the deformations are all given by extension classes of these two groups. It is thus possible to describe those in a very explicit manner. This is done in [**83**]. We next describe the final result.

4.2. Serre–Tate coordinates. Let A be an ordinary abelian variety over an algebraically closed field k of characteristic p. Let $T_p(A) = \varprojlim A[p^n]$.

THEOREM 4.2 (Serre–Tate Coordinates). (1) *We have an isomorphism of functors on the category* $\mathcal{C}_k$:

$$\widehat{\mathfrak{M}}_{A/k}(-) \cong \mathrm{Hom}_{\mathbb{Z}_p}\big((T_pA)(k) \otimes_{\mathbb{Z}_p} (T_pA^\vee)(k), \widehat{\mathbb{G}}_m(-)\big), \quad R \in \mathcal{C}_k. \tag{4.5}$$

(2) *Let* $R \in \mathcal{C}_k$, *and let* $\mathcal{A}/R$, $\mathcal{B}/R$ *be deformations of two ordinary abelian varieties* A/k, B/k, *respectively, with corresponding bilinear forms* $q_{\mathcal{A}}$, $q_{\mathcal{B}}$. *A homomorphism* $f\colon A \to B$ *extends to homomorphism* $\mathbf{f}\colon \mathcal{A} \to \mathcal{B}$ *if and only if*

$$q_{\mathcal{A}}\big(x, f^\vee(y)\big) = q_{\mathcal{B}}(f(x), y), \quad \forall x \in T_pA(k), y \in T_pB^\vee(k). \tag{4.6}$$

---

[5]This notation is intended to suggest that $\mathcal{O}$ is the completion of the local ring of the point corresponding to A/k in the moduli space $\mathcal{M}$ of abelian varieties, if such existed. In fact, if one rigidifies the situation by introducing more structure into the deformation problem (e.g., polarization, real multiplication) then the corresponding complete local ring that pro-represents the new functor (it would be a quotient of $\mathcal{O}$ by some ideal) is the completion of the local ring of the moduli space $\mathcal{M}$ parameterizing such an object at the point corresponding to A + the extra data.

As a *General Principle*, if one has a fine moduli space $\mathfrak{M}$ representing a functor $\mathcal{F}$, and $x \in \mathfrak{M}(k)$, then $\widehat{\mathfrak{M}}_x$—the completion of the local ring of $\mathfrak{M}$ at x—pro-represents the local deformation functor $\mathcal{F}|_{\mathcal{C}_k}$ with the initial data given by the object parameterized by x.

TABLE 4.1

	μ_{q^r}	$\mathbb{Z}/q^r\mathbb{Z}$
f	raising to q power; $\mathrm{Ker}(f) = \mu_q$	id
$f^\vee$	id	multiplying by q; $\mathrm{Ker}(f^\vee) = q^{r-1}\mathbb{Z}/q^r\mathbb{Z}$

We recall that if A is ordinary then $T_pA(k), T_pA^\vee(k)$ are free $\mathbb{Z}_p$-modules of rank g, and that $\widehat{\mathbb{G}}_m$ is defined by

$$\widehat{\mathbb{G}}_m = \varinjlim \mu_{p^n}. \tag{4.7}$$

EXERCISE 4.3. *Prove that for $R \in \mathcal{C}_k$, $\widehat{\mathbb{G}}_m(R) = 1 + \mathfrak{m}_R$. Thus* (4.5) *says that for all $R \in \mathcal{C}_k$:*

$$\mathcal{F}(R) = \widehat{\mathfrak{M}}_{A/k}(R) = \mathrm{Hom}_{\mathbb{Z}_p}(T_pA(k) \otimes_{\mathbb{Z}_p} T_pA^\vee(k), 1 + \mathfrak{m}_R). \tag{4.8}$$

Choosing coordinates for $T_pA(k), T_pA^\vee(k)$ as $\mathbb{Z}_p$ modules, we find

$$\widehat{\mathfrak{M}}_{A/k} \cong \mathbb{Z}_p^{g^2} \otimes_{\mathbb{Z}_p} \widehat{\mathbb{G}}_m \cong \widehat{\mathbb{G}}_m{}^{g^2}. \tag{4.9}$$

More importantly, for every $R \in \mathcal{C}_k$, the deformations of A over R, $\widehat{\mathfrak{M}}_{A/k}(R) = \mathcal{F}(R)$, correspond to $\mathbb{Z}_p$-linear maps:

$$\mathcal{A}/R \leftrightarrow \langle\ ,\ \rangle_{\mathcal{A}} \colon T_pA(k) \times T_pA^\vee(k) \to 1 + \mathfrak{m}_R. \tag{4.10}$$

A remarkable consequence of the existence of the Serre–Tate coordinates is the existence a natural group structure on $\mathfrak{M}_{A/k}$. This group law is simply given by "addition" of bilinear forms (N.B.: The bilinear forms have values in a multiplicative group. Thus the sum of bilinear forms is obtain by multiplying their values). In particular, it has a distinguished element, the identity of the group which is just the trivial bilinear pairing $\langle\ ,\ \rangle \equiv 1$.

DEFINITION 4.4. For every R, we let $\mathcal{A}^{\mathrm{can}}/R$ be the deformation corresponding to the pairing $\langle\ ,\ \rangle \equiv 1$. It is called the *canonical lift* of A to R.

By taking limits, we can also get a canonical lift to $W(k)$ (this uses results of Grothendieck on algebraization of formal schemes and uses implicitly the existence of a polarization on the "limit" object "$\varinjlim \mathcal{A}^{\mathrm{can}}/(W(k)/p^nW(k))$").

One sees from the criterion furnished by (4.6) that $\mathrm{End}_k(A)$ lifts to $\mathcal{A}^{\mathrm{can}}/W(k)$. In general, a deformation $\mathcal{A}/W(k)$ such that $\mathrm{End}_k(A)$ lifts to $\mathcal{A}/W(k)$ corresponds to a $W(k)^\times$-valued pairing satisfying

$$\langle x, f^\vee(y)\rangle = \langle f(x), y\rangle \quad \forall f \in \mathrm{End}(A), x \in T_pA(k), y \in T_pA^\vee(k). \tag{4.11}$$

If $k = \overline{\mathbb{F}_p}$, then this property *characterizes* the canonical lift. Essentially, because A is defined over some $\mathbb{F}_q$, the endomorphisms of A given by $f = \mathrm{Fr}_q$ and $f^\vee = \mathrm{Ver}_q$ behave very differently. The action of f and $f^\vee$ on the group schemes μ_{q^r} and $\mathbb{Z}/q^r\mathbb{Z}$ is given in Table 4.1

We note that over k, $A[q^r] \cong \mu_{q^r}^g \oplus (\mathbb{Z}/q^r\mathbb{Z})^g$. Assume that $\mathcal{A}/W(k)$ is a deformation corresponding to a bilinear form $\langle\ ,\ \rangle$ such that every endomorphism of A lifts to $\mathcal{A}$. We want to show that $\mathcal{A} = \mathcal{A}^{\mathrm{can}}$. Let U_1 denote the units of $W(k)$

that are congruent to 1 modulo p. It is enough to show that for every r the induced pairing

$$\langle\ ,\ \rangle\colon T_pA(k)/q^rT_pA(k) \times T_pA^\vee(k)/q^rT_pA^\vee(k) \to U_1/U_1^{q^r}, \tag{4.12}$$

is trivial. Apply f^r. Since f^r is an endomorphism of $\mathcal{A}$, we must have

$$\langle x, (f^\vee)^r y\rangle = \langle f^r(x), y\rangle, \quad \forall (x,y) \in T_pA(k) \times T_pA^\vee(k). \tag{4.13}$$

However, on $T_pA(k)/q^rT_pA(k) \cong A[q^r](k) \cong (\mathbb{Z}/q^r\mathbb{Z})^g$ the morphism f^r is an automorphism, while on $T_pA^\vee(k)/q^rT_pA^\vee(k) \cong A^\vee[q^r](k) \cong (\mathbb{Z}/q^r\mathbb{Z})^g$, the morphism $(f^\vee)^r$ is zero. This proves that (4.12) is trivial. Examining the argument one derives the following proposition.

PROPOSITION 4.5. *Let $A/\mathbb{F}_q$ be an abelian variety. Let $k = \overline{\mathbb{F}_q}$, $R \in \mathcal{C}_k$ and $\mathcal{A}/R$ a deformation of A to R. Then $\mathcal{A}$ is the canonical lift of A to R if and only if* Fr_q *extends to an endomorphism of $\mathcal{A}$.*

We further remark, that if λ is a principal polarization of A, and we use it to identify T_pA with $T_pA^\vee$, then the deformations of (A,λ) (as a principally polarized abelian variety) are given by the symmetric elements of

$$\mathrm{Hom}_{\mathbb{Z}_p}\big(T_pA(k) \otimes_{\mathbb{Z}_p} T_pA^\vee(k), \widehat{\mathbb{G}}_m(-)\big). \tag{4.14}$$

EXAMPLE 4.6. If $\mathfrak{M}$ is the moduli space of principally polarized abelian varieties (with level n structure, $n \geq 3$), $x \in \mathfrak{M}(k)$ a point parameterizing a principally polarized ordinary abelian variety (A,λ), then

$$\widehat{\mathfrak{M}}_x = \widehat{\mathfrak{M}}_{(A,\lambda)/k} \cong \widehat{\mathbb{G}}_m{}^{g(g+1)/2}, \tag{4.15}$$

where $\widehat{\mathfrak{M}}_x$ is the formal spectrum of the completion of the local ring of x in $\mathfrak{M}$, and where $\widehat{\mathfrak{M}}_{(A,\lambda)/k}$ is the complete local ring that pro-represents the functor of deformations of (A,λ) over $\mathcal{C}_k$. Thus, at least at ordinary points, we know that $\mathfrak{M}$ is non-singular of dimension $g(g+1)/2$ (though this is true at every point).

Following Chai–Norman [**11**] we give another application of the Serre–Tate coordinates.

DEFINITION 4.7. A $\Gamma_0(p)$*-level structure* is a pair of two principally polarized abelian varieties (A,λ_A) and (B,λ_B) of dimension g, and an isogeny:

$$(A,\lambda_A) \xrightarrow{f} (B,\lambda_B), \tag{4.16}$$

such that $f^*\lambda_B = p\lambda_A$.

EXERCISE 4.8. *Choose a line bundle $\mathcal{L}$ such that $\phi_{\mathcal{L}} = \lambda_A$. Prove that the group $K(\mathcal{L}^p)$ together with its Mumford pairing are independent of the choice of $\mathcal{L}$. Prove that to give a $\Gamma_0(p)$-level structure $f\colon (A,\lambda_A) \to (B,\lambda_B)$ is equivalent to giving a maximal isotropic subgroup of $K(\mathcal{L}^p)$. How many such subgroups are there if A is ordinary? if A is isomorphic to a product of supersingular elliptic curves with the product polarization?*

Let us compute the structure of the moduli space of $\Gamma_0(p)$-level structure at a formal neighborhood of an ordinary point. A deformation of $(A,\lambda_A) \xrightarrow{f} (B,\lambda_B)$ over $R \in \mathcal{C}_k$ is a diagram of polarized abelian schemes over R, $(\mathcal{A},\Lambda_{\mathcal{A}}) \xrightarrow{f_R} (\mathcal{B},\Lambda_{\mathcal{B}})$ such that $f_R^*\Lambda_{\mathcal{B}} = p\Lambda_{\mathcal{A}}$, together with identifications $\mathcal{A}\otimes k \cong A$ and $\mathcal{B}\otimes k \cong B$ such that under these identifications $\Lambda_{\mathcal{A}}$ induces λ_A, $\Lambda_{\mathcal{B}}$ induces λ_B and f_R induces f. Put $V = T_pA$, $W = T_pB$. Using Serre–Tate coordinates we see that the local deformation functor $\mathcal{C}_k \to$ **Sets** is given by:

$$R \mapsto \{\text{symmetric pairings } \langle\, , \rangle_V \colon V \otimes V \to \widehat{\mathbb{G}}_m(R), \\ \langle\, , \rangle_W \colon W \otimes W \to \widehat{\mathbb{G}}_m(R), \text{ such that} \\ \langle v, f^\vee(w)\rangle_V = \langle f(v), w\rangle_W, \text{ for } v \in V,\, w \in W\}. \tag{4.17}$$

By the Elementary Divisors Theorem, we may choose bases $(v_1, \dots, v_g)$ for V, and $(w_1, \dots, w_g)$ for W, such that

$$f(v_1) = w_1, \dots, f(v_a) = w_a; \quad f(v_{a+1}) = pw_{a+1}, \dots, f(v_g) = pw_g, \tag{4.18}$$

and hence

$$f^\vee(w_1) = pv_1, \cdots, f^\vee(w_a) = pv_a; \quad f^\vee(w_{a+1}) = v_{a+1}, \cdots, f^\vee(w_g) = v_g. \tag{4.19}$$

We have a $g(g+1)/2$-dimensional space of pairings such that:

$$\langle v_i, f^\vee(w_j)\rangle_V = \langle f(v_i), w_j\rangle_W, \quad \forall i, j. \tag{4.20}$$

More precisely, the conditions

$$\langle v_i, pv_j\rangle_V = \langle w_i, w_j\rangle_W, \quad 1 \le i \le a,\ 1 \le j \le a, \tag{4.21}$$

give an $a(a+1)/2$-dimensional space; the conditions

$$\langle v_i, v_j\rangle_V = \langle w_i, w_j\rangle_W, \quad 1 \le i \le a,\ a+1 \le j \le g, \tag{4.22}$$

give an $a(g-a)$-dimensional space; the conditions

$$\langle v_i, v_j\rangle_V = \langle pw_i, w_j\rangle_W, \quad a+1 \le i \le g,\ a+1 \le j \le g, \tag{4.23}$$

give a $(g-a)(g-a+1)/2$-dimensional space. Together we get $a(a+1)/2 + a(g-a) + (g-a)(g-a+1)/2 = g(g+1)/2$ dimensions. Therefore, the following theorem is proved.

THEOREM 4.9 (Chai–Norman). *The coarse moduli space $\mathcal{M}$ for $\Gamma_0(p)$-level structure at an ordinary point $f : (A, \lambda_A) \to (B, \lambda_B)$ satisfies*

$$\widehat{\mathfrak{M}}_{(f\colon A \to B)} \cong \mathbb{Z}_p^{g(g+1)/2} \otimes_{\mathbb{Z}_p} \widehat{\mathbb{G}}_m. \tag{4.24}$$

5. Abelian varieties with real multiplication over a general field

In this section we discuss certain features of abelian varieties with real multiplication defined over a general field. We provide some information on endomorphism and polarizations but mainly focus on two kinds of "regularity conditions" for such abelian varieties defined by Rapoport and Deligne–Pappas and compare them.

Let k be a field. Let L be a totally real field of degree g with discriminant d_L.

DEFINITION 5.1. An abelian variety with real multiplication by $\mathcal{O}_L$ (or, an "abelian variety with RM" for short) is an abelian variety A over k together with an embedding of rings

$$\iota\colon \mathcal{O}_L \hookrightarrow \mathrm{End}_k(A), \tag{5.1}$$

such that the Deligne–Pappas condition holds:

$$A \otimes_{\mathcal{O}_L} \mathcal{M}_A \cong A^\vee. \tag{DP}$$

A few remarks are in order. First, since $\mathcal{O}_L$ acts on A it also acts on $A^\vee$ by duality. One defines then the symmetric $\mathcal{O}_L$-linear homomorphisms

$$\mathcal{M}_A = \{\lambda\colon A \to A^\vee : \lambda \circ \iota(r) = \iota(r)^\vee \circ \lambda,\ \forall r \in \mathcal{O}_L\}. \tag{5.2}$$

The abelian variety $A \otimes_{\mathcal{O}_L} \mathcal{M}_A$ and the homomorphism $A \otimes_{\mathcal{O}_L} \mathcal{M}_A \to A^\vee$ are all determined uniquely by their behavior on k-algebras R:

$$(A \otimes_{\mathcal{O}_L} \mathcal{M}_A)(R) = A(R) \otimes_{\mathcal{O}_L} \mathcal{M}_A; \quad x \otimes \lambda \mapsto \lambda(x). \tag{5.3}$$

The module $\mathcal{M}_A$ is a projective $\mathcal{O}_L$-module of rank one endowed with a natural notion of positivity determined by the cone of polarizations in $\mathcal{M}_A$. Cf. Chapter 2, §2.2.

Second, one can show that if A satisfies condition (DP), so does $A^\vee$. To see that we cite the following useful lemma from Chai [**10**, Lemma 6], where the reader would find the proof.

LEMMA 5.2. *Let (A, ι) be an abelian variety with RM by $\mathcal{O}_L$ over an algebraically closed field k. Then A is isogenous to B^n for some simple abelian variety B/k. Let $D = \mathrm{End}^0(B)$ and therefore $\mathrm{End}^0(A) \cong M_n(D)$. There are four possibilities:*

Type I. *The algebra D is a totally real number field K_0 of degree $\dim(B)$. The field L contains K_0 and $[L : K_0] = n$. In this case, L is equal to its centralizer in $\mathrm{End}^0(A)$.*

Type II. *The algebra D is a totally indefinite quaternion division algebra over a totally real number field K_0 and $\dim(B) = 2[K_0 : \mathbb{Q}]$. The field L contains K_0 and $[L : K_0] = 2n$. In this case, L is equal to its centralizer in $\mathrm{End}^0(A)$.*

Type III. *The algebra D is a totally definite quaternion division algebra over a totally real number field K_0. There are two possibilities:*

(a) *The field k has characteristic p, $K_0 = \mathbb{Q}$ and B is a supersingular elliptic curve. Then D is the quaternion algebra $B_{p,\infty}$ over $\mathbb{Q}$ ramified precisely at p and ∞. The centralizer of L in $\mathrm{End}^0(A)$ is the quaternion algebra $B_{p,\infty} \otimes L$ over L.*

(b) *The abelian variety B is not an elliptic curve. Then $\dim(B) = 2[K_0 : \mathbb{Q}]$, the field L contains K_0 and $[L : K_0] = 2n$. In this case, L is equal to its centralizer in $\mathrm{End}^0(A)$.*

Type IV. *The algebra D is a central division algebra of dimension d^2 over a CM field K with a maximal totally real subfield K_0. Then $L \supseteq K_0$ and $[L : K_0] = nd$. The centralizer of L in $\mathrm{End}^0(A)$ is $L \cdot K$—the subalgebra of $\mathrm{End}^0(A)$ generated by L and K, which is isomorphic to $L \otimes_{K_0} K$. If k has characteristic zero then $d = 1$ and $D = K$.*

It follows from the lemma that taking any $\mathcal{O}_L$-linear polarization, the symmetric elements of the centralizer of L in $\mathrm{End}^0(A)$ under the Rosati involution of that polarization are just L. Consider therefore the natural pairing

$$\mathcal{M}_{A^\vee} \otimes_{\mathcal{O}_L} \mathcal{M}_A \to \mathcal{O}_L, \tag{5.4}$$

given by $(\lambda, \mu) \mapsto \lambda \circ \mu$. We use the following fact, left as an exercise to the reader.

EXERCISE★ 5.3. *Using condition (DP) prove that for every prime ℓ there exists an element $\lambda \in \mathcal{M}_A$ of degree not divisible by ℓ.*

Now, if $\lambda \in \mathcal{M}_A$ then $[\deg(\lambda)]\lambda^{-1} \in \mathcal{M}_{A^\vee}$. We conclude that for every prime ℓ the image of the pairing (5.4) contains elements of degree not divisible by ℓ and therefore that (5.4) is surjective. Hence,

$$\mathcal{M}_{A^\vee} \cong \mathcal{M}_A^{-1}, \tag{5.5}$$

and

$$A^\vee \otimes_{\mathcal{O}_L} \mathcal{M}_{A^\vee} \cong A \otimes_{\mathcal{O}_L} \mathcal{M}_A \otimes_{\mathcal{O}_L} \mathcal{M}_A^{-1} \cong A. \tag{5.6}$$

EXERCISE$^\bigstar$ 5.4. *The transformation of functors $(A, \iota) \mapsto (A^\vee, \iota^\vee)$ induces an automorphism of the coarse moduli space of abelian varieties with real multiplication by $\mathcal{O}_L$. Find its fixed points.*

Third, if $(\mathrm{char}(k), d_L) = 1$ or $\mathrm{char}(k) = 0$ then, as we shall see shortly, condition (DP) is equivalent to Rapoport's condition:

(R) $\mathfrak{t}^*_{A/k}$ is a free $\mathcal{O}_L \otimes_{\mathbb{Z}} k$-module of rank 1, and $\mathcal{M}_A$ is a projective $\mathcal{O}_L$-module of rank 1.

The condition (R) is heuristically easier to understand. We remark that for abelian varieties in characteristic zero, or for ordinary abelian varieties condition (R) holds automatically. See Chapter 2, Corollary 2.6, and Corollary 6.4 below. We also remark that since we are working over a field the requirement that $\mathcal{M}_A$ is a projective $\mathcal{O}_L$-module of rank 1 is superfluous. This always holds [**135**, Proposition 1.10].

Let us engage again in Kodaira–Spencer heuristic:

First, to obtain a good moduli problem one fixes representatives $\{(\mathfrak{a}, \mathfrak{a}^+)\}$ for $\mathrm{Cl}(L)^+$. One considers isomorphism classes of triples $(A, \iota, \lambda)/S$ where (A, ι) is an abelian scheme with RM over S, satisfying condition (DP) (or (R), see below), and λ is an isomorphism $(\mathcal{M}_A, \mathcal{M}_A^+) \to (\mathfrak{a}, \mathfrak{a}^+)$ (locally). Upon taking a rigid level structure (e.g., a full level N structure with $N \geq 3$, or a μ_N-level structure with $N \geq 4$—see Definitions 6.6, 6.8 below), one gets a fine moduli space.

We know that the moduli space of polarized complex abelian varieties with RM is g-dimensional. If a reasonable moduli scheme, say over $\mathrm{Spec}(\mathbb{Z})$, for abelian varieties with RM exists, we expect it then to be of relative dimension g. The Kodaira–Spencer method (see Chapter 1, Remark 6.10) allows us to compute the tangent space to the moduli problem. It is given by

$$\mathrm{Hom}_{\mathcal{O}_L}(\mathfrak{t}^*_A, \mathfrak{t}_{A^\vee}). \tag{5.7}$$

If condition (R) holds this is a g-dimensional space. Thus condition (R) is natural to impose. However, it turns out that the moduli scheme is then not proper over $\mathrm{Spec}(\mathbb{Z})$ not only because of lack of cusps (that are necessary because degenerations of abelian varieties with RM to semi-abelian varieties with RM do exist), but because of families of abelian varieties with RM satisfying (R) degenerating into abelian varieties with $\mathcal{O}_L$-action not satisfying (R). Deligne and Pappas discovered [**25**] that abelian varieties with RM satisfying (R) satisfy (DP) and, moreover, the property of (DP) is preserved under degenerations and specializations.

It also turns out, see [**25**] (see also [**3**]), that the moduli scheme is singular at all fibers dividing d_L. The singular locus is of codimension 2 and consists precisely of all points where condition (R) fails.

LEMMA 5.5. *Condition* (R) *always implies condition* (DP). *If $(p, d_L) = 1$, the conditions are equivalent.*

PROOF (Sketch). Assume that condition (R) holds. One first shows that A always has some $\mathcal{O}_L$-linear polarization. See [**135**, Proposition 1.10]. One then argues that $\mathcal{M}_A$ is a projective $\mathcal{O}_L$-module of rank 1. To prove that condition (DP) holds it is enough to show that for every prime ℓ there exists $\lambda \in \mathcal{M}_A$ of degree prime to ℓ (cf. Exercise 5.3).

One way (and, admittedly, not a very elegant way) to see that is the following. Using deformation theory of abelian varieties as in Chapter 6, one concludes that if condition (R) holds for (A, ι) then the functor of infinitesimal deformations of (A, ι) is formally smooth. That implies that (A, ι) admits a lift to characteristic zero. In characteristic zero we have the complex uniformization of Chapter 2, where one verifies easily that for every prime ℓ there exists an element of $\mathcal{M}_A$ of degree prime to ℓ (see Formula (2.31) in Chapter 2).

Assume now that condition (DP) holds. We assume that k has characteristic p, because in characteristic zero condition (R) always holds (by the Lefschetz principle it is enough to work over $\mathbb{C}$ and this is Corollary 2.6 in Chapter 2). One first shows that $H^1_{dR}(A)$ is a free $\mathcal{O}_L \otimes k$ module of rank 2. The module $H^1_{dR}(A)$ admits a filtration

$$0 \to H^0(A, \Omega_{A/k}) \to H^1_{dR}(A) \to H^1(A, \mathcal{O}_A) \to 0, \tag{5.8}$$

that may be rewritten as

$$0 \to \mathfrak{t}^*_{A/k} \to H^1_{dR}(A) \to \mathfrak{t}_{A^\vee} \to 0. \tag{5.9}$$

By Exercise 5.3 there exists a polarization of degree prime to p. It gives an isomorphism $\mathfrak{t}_{A^\vee} \cong \mathfrak{t}_A$ and thus we have an exact sequence

$$0 \to \mathfrak{t}^*_{A/k} \to H^1_{dR}(A) \to \mathfrak{t}_{A/k} \to 0. \tag{5.10}$$

If $p \nmid d_L$ this is a sequence of modules over the semisimple algebra $\mathcal{O}_L/(p)$ and one concludes that

$$(\mathcal{O}_L \otimes k)^2 \cong H^1_{dR}(A) \cong \mathfrak{t}^*_{A/k} \oplus \mathfrak{t}_{A/k} \cong (\mathfrak{t}^*_{A/k})^2, \tag{5.11}$$

and hence that $\mathcal{O}_L \otimes k \cong \mathfrak{t}^*_{A/k}$. □

Examples of abelian varieties with real multiplication can be constructed in characteristic zero by means of the complex uniformization given in Chapter 2, §2.2. One can then obtain examples in characteristic p by reducing modulo a suitable prime ideal.

Since the moduli space of polarized abelian varieties with RM and suitable level structure is smooth over $\mathbb{Z}[d_L^{-1}]$, every abelian variety in characteristic p, $(p, d_L) = 1$, can be lifted to characteristic zero together with its endomorphisms. The same holds for characteristic p such that p does divide d_L, but one needs a more elaborate argument, due essentially to Mumford (see [**127**, p. 249]). It is a general principle saying, roughly, that if (i) your object (in this case an abelian variety with RM) is a special fiber on a nice enough family, over an equi-characteristic integral scheme, of objects of the same kind, and if (ii) the generic fiber of such a family lifts to characteristic zero, then the special fiber lifts as well.

It is enough therefore, given an abelian variety with RM in characteristic p, represented by a point x in the moduli space, to construct a smooth curve in the moduli space modulo p that passes through x and whose generic fiber is a non-singular point of the ambient moduli space, namely, that the relative cotangent space is generically free. At least for $g = 2$ such a curve may be constructed using a Moret–Bailly family (See [**3**]). In general, such a curve exists because the moduli space is quasi-projective and the singularities are of codimension 2.

Another way to produce examples is to use the Honda–Tate Theorem (Theorem 3.2) to construct first isogeny classes having multiplication by some order in $\mathcal{O}_L$, and then finding particular representatives for this isogeny class that have multiplication by the whole of $\mathcal{O}_L$.

6. Irreducibility of the moduli space of μ_{p^∞}-level structure

Let L be a totally real field of degree $g = [L : \mathbb{Q}]$, with ring of integers $\mathcal{O}_L$, different $\mathcal{D}_L$, and discriminant d_L. We fix representatives $\{(\mathfrak{a}, \mathfrak{a}^+)\}$ to the ideal classes in $\mathrm{Cl}(L)^+$.

In this section, we follow Ribet's proof of the irreducibility of the moduli space of abelian varieties in characteristic p with RM and with μ_{p^n}-level structure when $(p, d_L) = 1$. The argument is essentially that of monodromy.

6.1. Moduli spaces for full and μ_N-level. We repeat more carefully the scheme theoretic analogue of our definition of abelian varieties with RM, generalizing the notion of abelian variety with real multiplication of Definition 5.1.

DEFINITION 6.1. Let S be a scheme in which d_L is invertible.[6] An *abelian scheme with real multiplication* by $\mathcal{O}_L$ is a couple $(A/S, \iota)$, where A/S is an abelian scheme and

$$\iota\colon \mathcal{O}_L \hookrightarrow \mathrm{End}(A/S) \tag{6.1}$$

is a ring injection such that $\mathfrak{t}_{A/S}$, the relative tangent sheaf (also denoted by $\mathrm{Lie}(A/S)$), is a locally free $\mathcal{O}_L \otimes_{\mathbb{Z}} \mathcal{O}_S$-module of rank 1, and such that the $\mathcal{O}_L$-module of symmetric $\mathcal{O}_L$-linear homomorphisms to the dual abelian scheme is a projective rank 1 module in the étale topology on S.

We say that (A, ι) is polarized if we are given locally étale on S an isomorphism $\lambda\colon (\mathcal{M}_A, \mathcal{M}_A^+) \to (\mathfrak{a}, \mathfrak{a}^+)$. We call (A, ι, λ) a polarized (or $(\mathfrak{a}, \mathfrak{a}^+)$-polarized) abelian scheme with RM.

REMARK 6.2. Let $A^\vee/S$ be the dual abelian scheme and define

$$\iota^\vee\colon \mathcal{O}_L \to \mathrm{End}(A^\vee/S), \quad \iota^\vee(m) = \big(\iota(m)\big)^\vee. \tag{6.2}$$

Then $(A^\vee/S, \iota^\vee)$ is again an abelian scheme with real multiplication.

Let us restrict our attention to the case of an algebraically closed field k of characteristic p not dividing the discriminant d_L. Let A/k be an ordinary abelian variety of dimension g with real multiplication by $\mathcal{O}_L$. Consider the Tate module $T_pA(k)$ ($\cong \mathbb{Z}_p^g$) as a module over $\mathbb{Z}_p \otimes \mathcal{O}_L$. Note that $\mathbb{Z}_p \otimes \mathcal{O}_L$ acts faithfully on $T_pA(k)$ because $\mathbb{Z}/p^n\mathbb{Z} \otimes \mathcal{O}_L$ acts faithfully on $A[p^n](k)$.

Indeed, using that $\mathbb{Z}/p^n\mathbb{Z} \otimes \mathcal{O}_L$ is a direct sum of local artinian rings, we may take $n = 1$. Now, one can show that A has a polarization of degree prime to p (see the proof of Lemma 5.5 for the idea, but use Serre–Tate coordinates, as below, to obtain a lift), hence $A[p]$ is a self-dual group scheme, since this polarization on A induces an isomorphism of $A[p]$ with $A^\vee[p] = A[p]^\vee$. So,

$$A[p] \cong (\mu_p)^g \oplus \underline{(\mathbb{Z}/p\mathbb{Z})}^g, \tag{6.3}$$

[6]That allows us to work with Condition (R). See Lemma 5.5.

and the faithfulness of the action of $\mathbb{Z}/p\mathbb{Z} \otimes \mathcal{O}_L$ on $(\mathbb{Z}/p\mathbb{Z})^g$ is equivalent to the faithfulness on μ_p^g. Thus, if $\iota(\ell)$ kills $A[p](k)$ it kills $A[p]$ and thus $\iota(\ell) = pm = \iota(p)m$ in $\mathrm{End}(A)$. It follows that ℓ is divisible by p in $\mathcal{O}_L$ and thus is zero in $\mathbb{Z}/p\mathbb{Z} \otimes \mathcal{O}_L$.

COROLLARY 6.3. *The $\mathbb{Z}_p \otimes \mathcal{O}_L$-module $T_pA(k)$ is free of rank* 1.

COROLLARY 6.4. *An ordinary abelian variety with RM satisfies condition* (R).

PROOF. The ring $\mathcal{O}_L \otimes k$ acts faithfully on $\mu_p^g \hookrightarrow A/k$ and hence acts faithfully on $\mathfrak{t}_{\mu_p^g} \cong \mathfrak{t}_{A/k}$. □

We now turn our attention to another question. We ask: *What are the deformations of $(A/k, \iota)$?*

Apply Serre–Tate coordinates: for $R \in \mathcal{C}_k$, we look for the pairings

$$(6.4) \qquad \langle\ ,\ \rangle \colon T_pA(k) \otimes T_pA^\vee(k) \to \widehat{\mathbb{G}}_m(R),$$

such that:

$$(6.5) \qquad \langle x, my\rangle = \langle mx, y\rangle, \quad \forall x \in T_pA(k),\ y \in T_pA^\vee(k),\ m \in \mathcal{O}_L.$$

Fix a generator y of $T_pA^\vee(k)$ as a $\mathbb{Z}_p \otimes \mathcal{O}_L$ module. Define

$$(6.6) \qquad \phi_y \colon T_pA(k) \to \widehat{\mathbb{G}}_m, \quad \phi_y(x) = \langle x, y\rangle.$$

Note that $\langle x, my\rangle = \phi_y(mx)$, and hence ϕ_y determines the pairing. Conversely, given a homomorphism $\phi : T_pA(k) \to \widehat{\mathbb{G}}_m$ define a pairing by $\langle x, my\rangle = \phi(mx)$. One checks that this is a well-defined bilinear pairing

$$(6.7) \qquad \langle\ ,\ \rangle \colon T_pA(k) \otimes T_pA^\vee(k) \to \widehat{\mathbb{G}}_m(R),$$

for which the action of $\mathcal{O}_L$ is self-adjoint. Thus, the $\mathcal{O}_L$-pairings are identified with $\mathrm{Hom}_{\mathbb{Z}_p}(T_pA(k), \widehat{\mathbb{G}}_m)$, and one concludes:

PROPOSITION 6.5. *There is an isomorphism of the universal local deformation ring $\widehat{\mathcal{M}}_{(A/k,\iota)}$ for $(A/k, \iota)$ with ${\widehat{\mathbb{G}}_m}^g$:*

$$(6.8) \qquad \widehat{\mathcal{M}}_{(A/k,\iota)} \cong {\widehat{\mathbb{G}}_m}^g.$$

DEFINITION 6.6. A *full level n structure* on an abelian scheme with real multiplication $(A/S, \iota)$ is an isomorphism of constant group schemes with $\mathcal{O}_L$-action

$$(6.9) \qquad \alpha \colon (\mathcal{O}_L/n\mathcal{O}_L)^2 \xrightarrow{\sim} A[n].$$

(Note that we do not take necessarily a symplectic level structure).

THEOREM 6.7 (Rapoport [**135**, Lemme 1.23, p. 267]). *Let $n \geq 3$ be an integer. Consider the functor over $\mathbb{Z}[1/(nd_L)]$-schemes:*

$$(6.10) \qquad S \to \{(A/S, \iota, \lambda, \alpha)\}/\cong,$$

associating to a $\mathbb{Z}[1/(nd_L)]$-scheme S the set of isomorphism class of abelian schemes $(A/S, \iota)$ over S with RM by $\mathcal{O}_L$, polarization $\lambda \colon (\mathcal{M}_A, \mathcal{M}_A^+) \to (\mathfrak{a}, \mathfrak{a}^+)$ (for a suitable representative $(\mathfrak{a}, \mathfrak{a}^+)$), and full level n structure $\alpha \colon (\mathcal{O}_L/n\mathcal{O}_L)^2 \xrightarrow{\sim} A[n]$. This functor is representable by a smooth morphism $\pi \colon \mathfrak{M} \to \mathbb{Z}[1/(nd_L)]$ of relative dimension g.

Furthermore, for $(p, n \cdot d_L) = 1$ the components of $\mathfrak{M} \otimes \overline{\mathbb{F}_p}$ are the same as the components of $\mathfrak{M} \otimes \mathbb{C}$.

See also [**135**, Théorème 1.28, p. 268, Variante 6.2, p. 325] for some more details. We remark that the complex variety $\mathfrak{M}(\mathbb{C})$ is complex analytically isomorphic to $\coprod_{(\mathfrak{a},\mathfrak{a}^+)}\coprod_{1\leq i\leq\phi(n)} G_{(\mathfrak{a},\mathfrak{a}^+)}\backslash\mathcal{H}^g$ where the group $G_{(\mathfrak{a},\mathfrak{a}^+)}$ is the principal congruence subgroup modulo n of $\mathrm{SL}(\mathcal{O}_L\oplus\mathfrak{a})$.

Another very useful version of moduli problems (and hence of moduli schemes) for abelian varieties with real multiplication is the following:

DEFINITION 6.8. A μ_N*-level structure* on an abelian scheme with RM by $\mathcal{O}_L$ $(A/S,\iota)$ is an $\mathcal{O}_L$-equivariant embedding

$$\mu_N\otimes_{\mathbb{Z}}\mathcal{D}_{L/\mathbb{Q}}^{-1}\to A[N]. \tag{6.11}$$

A μ_{p^∞}*-level structure* is a compatible sequence of μ_{p^n}-level structures for $n = 1,2,3,\ldots$.

We offer some words of explanation regarding the group scheme $\mu_N\otimes_{\mathbb{Z}}\mathcal{D}_{L/\mathbb{Q}}^{-1}$ (the reader may consult Appendix A for background on group schemes). The group scheme $\mu_N\otimes_{\mathbb{Z}}\mathcal{D}_{L/\mathbb{Q}}^{-1}$ is of course determined by its functor of points which is given by:

$$(\mu_N\otimes_{\mathbb{Z}}\mathcal{D}_{L/\mathbb{Q}}^{-1})(R)=\mu_N(R)\otimes_{\mathbb{Z}}\mathcal{D}_{L/\mathbb{Q}}^{-1}. \tag{6.12}$$

Thus, the group scheme $\mu_N\otimes_{\mathbb{Z}}\mathcal{D}_{L/\mathbb{Q}}^{-1}$ is isomorphic to μ_N^g (an isomorphism is given by choosing a $\mathbb{Z}$-basis to $\mathcal{D}_{L/\mathbb{Q}}^{-1}$). However, writing it as we do has the advantage of endowing it with a canonical $\mathcal{O}_L$ action coming from the action of $\mathcal{O}_L$ on the $\mathcal{O}_L$-module $\mathcal{D}_{L/\mathbb{Q}}^{-1}$.

Note that an abelian variety over a field of characteristic p with RM and a μ_{p^n}-level structure is ordinary.

Let N be prime to p and let $n\leq\infty$. As in the case of full level structure one considers the moduli problem of polarized abelian schemes with RM and with μ_{Np^n}-level structure. That is, we consider schemes over $\mathbb{Z}[d_L^{-1}]$ and the functor associating to such a scheme S the isomorphism classes of quadruples $(A,\iota,\lambda,\beta_{Np^n})/S$ consisting of a polarized abelian scheme with RM $(A,\iota,\lambda)/S$ and a μ_{Np^n} level structure given by $\beta_{Np^n}:\mu_{Np^n}\otimes\mathcal{D}_{L/\mathbb{Q}}^{-1}\to A$. For $Np^n\geq 4$ this is a rigid moduli problem—see Chapter 4, Lemma 3.1. The same methods giving Theorem 6.7 give also

THEOREM 6.9. *Let N be prime to p and $n\leq\infty$. The coarse moduli space of polarized abelian varieties with RM for schemes over $\mathbb{Z}[d_L^{-1}]$, together with level μ_{Np^n} exists; denote it by $\mathfrak{M}(\mu_{Np^n})\to\mathrm{Spec}(\mathbb{Z}[d_L^{-1}])$ (with $\mathfrak{M}=\mathfrak{M}(\mu_{p^0})$). If $Np^n\geq 4$ it is a fine moduli space. The morphism $\mathfrak{M}(\mu_{Np^n})\to\mathrm{Spec}(\mathbb{Z}[d_L^{-1}])$ is smooth over $\mathbb{Z}[(Npd_L)^{-1}]$.*

REMARK 6.10. Using condition (DP) instead of condition (R) for the moduli problems considered in Theorems 6.7, 6.9 one may obtain moduli schemes over $\mathbb{Z}[n^{-1}]$ and $\mathbb{Z}$, respectively. The moduli spaces are flat over $\mathbb{Z}[n^{-1}]$ and $\mathbb{Z}[(Np)^{-1}]$ respectively. Assume that $(n,d_L)=1$ and $(Np,d_L)=1$. If $g>1$ the fibers of the moduli spaces over primes ramified in $\mathcal{O}_L$ are singular, yet normal, varieties. The singular locus being of pure codimension 2. See [**25**].

For any integer $N\geq 1$ and prime to p, let $\mathfrak{M}(\mathbb{F}_p,\mu_{Np^n})$ be the reduction modulo p of $\mathfrak{M}(\mu_{Np^n})$ and let $\mathfrak{M}(\mathbb{F}_p,\mu_N)^{\mathrm{ord}}$ be the ordinary part of $\mathfrak{M}(\mathbb{F}_p,\mu_N)$. If $N=1$

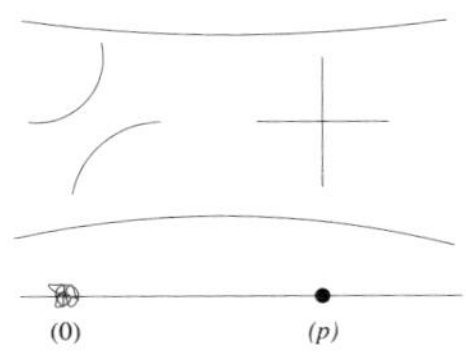

FIGURE 6.1. The curve $xy - p = 0$ over $\operatorname{Spec}(\mathbb{Z}_p)$.

we just write $\mathfrak{M}(\mathbb{F}_p)^{\mathrm{ord}}$. Then, for $N \geq 4$ the fibers of

$$f\colon \mathfrak{M}(\mathbb{F}_p, \mu_{Np^n}) \to \mathfrak{M}(\mathbb{F}_p, \mu_N)^{\mathrm{ord}} \tag{6.13}$$

are principal homogeneous spaces under $(\mathcal{O}_L/p^n\mathcal{O}_L)^\times$ (for $n = \infty$, take $(\mathcal{O}_L \otimes \mathbb{Z}_p)^\times$) and for any N the action of $(\mathcal{O}_L/p^n\mathcal{O}_L)^\times$ on the fibers is transitive.

We may ask how do level structures behave under reduction mod p when the prime p divides the level. After all, a regular scheme over $\operatorname{Spec}(\mathbb{Z}_p)$ with an irreducible generic fiber may very well have a reducible special fiber, e.g., $\operatorname{Spec}\big(\mathbb{Z}_p[x,y]/(xy - p)\big) \to \operatorname{Spec}(\mathbb{Z}_p)$ (Figure 6.1); see the reduction of $X_0(p)$ on p. 32; or see Kodaira's classification of the special fiber of the Néron model of an elliptic curve (e.g., [**157**, Chapter IV.8]). This is quite a subtle question and to study one such level structure, just for ordinary points, we would require Deligne's description of ordinary abelian varieties over finite fields.

6.2. Deligne's description of ordinary abelian varieties. Let A be a g-dimensional abelian variety over $\mathbb{F}_q$, $q = p^r$. Fix an algebraic closure $\overline{\mathbb{F}_q}$ and an embedding $W(\overline{\mathbb{F}_q}) \overset{\phi}{\hookrightarrow} \mathbb{C}$. Assume that A is ordinary; then, via ϕ, we get a canonical lifting $A_\mathbb{C}$ of A to $\mathbb{C}$: $A_\mathbb{C} := A^{\mathrm{can}} \otimes_{W(\overline{\mathbb{F}_q})} \mathbb{C}$. Put

$$T(A) = H_1(A_\mathbb{C}, \mathbb{Z}). \tag{6.14}$$

It is a lattice in $\mathbb{C}^g \cong H_1(A_\mathbb{C}, \mathbb{C})$ of rank $2g$ and $A_\mathbb{C} \cong H_1(A_\mathbb{C}, \mathbb{R})/H_1(A_\mathbb{C}, \mathbb{Z})$. Note that $\mathrm{Fr}_q \in \operatorname{End}_{W(\overline{\mathbb{F}_q})}(A) = \operatorname{End}_\mathbb{C}(A_\mathbb{C})$, hence it defines a linear operator $F \in \operatorname{End}\big(T(A)\big)$.

THEOREM 6.11 (Deligne [**21**]). *The functor*

$$\begin{aligned} \{\textit{ordinary abelian varieties over } \mathbb{F}_q\} &\to \{\textit{free even rank } \mathbb{Z}\textit{-lattices with an endomorphism } F \textit{ such that the conditions (1)–(4) hold}\}, \\ A &\mapsto (T(A), F) \end{aligned} \tag{6.15}$$

is an equivalence of categories. The conditions are the following:

(1) *F is a semisimple operator whose eigenvalues all have complex absolute value $\sqrt{q}$.*
(2) *Exactly half the roots of the characteristic polynomial of F are p-adic units.*
(3) *There exists a linear operator V such that $FV = q$.*
(4) *There exist free $\mathbb{Z}_p$-modules T'_p, T''_p such that;* (a) $T \otimes \mathbb{Z}_p = T'_p \oplus T''_p$, (b) $\operatorname{rank}_{\mathbb{Z}_p} T'_p = \operatorname{rank}_{\mathbb{Z}_p} T''_p$, *and* (c) *$F|_{T'_p}$ is invertible and $F|_{T''_p}$ is divisible by q.*

REMARK 6.12. If condition 1 holds, then $\{2,3\} \iff 4$.

REMARK 6.13. In fact, starting from $A/\mathbb{F}_q$, $T = T(A)$, then T'_p, T''_p are none-other than $T_pA(\overline{\mathbb{F}_q})$ and $\mathrm{Hom}_{\mathbb{Z}_p}\big(T_pA^\vee(\overline{\mathbb{F}_q}), \mathbb{Z}_p(1)\big)$.

COROLLARY 6.14. *Let $(A/\mathbb{F}_q, \iota)$ be an abelian variety with RM by $\mathcal{O}_L$. We get an induced embedding $T(\iota)\colon \mathcal{O}_L \to \mathrm{End}\big(T(A)\big)$ such that the image of $\mathcal{O}_L$ commutes with F. Thus,*

$$(A, \iota) \to \big(T(A), F, T(\iota)\big) \tag{6.16}$$

is an equivalence between ordinary abelian varieties with RM by $\mathcal{O}_L$ over $\mathbb{F}_q$ and free even rank $\mathcal{O}_L$-lattices with an $\mathcal{O}_L$-endomorphism F such that the conditions (1)–(4) *hold.*

Given $a \in \mathcal{O}_L$ such that $\big((a),(p)\big) = 1$, and n such that $a^2 - 4p^n \ll 0$, let

$$T = \mathcal{O}_L[x]/(x^2 - ax + p^n). \tag{6.17}$$

It is a $\mathbb{Z}$-lattice of rank $2g$ and a free $\mathcal{O}_L$-module (of rank 2). Let F be multiplication by x; then $\mathcal{O}_L$ acts on T as ordinary multiplication. Note that $T \otimes \mathbb{Q}$ is a CM field, hence F is semisimple.

$$|x| = \left|\frac{a \pm \sqrt{a^2 - 4p^n}}{2}\right| = \left|\frac{a + \sqrt{a^2 - 4p^n}}{2} \cdot \frac{a - \sqrt{a^2 - 4p^n}}{2}\right|^{1/2} = \sqrt{p^n}. \tag{6.18}$$

Mod p, x solves the equation $x(x-a) = 0$. This implies that half of the roots of the characteristic polynomial of F are p-adic units. Take $V = -(x-a)$ then $FV = p^n$.

Conclusion. The conditions imposed in Corollary 6.14 on (T, F, ι) are satisfied and (T, F, ι) corresponds to an ordinary abelian variety A of dimension g with RM by $\mathcal{O}_L$ over $\mathbb{F}_{p^n}$.

Remark that $T \otimes \mathbb{Z}_p = T'_p \oplus T''_p$ and F acts on T'_p by an invertible $\mathcal{O}_L$-linear endomorphism. We conclude that $F = \mathrm{Fr}_{p^n}$ acts on $T_pA(k)$ by the unique $\eta \in (\mathcal{O}_L \otimes \mathbb{Z}_p)^\times$ such that $\eta^2 - a\eta + p^n = 0$.

REMARK 6.15. $T \cong \mathcal{O}_L \oplus \mathcal{O}_L$ as $\mathcal{O}_L$-modules, hence the polarization module of the associated abelian variety with RM is just $\mathcal{D}_L^{-1}$. Replacing $\mathcal{O}_L$ by $\mathfrak{A}^{-1}$ in (6.17), we get the polarization module $\mathfrak{A}^2\mathcal{D}_L^{-1}$. It is easy to modify the construction to get any polarization module (and not just those of the form $\mathfrak{A}^2 \cdot \mathcal{D}_L^{-1}$). We will assume that henceforth. Thus, we have the following

COROLLARY 6.16. *If $\mathfrak{M}$ is the coarse moduli space of abelian varieties with RM then every component of $\mathfrak{M} \otimes \overline{\mathbb{F}_p}$ contains an ordinary point.*

This follows because Theorem 6.7 gives that the components of $\mathfrak{M} \otimes \overline{\mathbb{F}_p}$, with no level structure, are in bijection with classes $\mathfrak{A} \in \mathrm{Cl}(L)^+$. Ribet's construction gives an ordinary point on each component.

COROLLARY 6.17. *The ordinary locus is a dense open subset, hence it is irreducible in every component.*

Indeed, the Serre–Tate coordinates imply that the ordinary locus is open.

6.3. Irreducibility. Recall that $\mathfrak{M}(\mathbb{F}_p)$ and $\mathfrak{M}(\mathbb{F}_p, \mu_{Np^n})$ stand, respectively, for the coarse moduli spaces of abelian with RM and abelian varieties with RM and μ_{Np^n}-level structure (in the sense of §6.1), both in characteristic p. We want to prove that the geometric components of $\mathfrak{M}(\mathbb{F}_p, \mu_{Np^n})$ are the same as those of $\mathfrak{M}(\mathbb{F}_p)$. Namely, that if one fixes the polarization module then there is a unique component of $\mathfrak{M}(\mathbb{F}_p, \mu_{p^n})$ corresponding to this polarization module. In the case of elliptic curves, for $n = 1$, this means that the special fiber of $X_0(p)$, which is reducible(!), nonetheless has a unique component corresponding to μ_p-level structure (Figure 5.1). It is technically easier to deal with the moduli spaces $\mathfrak{M}(\mathbb{F}_p, \mu_{Np^n})$ and $\mathfrak{M}(\mathbb{F}_p, \mu_N)$, for some extraneous $N \geq 4$ and prime to p, because in this case the fibers of

$$f\colon \mathfrak{M}(\mathbb{F}_p, \mu_{Np^n}) \to \mathfrak{M}(\mathbb{F}_p, \mu_N)^{\mathrm{ord}} \tag{6.19}$$

are principal homogeneous spaces $(\mathcal{O}_L/p^n\mathcal{O}_L)^\times$ (for $n = \infty$, take $(\mathcal{O}_L \otimes \mathbb{Z}_p)^\times$). This implies the result for $N = 1$.

EXERCISE 6.18. *Prove that the fact that the components of* $\mathfrak{M}(\mathbb{F}_p, \mu_{Np^n})$ *are parameterized by classes in* $\mathrm{Cl}(L)^+$ *implies the same for* $\mathfrak{M}(\mathbb{F}_p, \mu_{p^n})$.

THEOREM 6.19 (Ribet [**138**]). *Let* $\mathfrak{A} \in \mathrm{Cl}(L)^+$, *let* $\mathfrak{B}$ *be the component of* $\mathfrak{M}(\mathbb{F}_p, \mu_N)^{\mathrm{ord}}$ *corresponding to* $\mathfrak{A}$, *and let* $\mathfrak{N}$ *be the preimage of* $\mathfrak{B}$ *under the morphism* $f\colon \mathfrak{M}(\mathbb{F}_p, \mu_{Np^n}) \to \mathfrak{M}(\mathbb{F}_p, \mu_N)^{\mathrm{ord}}$. *Then,* $\mathfrak{N}$ *is geometrically irreducible.*

PROOF. As explained above we may take $N \geq 4$. To simplify the exposition, assume that $\mathfrak{B}$ corresponds to the polarization class $\mathcal{D}_L^{-1}$. It is enough to prove the claim for $n = \infty$.

The morphism

$$f\colon \mathfrak{N} \to \mathfrak{B} \tag{6.20}$$

is étale with fibers being principal homogeneous spaces under $(\mathcal{O}_L \otimes \mathbb{Z}_p)^\times$. It gives a p-adic character

$$\chi\colon \pi_1(\mathfrak{B} \otimes \overline{\mathbb{F}_p}) \to (\mathcal{O}_L \otimes \mathbb{Z}_p)^\times. \tag{6.21}$$

The preimage $\mathfrak{N}$ is irreducible if and only if χ is surjective. Now,

$$\pi_1(\mathfrak{B} \otimes \overline{\mathbb{F}_p}) = \bigcap_n \pi_1(\mathfrak{B} \otimes \mathbb{F}_{p^n}), \tag{6.22}$$

and

$$\mathcal{O}_L \otimes \mathbb{Z}_p = \varprojlim \mathcal{O}_L \otimes (\mathbb{Z}/p^k\mathbb{Z}). \tag{6.23}$$

Therefore, it is enough to prove that *for every* k *there exists an* N *such that for every* $n \geq N$ *the homomorphism* $\pi_1(\mathfrak{B} \otimes \mathbb{F}_{p^n}) \xrightarrow{\chi} (\mathcal{O}_L/p^k\mathcal{O}_L)^\times$ *is surjective.*

Given $\alpha \in (\mathcal{O}_L/p^k\mathcal{O}_L)^\times$, choose $a \in \mathcal{O}_L$, such that $a \equiv \alpha \mod p^k$, and choose $N > k$ such that $a^2 - 4p^n \ll 0$ for $n \geq N$. Consider the ordinary abelian variety (A, ι) over $\mathbb{F}_{p^n}$ constructed as in (6.17) from $\mathcal{O}_L, a, n$.

Let $z \in \mathfrak{B}(\mathbb{F}_{p^n})$ be the point corresponding to (A, ι) with residue field $k(z)$. Let χ_0 be the composition:

$$\pi_1\Big(\mathrm{Spec}\big(k(z)\big)\Big) \to \pi_1(\mathfrak{B} \otimes \mathbb{F}_{p^n}) \xrightarrow{\chi} (\mathcal{O}_L/p^k\mathcal{O}_L)^\times. \tag{6.24}$$

But

$$\pi_1\Big(\mathrm{Spec}\big(k(z)\big)\Big) = \pi_1\big(\mathrm{Spec}(\mathbb{F}_{p^n})\big) \cong \widehat{\mathbb{Z}} \tag{6.25}$$

and $\langle \mathrm{Fr}_{p^n} \rangle$ is a dense subgroup of $\pi_1\big(\mathrm{Spec}(k(z))\big)$ that provides the isomorphism with $\widehat{\mathbb{Z}}$. The value of the character χ_0 on $\mathrm{Fr}_{p^n}$is just the action of Fr_{p^n} on the fiber of $f\colon \mathfrak{N} \to \mathfrak{B}$ over z. That is, it is the action of Fr_{p^n} on $\mu_{p^k} \otimes \mathcal{O}_L \hookrightarrow A[p^k]$. By duality, this is the action of Fr_{p^n} on $T_p(A/\overline{\mathbb{F}_{p^n}}) \mod p^k$. This was given by

$\eta \in (\mathcal{O}_L \otimes \mathbb{Z}_p)^\times$ such that $\eta^2 - a\eta + p^n = 0$. We may read this equation mod p^k obtaining $\eta(\eta - \alpha) = 0$. We get that $\chi_0(\mathrm{Fr}_{p^n})$ acts by $\eta \equiv \alpha \bmod p^k$ viewed as in $(\mathcal{O}_L/p^k\mathcal{O}_L)^\times$. □

CHAPTER 4

p-adic Elliptic Modular Forms

1. Introduction

In its simplest incarnation, as defined by Serre [**146**], a p-adic modular form is a power series

$$\sum_{n=0}^{\infty} a_n q^n, \quad a_n \in \mathbb{Q}_p, \tag{1.1}$$

which is a limit, in a suitable sense, of classical (i.e., usual) modular forms with coefficients in $\mathbb{Q}$. Another approach is due to Dwork [**36**], who used p-adic analytic functions on modular curves, endowed with the action of the U-operator. Those notions were superseded and generalized by Katz, who gave a more conceptual definition of such a form as a section of a line bundle over open sets of the moduli space of elliptic curves with level structure.

In the next two sections we shall describe briefly two sources of motivation for the study of p-adic modular forms: first, p-adic L-functions, and second, deformations of Galois representations. The connection between p-adic modular forms and p-adic L-functions is through q-expansions on the modular side, and special values on the L-functions side: values of p-adic L-functions at negative integers appear (up to simple factors) as constant coefficients in the q-expansion of Eisenstein series. The connection between modular forms and Galois representations is through the q-expansion of newforms and traces of Frobenius elements: given a newform f of weight k and level $\Gamma_0(N)$ (say) $f = \sum a_n q^n$ there is a two-dimensional continuous ℓ-adic Galois representation $\rho\colon G_{\mathbb{Q}} \to \mathrm{GL}_2(\mathbb{Q}_\ell)$ (for any ℓ) such that for almost every p the representation is unramified at p and the trace and determinant of the corresponding Frobenius element are $\mathrm{Tr}\big(\rho(\mathrm{Fr}_p)\big) = a_p$ and $\det\big(\rho(\mathrm{Fr}_p)\big) = p^{k-1}$. (Note that since the representation is semi-simple and the Frobenius elements are dense in $G_{\mathbb{Q}}$ the coefficients a_p determine the representation uniquely).

In Section 2 we discuss congruences modulo p between modular forms on $\mathrm{SL}_2(\mathbb{Z})$ with p-integral q-expansion. We take a very direct approach, following Swinnerton-Dyer [**158**] and Serre [**145**]. This approach uses much specific information on the graded ring of modular forms on $\mathrm{SL}_2(\mathbb{Z})$ and is thus not suitable for generalizations. In Section 3 we approach the same subject from a more "highbrow" perspective in which modular forms are sections of line bundles and Hecke (and other) operators are obtained from correspondences (and other natural geometric constructions). Still, in §§3.3, 3.4, and 4 we use the direct approach to discuss systems of eigenvalues, congruences between modular forms modulo p^m and the applications to p-adic zeta functions.

In Section 5 we take a geometric approach to congruences (using the language developed in §3). We find the ideal of congruences for modular forms of level μ_N

modulo p, for any N prime to p: If H is the Hasse invariant (§5.1) then the ideal of congruences for modular forms modulo p is generated by $H-1$. The method of proof is geometric and relates modular forms of level μ_N with functions of level μ_{Np}. This comparison leads to a certain dictionary between operations on modular forms and operations on functions. This is taken up in §5.3. It allows us to give, in particular, a natural interpretation for the operator θ and derive easily some of its properties.

Section 6 discusses p-adic modular forms over a p-adic base B, following N. Katz [**79**]. After defining p-adic modular forms using test objects, we determine them in the case p is nilpotent in B (§6.4) and in the case of a growth coefficient which is a unit in B (§6.5). One important result is the Katz expansion of a p-adic modular form discussed in §6.6. It allows us to show that (i) the q-expansion map on p-adic modular forms is injective (Corollary 6.20), and that (ii) Serre's p-adic modular forms (defined as "p-adic uniform limits" of classical modular forms) are the same as Katz's p-adic modular forms of growth condition 1 (Corollary 6.20).

The last section discusses the ring of divided congruences. Again the results are due to Katz. The main result from our perspective is Corollary 7.9 that relates sums of modular forms such that the sum of their q-expansions is p-integral, and thus is a classical object, with the ring of functions over the p-adic analytic space consisting of the complement of the supersingular discs, which is a p-adic object.

1.1. p-adic L-functions. Let K be a number field of degree $g=[K:\mathbb{Q}]$, ring of integers $\mathcal{O}_K$, discriminant d_K, r_1 real embeddings and $2r_2$ complex embeddings. Its zeta function

$$\zeta_K(s) = \sum_{\mathfrak{a}\subset\mathcal{O}_K} \frac{1}{\operatorname{Norm}(\mathfrak{a})^s} = \prod_{\substack{\mathfrak{p}\subset\mathcal{O}_K\\ \mathfrak{p}\text{ prime}}} \frac{1}{1-\operatorname{Norm}(\mathfrak{p})^{-s}}, \qquad \Re(s)>1, \tag{1.2}$$

can be continued meromorphically to the complex plane and satisfies the following functional equation:

$$A^s\Gamma\left(\frac{s}{2}\right)^{r_1}\Gamma(s)^{r_2}\zeta_K(s) = A^{1-s}\Gamma\left(\frac{1-s}{2}\right)^{r_1}\Gamma(1-s)^{r_2}\zeta_K(1-s), \qquad A = 2^{-r_2}\pi^{-g/2}\sqrt{d_K}. \tag{1.3}$$

Here Γ is the complex gamma function interpolating the values $\Gamma(n+1)=n!$ for a non-negative integer n. Note that the product in (1.2), called an *Euler product*, shows that ζ_K does not vanish for $\mathrm{Re}(s)>1$.

The set of poles of the gamma function is precisely the set of non-positive integers. The gamma function does not vanish on the real line. Therefore, if K is not totally real (i.e. $r_2>0$), then $\zeta_K(1-m)=0$ for every integer $m>1$; if K is totally real, then $\zeta_K(1-m)=0$ for every odd integer $m>1$. Thus, we are interested in the numbers $\zeta_K(1-m)$, where m is a positive even integer, and K is totally real.[1]

Recall that $\zeta_K(1-m)$, $m\geq 2$, is a rational number (See Chapter 3, §4.1). As a further indication that there is bound to be interesting information encoded

[1]These special values of the zeta function are related to ratios of the orders of the torsion part of the K-groups of $\mathcal{O}_K$. Lichtenbaum's conjecture [**92**] predicts that even when $\zeta_K(1-m)=0$ the leading coefficient of the Taylor expansion around that point retains the K-theoretic interpretation. See [**87, 88**].

TABLE 1.1. Values of Riemann's zeta function at negative integers.

k	$\zeta_{\mathbb{Q}}(1-k)$	k	$\zeta_{\mathbb{Q}}(1-k)$
2	$\frac{-1}{2^2\cdot 3}$	20	$\frac{283\cdot 617}{2^3\cdot 3\cdot 5^2\cdot 11}$
4	$\frac{1}{2^3\cdot 3\cdot 5}$	22	$\frac{-131\cdot 593}{2^2\cdot 3\cdot 23}$
6	$\frac{-1}{2^2\cdot 3^2\cdot 7}$	24	$\frac{103\cdot 2294797}{2^4\cdot 3^2\cdot 5\cdot 7\cdot 13}$
8	$\frac{1}{2^4\cdot 3\cdot 5}$	26	$\frac{-657931}{2^2\cdot 3}$
10	$\frac{-1}{2^2\cdot 3\cdot 11}$	28	$\frac{9349\cdot 362903}{2^3\cdot 3\cdot 5\cdot 29}$
12	$\frac{691}{2^3\cdot 3^2\cdot 5\cdot 7\cdot 13}$	30	$\frac{-1721\cdot 1001259881}{2^2\cdot 3^2\cdot 7\cdot 11\cdot 31}$
14	$\frac{-1}{2^2\cdot 3}$	32	$\frac{37\cdot 683\cdot 305065927}{2^6\cdot 3\cdot 5\cdot 17}$
16	$\frac{3617}{2^5\cdot 3\cdot 5\cdot 17}$	34	$\frac{-151628697551}{2^2\cdot 3}$
18	$\frac{-43867}{2^2\cdot 3^3\cdot 7\cdot 19}$	36	$\frac{26315271553053477373}{2^3\cdot 3^3\cdot 5\cdot 7\cdot 13\cdot 19\cdot 37}$

in these values, we recall that the ubiquitous Bernoulli numbers appear as special values of the Riemann zeta function $\zeta_{\mathbb{Q}}$:

$$\zeta_{\mathbb{Q}}(1-k) = -\frac{B_k}{k}, \quad k \geq 2. \tag{1.4}$$

Here B_k is the k-th Bernoulli number, which is zero for k odd. See Table 1.1 of some values of $\zeta_{\mathbb{Q}}$.

Say K is a totally real number field. Another motivation to study special values of ζ_K is Siegel's formula for the hyperbolic volume of $\mathrm{SL}_2(\mathcal{O}_K)\backslash\mathcal{H}^g$:

$$\int_{\mathrm{SL}_2(\mathcal{O}_K)\backslash\mathcal{H}^g} (-1)^g \frac{1}{(2\pi)^g} \frac{dx_1 \wedge dy_1}{y_1^2} \wedge \cdots \wedge \frac{dx_g \wedge dy_g}{y_g^2} = 2\zeta_K(-1). \tag{1.5}$$

This formula was generalized to other algebraic groups. In these more general situations one obtains a formula for the volume of a bounded symmetric domain in terms of products of values of zeta functions. A good source is [**148**]. Similar products appear in calculating the number of strata in moduli spaces of abelian varieties. See [**45**].

Let p be a prime. Let $g = [K : \mathbb{Q}]$ and $g_p = [K \cap \mathbb{Q}(\mu_{p^\infty}) : \mathbb{Q}]$, where μ_{p^∞} is the group of p-power roots of unity in $\mathbb{C}$. We have the following congruence relations:

THEOREM 1.1[2] (Kummer's congruences [**145**, §4], [**146**, §5.3]).

(1) *Assume* $p \geq 3, m > 0$ *even.*
  (a) *If* $g_p m \not\equiv 0 \bmod p-1$ *then* $\mathrm{val}_p\big(\zeta_K(1-m)\big) \geq 0$.
  (b) *If* $gm \equiv 0 \bmod p-1$ *then* $\mathrm{val}_p\big(\zeta_K(1-m)\big) \geq -1 - \mathrm{val}_p(gm)$.

(2) $\mathrm{val}_2\big(\zeta_K(1-m)\big) \geq g - 2 - \mathrm{val}_2(gm)$.

[2]In fact this theorem can be improved, using Hilbert modular forms. See Chapter 5, § 5 and [**49**]. It can also be formulated for the prime 2.

(3) *If* $m \equiv m' \bmod p^n(p-1)$, *and* $gm \not\equiv 0 \bmod p-1$, *then*

$$\prod_{\mathfrak{p}|p}(1-\mathrm{Norm}(\mathfrak{p})^{m-1})\zeta_K(1-m) \equiv \prod_{\mathfrak{p}|p}(1-\mathrm{Norm}(\mathfrak{p})^{m'-1})\zeta_K(1-m') \bmod p^{n+1} \tag{1.6}$$

We shall prove certain parts of this theorem in Corollary 4.9 and in Corollary 5.4 in Chapter 5. The link to modular forms is that one can find special values, say $\zeta_K(1-m)$ or $L(1-m,\chi)$ where χ is a Dirichlet character, as leading coefficients of modular forms whose higher coefficients are *integral.* We already saw and used that for the zeta function in Chapter 2, §§4.1, 6. More general constructions appear in [**27**]. See also [**165**].

One of the reasons for interest in this theorem is that it provides the p-adic interpolation of the values $\zeta_K(1-m)$ ($m \geq 2$ even), modified by a suitable Euler factor, by a continuous function whose domain is $\mathbb{Z}_p$. The group $\mathbb{Z}_p$ plays an essential role in Galois theory (e.g., via the Kronecker–Weber Theorem) and of course in Iwasawa theory [**74**], [**167**, Chapters 7, 13].

A more subtle link comes as follows: Recall that p divides $\sharp\,\mathrm{Cl}(\mathbb{Q}(\zeta_p))$ iff p divides the numerator of some Bernoulli number B_k, for $k = 2, 4, \dots, p-3$ ([**167**, Theorem 5.16]); in that case, p is called an *irregular prime*.

Let $G = \mathrm{Gal}(\mathbb{Q}(\zeta_p)/\mathbb{Q})$. Let A denote the p-Sylow subgroup of $\mathrm{Cl}(\mathbb{Q}(\zeta_p))$. The abelian group A can be considered as a $\mathbb{Z}_p[G]$-module. The group G is provided with an isomorphism to $(\mathbb{Z}/p\mathbb{Z})^\times$, a residue class n operating by $\zeta_p \mapsto \zeta_p^n$. Assume that p is odd. Let ω be the Teichmüller character; it is the unique character of conductor p with values in $\mu_{p-1} \subset \mathbb{Z}_p$ such that

$$\omega(n) \equiv n \bmod p, \quad n \not\equiv 0 \bmod p. \tag{1.7}$$

EXERCISE 1.2. *Find the first few p-adic digits of the values of ω for $p = 5$.*

The character ω is a generator for the character group $\widehat{G}$ of G,

$$\widehat{G} = \{\omega^i : 0 \leq i \leq p-2\}. \tag{1.8}$$

We decompose A in ω^i-eigenspaces:

$$A = \bigoplus_{i=0}^{p-2} A_i. \tag{1.9}$$

We have $A_0 = 0$ (trivially), and $A_1 = 0$ (this needs proof). For every $i = 1, 2, \dots, p-2$, it is known [**167**, Prop. 6.16] that A_i is killed by the generalized Bernoulli number $B_{1,\omega^{-i}}$ that has the property

$$B_{1,\omega^{-i}} \equiv \frac{B_{p-i}}{p-i} \bmod p. \tag{1.10}$$

We recall that for every Dirichlet character χ of conductor f, the generalized Bernoulli numbers $B_{n,\chi}$ are defined via the expansion of the following function in one variable:

$$F_\chi(t) = \sum_{a=1}^{f} \frac{\chi(a)te^{at}}{e^{ft}-1} = \sum_{n=0}^{\infty} B_{n,\chi}\frac{t^n}{n!}. \tag{1.11}$$

The formula (1.4) for the values $\zeta_{\mathbb{Q}}(1-m)$ can be generalized:

$$L(1-m,\chi) = -\frac{B_{m,\chi}}{m}, \quad m \geq 1. \tag{1.12}$$

The analogue of Theorem 1.1 enables one to prove:

THEOREM 1.3 (Kubota–Leopoldt). *Let χ be a Dirichlet character. There exists a unique p-adic meromorphic function $L_p(s,\chi)$ with the following properties:*

(1)

$$\begin{aligned} L_p(s,\chi) &= \frac{a_{-1}}{s-1} + \sum_{n=0}^{\infty} a_n (s-1)^n, \\ a_n \in \mathbb{Q}(\chi) := \mathbb{Q}(\{\chi(a) : a \in \mathbb{Z}\}), &\quad a_{-1} = \begin{cases} 1-1/p, & \chi = 1 \\ 0, & \chi \neq 1. \end{cases} \end{aligned} \tag{1.13}$$

The series converges in the disc $\{s \in \mathbb{C}_p : |s-1| < r\}$, $r = |p|^{1/(p-1)}|q|^{-1} > 1$ (if $p \neq 2$, put $q = p$; otherwise, put $q = 4$).

(2) *For a positive integer n, we have*

$$L_p(1-n,\chi) = (1-(\chi\cdot\omega^{-n})(p)p^{n-1})\cdot L(1-n,\chi\cdot\omega^{-n}). \tag{1.14}$$

For the proof see [**74**, Chapter 3], or [**167**, Chapter 5.2]; $\mathbb{C}_p$ is a completion of an algebraic closure of $\mathbb{Q}_p$. If $p = 2$, the Teichmüller character ω is the unique non-trivial character of $\mathbb{Z}_2^\times$ of conductor 4 with values in $\{\pm 1\}$.

1.2. Deformation of Galois representations. Let K be a field. We denote by G_K the absolute Galois group of K. For every prime p we choose an embedding $\overline{\mathbb{Q}} \hookrightarrow \overline{\mathbb{Q}_p}$ and denote by D_p and I_p the corresponding decomposition and inertia subgroups of $G_{\mathbb{Q}}$. Fix a prime ℓ. Let ϵ_ℓ be the $\mathbb{Z}_\ell^\times$-valued character giving the action of $G_{\mathbb{Q}}$ on ℓ-th powers roots of 1: $\sigma(\zeta) = \zeta^{\epsilon_\ell(\sigma)}$ if $\zeta^{\ell^n} = 1$ for some n. Let $\mathbb{F}$ be a finite field of characteristic ℓ and let

$$\bar\rho\colon G_{\mathbb{Q}} \to \mathrm{GL}_2(\mathbb{F}) \tag{1.15}$$

be a two-dimensional representation with determinant ϵ_ℓ. Let S be a finite set of primes of $\mathbb{Q}$. We shall assume that

- $\bar\rho$ is *absolutely irreducible*. That is, for every extension $\mathbb{F}'$ of $\mathbb{F}$ the representation space $\mathbb{F}'^2$ associated to $\bar\rho\otimes_{\mathbb{F}}\mathbb{F}'$ has no proper subspace invariant under the action of $G_{\mathbb{Q}}$.
- *Unramified outside S*. That is, if $p \notin S$ then $\bar\rho|_{I_p}$ is trivial. Note that this does not depend on the choice of the embedding $\overline{\mathbb{Q}} \hookrightarrow \overline{\mathbb{Q}_p}$.
- *Semistable at every prime different from ℓ*. That is, for every prime $p \neq \ell$ the representation $\bar\rho|_{I_p} \sim \left(\begin{smallmatrix} 1 & * \\ 0 & 1 \end{smallmatrix}\right)$. Note that $\bar\rho$ is unramified at p iff $* = 0$.
- *$\bar\rho$ is ordinary at ℓ*. That is, the dimension of its coinvariants is one.

One of the most important examples of such a representation is the following. Let A be an abelian variety over $\mathbb{Q}$ with real multiplication by $\mathcal{O}_L$, which is semistable at all primes. That is, for every prime p there exists an abelian scheme $\mathcal{A} \to \mathrm{Spec}(\mathbb{Z}_p)$ such that $\mathcal{A}\otimes_{\mathbb{Z}_p}\mathbb{Q}_p \cong A\otimes_{\mathbb{Z}}\mathbb{Q}_p$ and $\mathcal{A}\otimes\mathbb{F}_p$ is a semi-abelian variety. For example, if A is an elliptic curve, that is the case iff the conductor N_A is square free.[3] Assume that A has ordinary reduction at ℓ in the sense that the abelian part of $\mathcal{A} \otimes \mathbb{F}_\ell$ is an ordinary abelian variety. Then, the Galois representation

$$\bar{\rho}\colon G_\mathbb{Q} \to \mathrm{Aut}_{\mathcal{O}_L}\big(A[\ell](\overline{\mathbb{Q}})\big), \tag{1.16}$$

if absolutely irreducible, is a representation satisfying the assumptions above for $S = \{\ell, \infty\} \cup \{p : A \text{ has bad reduction at } p\}$.

EXERCISE 1.4. *Prove that if A is an elliptic curve over $\mathbb{Q}$ then A has ordinary reduction at ℓ (in the sense above) iff $a_\ell \neq 0$, where a_ℓ is the ℓ-th coefficient of the L-function of A.*

Let $\widehat{\mathcal{C}}_\mathbb{F}$ be the category of complete Noetherian local rings $(R, \mathfrak{m}_R)$ with a given isomorphism $R/\mathfrak{m}_R \cong \mathbb{F}$. Morphisms in this category are local homomorphisms. That is, homomorphisms $R_1 \to R_2$ such that the induced map $\mathrm{Spec}(R_2) \to \mathrm{Spec}(R_1)$ takes the unique closed point of $\mathrm{Spec}(R_2)$ to the unique closed point of $\mathrm{Spec}(R_1)$.

Given $R \in \widehat{\mathcal{C}}_\mathbb{F}$, we say that a continuous homomorphism

$$\rho\colon G_\mathbb{Q} \to \mathrm{GL}_2(R) \tag{1.17}$$

is a *deformation of* $\bar{\rho}$ if the composition with the projection to $\mathbb{F}$ is $\bar{\rho}$. Two deformations ρ_1, ρ_2 over R are *equivalent* if they are conjugate by a matrix in $\mathrm{GL}_2(R) \cap \big(1 + M_2(\mathfrak{m}_R)\big)$, i.e., by a matrix in the kernel of the projection $\mathrm{GL}_2(R) \to \mathrm{GL}_2(\mathbb{F})$. Instead of using the language of matrices, one can use the language of modules. Then, a representation is a free R-module of rank 2, which is equipped with a continuous $G_\mathbb{Q}$ action, and the notion of deformation corresponds to the notion of being "isomorphic upon reduction."

The study of deformations of Galois representations was initiated by B. Mazur in [**95**]. At the time one could have considered it as a natural offspring of Hida's theory, as a natural analog of the deformation theory of varieties, and as a natural application of Schlessinger's theory [**142**]. However, the real importance of this theory became clear only after later developments. Most notably, Wiles applied this theory in his proof of the Shimura–Taniyama–Weil conjecture for semi-stable elliptic curves and hence to Fermat's Last Theorem [**164, 173**]. Since the application of Mazur's theory by Wiles is relevant to the following, we recall the gist of Wiles's argument for establishing Shimura–Taniyama–Weil for semi-stable elliptic curves. See [**37, 121, 151**] for a more detailed survey, and [**19, 20**] for text books.

Wiles translates the problem of proving that a given semi-stable elliptic curve E is modular to proving that the associated ℓ-adic representation

$$\rho_{E,\ell} \to \mathrm{Aut}\big(T_\ell(E)\big) \cong \mathrm{GL}_2(\mathbb{Z}_\ell) \tag{1.18}$$

is modular, in the sense that it is the two-dimensional Galois representation associated to a weight 2 newform (of level N_E). In order to do so, Wiles considers the mod ℓ representation

$$\bar{\rho}\colon G_\mathbb{Q} \to \mathrm{Aut}(E[\ell]) \cong \mathrm{GL}_2(\mathbb{F}_\ell). \tag{1.19}$$

[3] There are similar results for every abelian variety over a dvr R of mixed characteristic. Let K be the quotient field of R. Let p be the residue characteristic. For A/K one knows that A has semi-stable reduction over R if and only if there exists a submodule $U_0 \subseteq U = T_\ell(A/K)$ ($\ell \neq p$ a prime) such that the inertia group I_p at p acts trivially on U_0 and on U/U_0. Moreover, A has good reduction over R if and only if one can choose $U_0 = U$. One may relate these results to the Artin and Swan conductors of the representation $G_{\overline{K}} \to \mathrm{Aut}(U)$. See [**66**] Cor. 2.2.9, Prop. 3.5. and §4, especially (4.6.3).

Wiles argues that after passing to an auxiliary elliptic curve one may assume that $\bar{\rho}$ enjoys many wonderful properties. In fact, one may assume that ℓ is odd and

- $\bar{\rho}$ is *modular*, i.e., arises from the reduction mod ℓ of the representation associated to a weight 2 newform. (This is the hardest condition to verify).
- $\bar{\rho}$ is continuous, has determinant ϵ_ℓ, and is absolutely irreducible, even after restriction to the quadratic subfield of $\mathbb{Q}(\zeta_\ell)$.
- $\bar{\rho}$ is semi-stable at every prime $p \neq \ell$.
- $\bar{\rho}$ is either ordinary or Barsotti–Tate at ℓ.

The last condition means that either $\bar{\rho}|_{I_\ell}$ has coinvariants of dimension one, or that there exists a finite flat group scheme $\mathcal{G}$ over $\mathbb{Z}_\ell$ such that $\bar{\rho}$ is the Galois representation of $\mathcal{G}(\overline{\mathbb{Q}_\ell})$.

Instead of considering the representation $\rho_{E,\ell}$ alone, Wiles considers two sets of representations into rings in $\widehat{\mathcal{C}}_{\mathbb{F}}$, where $\mathbb{F}$ is a suitable finite extension of $\mathbb{F}_\ell$. One set is a set of representations arising from modular forms. The other set is the set of all "reasonable" Galois representations lifting $\bar{\rho}$. To be precise, Wiles chooses a set Σ of primes that are unramified for $\bar{\rho}$ and considers representations

$$\rho\colon G_{\mathbb{Q}} \to \mathrm{GL}_2(R), \quad R \in \mathrm{Ob}\,\widehat{\mathcal{C}}_{\mathbb{F}}, \tag{1.20}$$

which are of "type Σ." This is a set of technical conditions and we refer the reader to any of the above references for details, being content to say that it means essentially that ρ is continuous, of determinant ϵ_ℓ, unramified outside Σ where $\bar{\rho}$ is unramified, "minimally ramified" in general and is ordinary or Barsotti–Tate at ℓ.

Mazur [**95**] proved the following fundamental theorem:

THEOREM 1.5 (Mazur). *Let $\bar{\rho}$ be an absolutely irreducible representation. Let S be a finite set of primes containing the primes ramified in $\bar{\rho}$. There exists a universal deformation*

$$\rho^U : G_{\mathbb{Q}} \to \mathrm{GL}_2(R^U) \tag{1.21}$$

for representations unramified outside S. That is, there exists a ring R^U in $\widehat{\mathcal{C}}_{\mathbb{F}}$ and a representation $\rho^U : G_{\mathbb{Q}} \to \mathrm{GL}_2(R^U)$ unramified outside S, such that for every ring R in $\widehat{\mathcal{C}}_{\mathbb{F}}$ and a representation $\rho\colon G_{\mathbb{Q}} \to \mathrm{GL}_2(R)$ which is unramified outside S, there exists a unique homomorphism $\varphi\colon R^U \to R$ such that the representation $\varphi_*\rho^U$ (obtained by composing ρ^U with $\varphi\colon \mathrm{GL}_2(R^U) \to \mathrm{GL}_2(R)$) is equivalent to the representation ρ.

Similarly, one proves (following [**134**]) that there exists a universal deformation of type Σ:

$$\rho^U_\Sigma : G_{\mathbb{Q}} \to \mathrm{GL}_2(R^U_\Sigma). \tag{1.22}$$

On the other hand, Wiles considers all newforms f of weight 2 and trivial character such that ℓ^2 does not divide the conductor of f and the associated Galois representation ρ_f is a lift of $\bar{\rho}$ of type Σ. Moreover, he constructs (out of Hecke algebras) a ring $\mathbb{T}_\Sigma$ and a "universal modular deformation of type Σ." That is, the ring $\mathbb{T}_\Sigma$ is finite flat $\mathbb{Z}_\ell$-algebra and there exists a representation of type Σ

$$\rho^M_\Sigma : G_{\mathbb{Q}} \to \mathrm{GL}_2(\mathbb{T}_\Sigma) \tag{1.23}$$

such that all the representations ρ_f (after extending scalars) are specializations of ρ^M_Σ and, in fact, any such specialization is coming from a modular form. One also gets easily a surjective homomorphism $R^U_\Sigma \to \mathbb{T}^U_\Sigma$ and the whole point and difficulty is to prove that this homomorphism is an isomorphism. Then the representation $\rho_{E,\ell}$, initially a specialization of ρ^U_Σ, is also a specialization of ρ^M_Σ and therefore modular.

One of the obstacles in Wiles's approach and its further generalizations [**7, 17, 33**] is getting to a situation where a mod ℓ representation is *known* to be modular. The reduction to this case uses a deep theorem of Langlands and Tunnell on modularity of certain Artin representations. In this context we recall the celebrated Serre conjecture. Let $\ell \geq 3$.

CONJECTURE 1.6 (Serre). *Let $\rho\colon G_{\mathbb{Q}} \to \mathrm{GL}_2(\mathbb{F})$ be an irreducible, odd, two-dimensional representation. There exists a normalized eigenform f of level $N(\rho)$, weight $k(\rho)$ and character $\epsilon(\rho)$ such that:*

$$\rho \sim \rho_f. \tag{1.24}$$

One important point is that the weight, level, and character are all specified. See [**150**, §§1–2], [**139**].

Coming back to our discussion of deformations of Galois representations, one could say that Wiles artfully defines a subscheme of $\mathrm{Spec}(R^U)$ (namely $\mathrm{Spec}(R^U_\Sigma)$) and a suitable modular candidate for it, namely $\mathrm{Spec}(\mathbb{T}_\Sigma)$, and shows they are isomorphic. All further strengthening of Wiles's methods and applications to the Shimura–Taniyama–Weil conjecture, carried out by various subsets of C. Breuil, B. Conrad, F. Diamond and R. Taylor [**7, 17, 33**], use the same general approach and technique of singling out subschemes of $\mathrm{Spec}(R^U)$ on the one hand, and on the other hand defining a correct modular candidate by increasingly convoluted procedures.

Taken in this light, Mazur's conjecture—proven under some restrictions by Gouvea and Mazur—as described below is conceptually a very important insight on deformations of Galois representations.

Let N be prime to ℓ. Recall the moduli space $\mathfrak{M}(\mathbb{F}_\ell, \mu_{N\ell^\infty})$ of elliptic curves over $\mathbb{F}_\ell$-algebras with $\Gamma_1(N)$-structure and $\Gamma_1(\mu_{\ell^\infty})$-structure ($\mu_{\ell^n} \hookrightarrow E, \forall n$). Let R be a ring in $\widehat{\mathcal{C}}_{\mathbb{F}}$ with residue field $\mathbb{F}$. One lets $V_{1,\infty}$ be (essentially) the space of all functions on $\mathfrak{M}(R, \mu_{N\ell^\infty})$. It is a space in which every space of classical modular forms $S_k(\Gamma_1(N\ell^\nu), R)$ injects, and their union is dense in $V_{1,\infty}$. There is a certain Hecke algebra $\mathbb{T}$ acting on $V_{1,\infty}$ and an eigenvector f for $\mathbb{T}$ corresponds to a homomorphism $\mathbb{T} \xrightarrow{\phi_f} R$. Let m_f be the kernel of the composition $\mathbb{T} \xrightarrow{\phi_f} R \to \mathbb{F}$; it is associated to $\bar{f}$—the reduction of f modulo ℓ. We call the completion of $\mathbb{T}$ at m_f the universal modular deformation ring $R^M(\bar{f})$. To $\bar{f}$ there is associated a two-dimensional Galois representation $\bar{\rho}\colon G_{\mathbb{Q}} \to \mathrm{GL}_2(\mathbb{F})$. We assume that the conductor of $\bar{\rho}$ is exactly N.

Let S be the set consisting of the primes dividing N, of ℓ and of ∞. The following theorem is due to Gouvea and Hida. See [**53**, Theorem 6], [**52**].

THEOREM 1.7. *Suppose that $\ell \geq 5$. There exists a representation*

$$\rho^M : G_{\mathbb{Q}} \to \mathrm{GL}_2(R^M(\bar{f})), \tag{1.25}$$

which is unramified outside S and satisfies for any prime $p \notin S$

$$\det(\rho^M(\mathrm{Fr}_p)) = p^{-1}\langle p\rangle, \quad \mathrm{Tr}(\rho^M(\mathrm{Fr}_p)) = T_p, \tag{1.26}$$

where T_p is the Hecke operator at p and $\langle p\rangle$ is the diamond operator at p.

Suppose $\bar{\rho}$ is absolutely irreducible. We let S_0 be the subset of S consisting of primes p dividing N at which $\bar{\rho}$ is ordinary (semi-stable and ramified). One may consider the functor of deformations of $\bar{\rho}$ into deformations unramified outside S and ordinary at S_0. The representation ρ_f associated to f is an example of such a representation. There exists a universal deformation $\rho^U : G_{\mathbb{Q}} \to \mathrm{GL}_2(R^U(\bar{f}))$ unramified outside S and semi-stable at S_0. By construction one obtains a homomorphism

$$R^U(\bar{f}) \to R^M(\bar{f}). \tag{1.27}$$

CONJECTURE 1.8 (Mazur, Gouvea–Mazur [**55**]). *The universal deformation ring and the universal modular deformation ring associated to $\bar{f}$ are isomorphic:*

$$R^U(\bar{f}) \cong R^M(\bar{f}). \tag{1.28}$$

That is, all the Galois deformations of a modular residual representation (with suitable conditions at S_0 and outside S) are (ℓ-adically) modular.

See [**53, 54, 95**] for a more complete introduction to deformations of Galois representations. As we mentioned Gouvea–Mazur [**55**] proved the conjecture under a technical condition. See also recent work by G. Böckle.

2. Congruences between modular forms mod p

In this section, we explore some classical congruences involving coefficients of q-expansions of modular forms. We follow [**158**] and begin our discussion trying to stay very concrete and down to earth, so that a story emerges out of the facts.

To begin with, there are striking congruences for the Fourier coefficients $\tau(n)$ of Δ:

$$\Delta(q) = q\prod_{n=1}^{\infty}(1-q^n)^{24} = \sum_{n=1}^{\infty}\tau(n)q^n. \tag{2.1}$$

We recall that Δ (discussed in Chapter 1, §5) is the unique normalized cusp form of weight 12 for $\mathrm{SL}_2(\mathbb{Z})$, hence an eigenform for all the Hecke operators. This already implies the relations:

$$\begin{aligned} \tau(mn) &= \tau(m)\tau(n), \quad (m,n)=1; \\ \tau(p^{n+1}) &= \tau(p)\tau(p^n) - p^{11}\tau(p^{n-1}), \quad p \text{ prime}. \end{aligned} \tag{2.2}$$

Perhaps the most elegant of the congruences satisfied by Δ, found by Ramanujan, is:

$$\tau(n) \equiv \sigma_{11}(n) \bmod 691. \tag{2.3}$$

See Table 2.1 for some values of $\tau(n)$. See [**158**] for some striking congruences between arithmetic functions.

EXERCISE* 2.1. *What can you say on the sign of $\tau(n)$?*

Ramanujan's τ function played a very important role in the development of the theory of modular forms. We just mention the consequence of Deligne's deep work [**22, 23**] on the Weil conjectures:

$$|\tau(n)| \leq \sigma_0(n)n^{11/2}, \tag{2.4}$$

(called "Ramanujan's conjecture") and the intriguing

TABLE 2.1. Values of Ramanujan's tau function.

n	$\tau(n)$
1	1
2	−24
3	252
4	−1472
5	4830
6	−6048
7	−16744
8	84480
9	−113643
10	−115920
11	534612
12	−370944

n	$\tau(n)$
13	−577738
14	401856
15	1217160
16	987136
17	−6905934
18	2727432
19	10661420
20	−7109760
21	−4219488
22	−12830688
23	18643272
24	21288960

CONJECTURE 2.2 (Lehmer). $\tau(n) \neq 0$, $\forall n \geq 1$.

Recall that the Eisenstein series $E_m = E_m^{\mathbb{Q}}$ for an even integer $m \geq 4$ are modular forms of level 1 and weight m (Chapter 1, §5) and have the following q-expansion

$$E_m^{\mathbb{Q}} = 1 + \frac{2}{\zeta_{\mathbb{Q}}(1-m)} \sum_{n=1}^{\infty} \sigma_{m-1}(n) q^n = 1 - \frac{2m}{B_m} \sum_{n=1}^{\infty} \sigma_{m-1}(n) q^n. \tag{2.5}$$

We also define

$$\begin{aligned} E_m^* = E_m^{\mathbb{Q},*} &= \frac{\zeta_{\mathbb{Q}}(1-m)}{2} E_m^{\mathbb{Q}} \\ &= \frac{\zeta_{\mathbb{Q}}(1-m)}{2} + \sum_{n=1}^{\infty} \sigma_{m-1}(n) q^n. \end{aligned} \tag{2.6}$$

Put

$$P = E_2 = 1 - 24 \sum \sigma_1(n) q^n, \tag{2.7}$$

$$Q = E_4 = 1 + 240 \sum \sigma_3(n) q^n, \tag{2.8}$$

$$R = E_6 = 1 - 504 \sum \sigma_5(n) q^n. \tag{2.9}$$

Ramanujan's congruence

$$\tau(n) \equiv \sigma_{11}(n) \bmod 691 \tag{2.10}$$

is equivalent to

$$\Delta \equiv E_{12}^* \bmod 691. \tag{2.11}$$

To see that the last congruence holds, recall that

$$\Delta = \frac{1}{1728}(Q^3 - R^2), \tag{2.12}$$

because the right-hand side is a normalized cusp form of weight 12. Granting ourselves the expression

$$E_{12}^* = \frac{1}{65520}(441Q^3 + 250R^2), \tag{2.13}$$

and noting that $441 \cdot 1728 \equiv 65520 \bmod 691$, and $250 \equiv -441 \bmod 691$, the congruence (2.11) follows, because the polynomials in Q, R expressing both sides are congruent mod 691. We also note that $E_{690} \equiv 1 \bmod 691$. We shall see that all the congruences are of these types. That is, congruences, say modulo the prime $p = 691$, are either coming from p-integral polynomial expressions being congruent modulo 691, or from $E_{690} \equiv 1 \bmod 691$.

The expression (2.13) follows from the following

LEMMA 2.3. *Put $F_m = E_m^*/(m-2)!$. Then, for $m \geq 4$ even,*

$$\frac{(m-2)(m+5)}{12} F_{m+4} = F_4 F_m + F_6 F_{m-2} + \cdots + F_m F_4. \tag{2.14}$$

PROOF. The Weierstrass $\wp$ function is given by

$$\wp(z,\tau) = \frac{1}{z^2} + 2 \sum_{m \in 2\mathbb{N}}^{\infty} (-1)^{(m+2)/2} (2\pi)^{m+2} F_{m+2} z^m, \tag{2.15}$$

and it satisfies the differential equation

$$\wp'(z,\tau)^2 = 4\wp(z,\tau)^3 - g_2(\tau)\wp(z,\tau) - g_3(\tau), \tag{2.16}$$

where $g_2(\tau)/E_4(\tau)$ and $g_3(\tau)/E_6(\tau)$ are constant and the derivatives are taken with respect to the variable z. The second derivative is:

$$\wp''(z,\tau) = 6\wp(z,\tau)^2 - g_2(\tau)/2. \tag{2.17}$$

Compare both sides of (2.17): The coefficient of z^m on the left-hand side is

$$2 \cdot (-1)^{(m+4)/2} (2\pi)^{m+4} (m+1)(m+2) F_{m+4}, \tag{2.18}$$

while on the right-hand side we have:

$$6 \cdot 4 \cdot \sum_{\substack{l+k=m, \\ l,\, k \text{ even}}} (-1)^{(l+2+k+2)/2} (2\pi)^{l+2+k+2} F_{k+2} F_{l+2} + 6 \cdot 4 \cdot (-1)^{(m+4)/2} (2\pi)^{m+4} F_{m+4}. \tag{2.19}$$

The lemma follows. □

Some explicit expressions are given in Table 2.2

Let $\mathcal{M}(\mathbb{C}, \mathrm{SL}_2(\mathbb{Z})) = \bigoplus_k \mathcal{M}(\mathbb{C}, k, \mathrm{SL}_2(\mathbb{Z}))$ be the ring of complex holomorphic modular forms on $\mathrm{SL}_2(\mathbb{Z})$. For every ring $B \supset \mathbb{Z}$ we let

$$\mathcal{M}(B, \mathrm{SL}_2(\mathbb{Z})) = \bigoplus_k \mathcal{M}(B, k, \mathrm{SL}_2(\mathbb{Z})) \tag{2.20}$$

TABLE 2.2. Polynomial expressions for Eisenstein series.

$E_2 = P$	$E_8 = Q^2$
$E_4 = Q$	$E_{10} = QR$
$E_6 = R$	$E_{12} = \frac{1}{691}(441Q^3 + 250R^2)$
	$E_{14} = Q^2 R$

be the subring of modular forms whose Fourier coefficients belong to B. Let $\mathcal{S} = \bigoplus_k \mathcal{S}(B, k, \mathrm{SL}_2(\mathbb{Z}))$ be the ideal of cusp forms. Recall that

$$\mathcal{M}(\mathbb{C}, \mathrm{SL}_2(\mathbb{Z})) = \mathbb{C}[Q, R] \tag{2.21}$$

is a free polynomial ring in two variables Q and R of weights 4 and 6 respectively, and that $f \mapsto f \cdot \Delta$ yields an isomorphism

$$\mathcal{M}(B, k, \mathrm{SL}_2(\mathbb{Z})) \cong \mathcal{S}(B, k+12, \mathrm{SL}_2(\mathbb{Z})) \tag{2.22}$$

for every ring B.

LEMMA 2.4. *Let f be a modular form, and*

$$\mathcal{O}_f = \mathbb{Z}[a_0(f), a_1(f), \dots]. \tag{2.23}$$

Then f has a unique expression as an isobaric element of

$$\mathcal{O}_f[\Delta, Q] \oplus R\mathcal{O}_f[\Delta, Q]. \tag{2.24}$$

PROOF. The uniqueness is easy and is left as an exercise to the interested reader. We prove the existence by induction on the weight. For $k < 12$, it is clear. For $k \geq 12$, $k = 4a + 6b$, we obtain the cusp form $f - a_0(f)Q^aR^b = \Delta \cdot g$. We have $\mathcal{O}_g \subset \mathcal{O}_f$ (because $q^{-1}\Delta \in \mathbb{Z}[\![q]\!]^\times$), and the result follows by induction. □

EXERCISE 2.5. *Prove that $\mathcal{O}_f$ is finitely generated over $\mathbb{Z}$.*

Before addressing the issue of determining all the congruences between modular forms on $\mathrm{SL}_2(\mathbb{Z})$, we discuss a certain derivation operator θ due to Ramanujan.

Let θ be the derivation

$$\theta = q\frac{d}{dq} = \frac{1}{2\pi i}\frac{d}{d\tau}. \tag{2.25}$$

Given a q-expansion $f = \sum a_n q^n$, we obtain a new q-expansion $\theta f = \sum n a_n q^n$. We shall use the following

FACT 2.6. We have

$$P(\tau) = P(\tau+1), \quad P\left(-\frac{1}{\tau}\right) = \tau^2 P(\tau) + \frac{12\tau}{2\pi i}. \tag{2.26}$$

THEOREM 2.7 (Ramanujan). *If $f \in \mathcal{M}(B, k, \mathrm{SL}_2(\mathbb{Z}))$, then*

$$\delta_k f := 12\theta f - kPf \in \mathcal{M}(B, k+2, \mathrm{SL}_2(\mathbb{Z})). \tag{2.27}$$

PROOF. We want to show: $(\delta_k f)(-1/\tau) = \tau^{k+2}\delta_k f(\tau)$.

$$\begin{aligned} \tau^{k+2}\delta_k f(\tau) &= 12\tau^{k+2}\theta f(\tau) - k\left(\tau^2 \cdot P(\tau) + \frac{12\tau}{2\pi i}\right)\tau^k f(\tau) + \frac{12k}{2\pi i}\tau^{k+1} f(\tau) \\ &= -k(Pf)\left(-\frac{1}{\tau}\right) + \frac{12k}{2\pi i}\tau^{k+1} f(\tau) + 12\tau^{k+2}\theta f(\tau). \end{aligned} \tag{2.28}$$

On the other hand

$$\begin{aligned} (\theta f)(\tau) &= \frac{1}{2\pi i}\frac{d}{d\tau}\left(\frac{1}{\tau^k} f\left(-\frac{1}{\tau}\right)\right) \\ &= -\frac{k}{2\pi i}\tau^{-(k+1)} f\left(-\frac{1}{\tau}\right) + \tau^{-(k+2)}\theta f\left(-\frac{1}{\tau}\right). \end{aligned} \tag{2.29}$$

Replacing in (2.29) τ by $-1/\tau$, using k even and (2.28), we get:

$$(2.30)\quad 12(\theta f)\Big(-\frac{1}{\tau}\Big) = \frac{12k}{2\pi i}\tau^{k+1}f(\tau)+12\tau^{k+2}(\theta f)(\tau) = \tau^{k+2}\delta_k f(\tau)+kPf\Big(-\frac{1}{\tau}\Big),$$

and (2.27) follows. □

DEFINITION 2.8. Define $\delta\colon \mathcal{M}\big(B,\mathrm{SL}_2(\mathbb{Z})\big) \to \mathcal{M}\big(B,\mathrm{SL}_2(\mathbb{Z})\big)$, $\delta = \bigoplus \delta_k$. If $f \in \mathcal{M}\big(B,k,\mathrm{SL}_2(\mathbb{Z})\big)$ then $\delta f = \delta_k f = 12\theta f - kPf$.

EXERCISE 2.9. *Verify the following identities:*

$$(2.31)\qquad \delta Q = -4R, \quad \delta R = -6Q^2, \quad \delta\Delta = 0, \quad \delta P = -P^2 - Q.$$

COROLLARY 2.10. (1) *δ is the unique derivation of $\mathcal{M}\big(\mathbb{C},\mathrm{SL}_2(\mathbb{Z})\big)$ such that*

$$(2.32)\qquad \delta Q = -4R, \quad \delta R = -6Q^2.$$

(2) *$\mathbb{C}[P,Q,R]$ is stable under δ.*

PROOF. The only thing to prove is that δ is a derivation. Indeed,

$$(2.33)\qquad \begin{aligned}\delta(f_1f_2) &= 12\theta(f_1\cdot f_2) - (k_1+k_2)Pf_1f_2\\ &= (12\theta f_1 - k_1Pf_1)f_2 + (12\theta f_2 - k_2Pf_2)f_1\\ &= \delta(f_1)f_2 + \delta(f_2)f_1.\end{aligned}$$

□

We now begin to study modular forms modulo p, and we look first at the q-expansion map modulo p. While in characteristic zero the q-expansion map

$$(2.34)\qquad \mathcal{M}\big(\mathbb{C},\mathrm{SL}_2(\mathbb{Z})\big) = \bigoplus_k \mathcal{M}\big(\mathbb{C},k,\mathrm{SL}_2(\mathbb{Z})\big) \to \mathbb{C}[\![q]\!],$$

is an injective ring homomorphism, this is not true in characteristic p. As we shall see, first for $\mathrm{SL}_2(\mathbb{Z})$ (by "naive" methods) and then in greater generality, the q-expansion map is injective on each $\mathcal{M}\big(\overline{\mathbb{F}_p},k,\mathrm{SL}_2(\mathbb{Z})\big)$ but has a kernel on $\mathcal{M}\big(\overline{\mathbb{F}_p},\mathrm{SL}_2(\mathbb{Z})\big)$.

DEFINITION 2.11. Let $\mathcal{N}_k = \mathcal{M}\big(\mathbb{Z}_{(p)},k,\mathrm{SL}_2(\mathbb{Z})\big)$. We identify $\mathcal{N}_k$ with its image under the q-expansion map. Thus $\mathcal{N}_k$ consists of all q-expansions of modular forms of weight k on $\mathrm{SL}_2(\mathbb{Z})$ whose coefficients are p-integral. We let

$$(2.35)\qquad \mathcal{N} = \bigoplus_k \mathcal{N}_k \hookrightarrow \mathbb{Z}_{(p)}[\![q]\!].$$

REMARK 2.12. The ring homomorphism $\mathcal{N} \to \mathbb{Z}_p[\![q]\!]$ is indeed injective. While this may be verified using characteristic 0 techniques, it also follows from our description of $\widetilde{\mathcal{N}}$ below.

Let $\widetilde{\mathcal{N}}$ be the reduction of $\mathcal{N} \subset \mathbb{Z}_{(p)}[\![q]\!]$ modulo p, so $\widetilde{\mathcal{N}} \subset \mathbb{F}_p[\![q]\!]$.

THEOREM 2.13. *Let $p \geq 5$.*

(1) $\mathcal{N} = \mathbb{Z}_{(p)}[Q,R]$.

(2) *Let A be the polynomial such that $A(Q,R) = E_{p-1}$. Then, via the q-expansion map,*[4]

$$\widetilde{\mathcal{N}} \cong \mathbb{F}_p[Q,R]/(\tilde{A}-1). \tag{2.36}$$

PROOF. Lemma 2.4 gives $\mathcal{N} = \mathbb{Z}_{(p)}[Q,\Delta] \oplus R\mathbb{Z}_{(p)}[Q,\Delta]$. Note that $\Delta = (Q^3 - R^2)/1728 \in \mathbb{Z}_{(p)}[Q,R]$. Thus the first assertion follows.

Let

$$\mathfrak{A} = \operatorname{Ker}(\mathbb{F}_p[Q,R] \to \widetilde{\mathcal{N}}). \tag{2.37}$$

The ideal $\mathfrak{A}$ is a prime ideal that cannot be maximal, else both $\widetilde{R}-1$ and $\widetilde{Q}-1$ would be algebraic over $\mathbb{F}_p$. But $\gcd(240, 504, p) = 1$, and therefore the coefficient of q in either $\widetilde{Q}-1$ or $\widetilde{R}-1$ is not zero, contradiction. By the Clausen–von Staudt Theorem $2\zeta(2-p)^{-1}$ is zero modulo p. Thus, E_{p-1} has q-expansion congruent to 1 modulo p. Therefore $\tilde{A}-1 \in \mathfrak{A}$ and since $\dim(\mathbb{F}_p[Q,R]) = 2$, we conclude that it is enough to show that $\tilde{A}-1$ is irreducible. We shall use the following

FACT 2.14 (Igusa [**72**]). The polynomial $\tilde{A}$ has simple roots.

We will prove this fact later, in much greater generality in fact. See Chapter 5, Theorem 3.6.

Suppose $\tilde{A}-1$ is reducible, then it has a factor of the form

$$\Phi(Q,R) = \Phi_n(Q,R) + \cdots + 1, \tag{2.38}$$

($n < p-1$, Φ_i = weight i part). The group $\mathbb{F}_p^\times$ acts on the graded ring $\mathbb{F}_p[Q,R]$ (where Q has weight 4 and R has weight 6): If $\zeta \in \mathbb{F}_p^\times$ and g is homogeneous of weight n then $[\zeta]g = \zeta^n g$. We note that $\tilde{A}-1$ is *fixed* by this action. Therefore,

$$[\zeta]\Phi(Q,R) = \zeta^n\Phi_n(Q,R) + \cdots + 1 \tag{2.39}$$

is a *distinct* factor of $\tilde{A}-1$, and hence $\Phi(Q,R)\cdot[\zeta]\Phi(Q,R)$ divides $(\tilde{A}-1)$. By weight considerations, $\Phi_n(Q,R)\cdot\zeta^n\Phi_n(Q,R)$ divides $\tilde{A}$, hence $\tilde{A}$ has a repeated factor, contradicting Fact 2.14. $\square$

COROLLARY 2.15. (1) *$\widetilde{\mathcal{N}}$ has a natural $\mathbb{Z}/(p-1)\mathbb{Z}$ grading,*

$$\widetilde{\mathcal{N}} = \bigoplus_{\alpha\in\mathbb{Z}/(p-1)\mathbb{Z}} \widetilde{\mathcal{N}}^\alpha, \quad \widetilde{\mathcal{N}}^\alpha = \bigcup_{k\equiv\alpha \bmod p-1} \widetilde{\mathcal{N}_k}. \tag{2.40}$$

(2) *Let $f \in \mathcal{N}_k$, $f' \in \mathcal{N}_{k'}$ and $f \not\equiv 0 \bmod p$. Then $f \equiv f' \bmod p$ implies that $k \equiv k' \bmod p-1$.*

Let $A(Q,R)$, $B(Q,R)$ be the polynomials such that $A(Q,R) = E_{p-1}$ and $B(Q,R) = E_{p+1}$. Examples for $p = 5, 7, 11$ and 13 are provided in Table 2.2.

THEOREM 2.16. (1) $\tilde{A}(\widetilde{Q},\widetilde{R}) = 1$, $\widetilde{B}(\widetilde{Q},\widetilde{R}) = \widetilde{P}$.

(2) $\delta\tilde{A} = \widetilde{B}$, $\delta\widetilde{B} = -Q\tilde{A}$.

(3) *The polynomials $\tilde{A}$ and $\widetilde{B}$ are relatively prime.*

---

[4]We use $\tilde{}$ to denote reduction modulo p. When we write $\widetilde{Q}$ etc., we mean the reduction of the q-expansion of Q modulo p. If A, B are polynomials (say in variables Q, R) with $\mathbb{Z}_p$ coefficients, we let $\tilde{A}$, $\widetilde{B}$ denote the polynomials obtained from A, B respectively by reducing their coefficients modulo p. Thus $\tilde{A}(Q,R)$ is a polynomial modulo p in the variables Q, R, while $\tilde{A}(\widetilde{Q},\widetilde{R})$ is the q-expansion modulo p obtained by substituting the q-expansions modulo p of Q, R for the variables Q, R in the polynomial $\tilde{A}$.

(4) *The algebra* $\widetilde{\mathcal{N}}$ *is stable under the derivation* θ.

PROOF. First, recall that

$$E_{p+1} = 1 - \frac{2(p+1)}{B_{p+1}} \sum \sigma_p(n) q^n. \tag{2.41}$$

However, by the Kummer congruences (Theorem 1.1)

$$\frac{B_{p+1}}{p+1} \equiv \frac{B_2}{2} \mod p, \tag{2.42}$$

and obviously $\sigma_p(n) \equiv \sigma_1(n) \bmod p$ for all n. Hence,

$$E_{p+1} \equiv P \bmod p. \tag{2.43}$$

We now compute the action of δ:

$$\begin{aligned} \delta\tilde{A}(\widetilde{Q}, \widetilde{R}) &= 12\theta\tilde{A}(\widetilde{Q}, \widetilde{R}) - (p-1)\widetilde{P}\tilde{A}(\widetilde{Q}, \widetilde{R}) \\ &= \widetilde{P} \\ &= \widetilde{B}(\widetilde{Q}, \widetilde{R}). \end{aligned} \tag{2.44}$$

Therefore, $\delta A(Q,R) - B(Q,R)$ has q-expansion with coefficients in $p\mathbb{Z}_{(p)}$. Hence by Lemma 2.4, $\delta A(Q,R) - B(Q,R)$ is in $p\mathbb{Z}_{(p)}[Q,\Delta] \oplus Rp\mathbb{Z}_{(p)}[Q,\Delta] = p\mathbb{Z}_{(p)}[Q,R]$. Thus, $\delta A - B = 0 \bmod p$.

We use a similar argument for B:

$$\begin{aligned} \delta\widetilde{B}(\widetilde{Q}, \widetilde{R}) &= 12\theta\widetilde{B}(\widetilde{Q}, \widetilde{R}) - (p+1)\widetilde{P}\widetilde{B}(\widetilde{Q}, \widetilde{R}) \\ &= 12\theta\widetilde{P} - \widetilde{P}^2 \\ &= \delta\widetilde{P} + \widetilde{P}^2 \\ &= -\widetilde{Q} \\ &= -\widetilde{Q}\tilde{A}(\widetilde{Q}, \widetilde{R}) \end{aligned} \tag{2.45}$$

(we used the identities (2.31)). Thus, $\delta\widetilde{B}(\widetilde{Q}, \widetilde{R}) = -\widetilde{Q}\tilde{A}(\widetilde{Q}, \widetilde{R})$, whence $\delta\widetilde{B} = -Q\tilde{A}$.

We next prove that the polynomials $\tilde{A}$ and $\widetilde{B}$ are relatively prime. Say

$$\tilde{A} = \Phi(Q,R) \cdot \Psi(Q,R), \tag{2.46}$$

where necessarily $(\Phi, \Psi) = 1$. Then

$$\widetilde{B} = \delta\tilde{A} = \delta\Phi \cdot \Psi + \delta\Psi \cdot \Phi. \tag{2.47}$$

If $\Phi \mid \widetilde{B}$ then $\Phi \mid \delta\Phi$ ($\tilde{A}$ has no repeated factor). But $\mathrm{wt}(\delta\Phi) = \mathrm{wt}(\Phi) + 2$ and we get a contradiction, because $\mathbb{F}_p[Q,R]$ has no elements of weight 2.

Finally, we see that $\widetilde{P} \in \widetilde{\mathcal{N}}$, hence

$$\theta f = \frac{1}{12}(\delta f + kPf) \in \widetilde{\mathcal{N}}. \tag{2.48}$$

□

3. Operators and systems of eigenvalues

3.1. A highbrow view of modular forms in characteristic p. So far we discussed modular forms in characteristic p as the reduction modulo p of modular forms in characteristic zero, and that from two perspectives: first, as the reduction of the algebra $\mathbb{Z}_{(p)}[Q, R]$ modulo p; second, as the reduction of q-expansions $\sum a_n q^n$ of modular forms such that $a_n \in \mathbb{Z}_{(p)}$.

Those points of view are clearly very restricted and artificial. Moreover, the methods use very heavily the description of modular forms on a very particular group: $\mathrm{SL}_2(\mathbb{Z})$. One would like to consider modular forms in a more intrinsic way, along the lines hinted in Chapter 1, §4. This approach is due to Katz and uses in an essential way the fact that modular forms are "living" over moduli spaces.

Recall the definition of an abelian variety with RM by $\mathcal{O}_L$ (L a totally real field) with μ_N level structure: Let S be a scheme over which d_L is invertible. Then we consider triples $(A, \iota, \beta)_{/S}$, where A/S is an abelian scheme, $\iota\colon \mathcal{O}_L \to \mathrm{End}_S(A)$ is an embedding of rings making the relative tangent space $\mathfrak{t}_{A/S}$ into a locally free $\mathcal{O}_L \otimes \mathcal{O}_S$ module of rank 1, and $\beta\colon \mu_N \otimes \mathcal{D}_L^{-1} \to A$ is an $\mathcal{O}_L$-equivariant homomorphism. Moreover, the module of symmetric $\mathcal{O}_L$-homomorphisms $A \to A^t$ is a projective $\mathcal{O}_L$-module of rank 1 in the étale topology.

The existence of a μ_N structure implies that the fiber of A/S, over every point of S with residue field of characteristic dividing N, is ordinary.

LEMMA 3.1. *Let $N \geq 4$. The moduli problem of abelian varieties with RM and μ_N-level is rigid, in the sense that an object (A, ι, β) parameterized by this moduli problem has no automorphisms besides the identity.*

PROOF. Let $\underline{A} = (A, \iota, \beta)_{/S}$ be an abelian variety with RM and μ_N-level. We may assume S is the spectrum of an algebraically closed field. Let D be the centralizer of L in $\mathrm{End}(A) \otimes \mathbb{Q}$. It is known that D is either L, a CM field such that $D^+ = L$, or a quaternion algebra over L that is ramified everywhere at ∞. See Chapter 3, Lemma 5.2.

Let $\mathcal{O}_D = D \cap \mathrm{End}(\underline{A})$. If $\xi \in \mathcal{O}_D$ is an automorphism of A preserving the polarization, then $\xi\xi^* = 1$, where $*$ is the unique positive involution of D. Hence, ξ is of finite order. It follows that the field $L(\xi)$ is either L, or a CM field whose totally real subfield is L, and that ξ is a root of unity of order n. The case of $L(\xi) = L$ is just the case of $\xi = \pm 1$ and is easily dispensed with. We assume that $L(\xi) \neq L$. Hence, $[L(\xi) : \mathbb{Q}] = 2g$. Equivalently, $1 < \phi(n)$, $\phi(n) \mid 2g$ and $L \cap \mathbb{Q}(\xi) = \mathbb{Q}(\xi)^+$.

If ξ preserves a μ_N-level structure, it follows that $N^g \mid \deg(1 - \xi)$. Hence, n is a prime power. Say $n = \ell^r$, ℓ a prime. Then $\deg(1 - \xi) = \ell^{2g/\phi(n)}$. Since $\phi(n) > 1$, this is divisible by a g-th power if and only if $\phi(n) = 2$. On the other hand, $\phi(n) = \ell^{r-1}(\ell - 1)$. This implies $r = 1$ and $\ell = 3$, or $r = 2$ and $\ell = 2$. Both imply $N < 4$. □

It follows from the general theory of moduli spaces of abelian varieties that the moduli problem of μ_N-level is representable by a scheme $\mathfrak{M}(\mu_N)$ over $\mathbb{Z}$. The morphism $\mathfrak{M}(\mu_N) \to \mathbb{Z}[(Nd_L)^{-1}]$ is smooth of relative dimension g. There exists a universal object $\underline{A}^U = (A^U, \iota^U, \beta^U) \to \mathfrak{M}(\mu_N)$.

We now restrict our attention to the case of elliptic curves. That is $L = \mathbb{Q}$. We shall later on (Chapter 5) lift the general discussion from the point we leave it.

DEFINITION 3.2. Let $N \geq 4$. Let $\mathfrak{M}(B, \mu_N)$ denote the base change $\mathfrak{M}(\mu_N) \times \mathrm{Spec}(B)$. It represents the functor of isomorphism classes of elliptic curves with a μ_N-level structure over bases $S \to \mathrm{Spec}(B)$. Let

$$\underline{A}^B = (A^B, \beta^B) \to \mathfrak{M}(B, \mu_N) \tag{3.1}$$

denote its universal object (obtained by base change from $(A^U, \beta^U) \to \mathfrak{M}(\mu_N)$).

A *modular form* f over B of weight k and μ_N-level is a section of the k-th tensor power of the relative cotangent space at the origin: $(\mathfrak{t}^*_{\underline{A}^B/\mathfrak{M}(B,\mu_N)})^{\otimes k}$. We denote the B-module of modular forms over B of weight k and level μ_N by

$$\mathcal{F}(B, k, \mu_N). \tag{3.2}$$

Using the property of the moduli space as classifying couples $(A, \beta)/S/B$, and the fact that if $(A, \beta)/S$ is parameterized by a morphism $\phi\colon S \to \mathfrak{M}(B, \mu_N)$ of schemes over B then $\phi^*(\mathfrak{t}^*_{\underline{A}^B/\mathfrak{M}(B,\mu_N)})^{\otimes k} = (\mathfrak{t}^*_{A/S})^{\otimes k}$ we find the following reformulation of Definition 3.2:

DEFINITION 3.3. Let $N \geq 4$. A *modular form* over B of weight k and μ_N-level is a rule f associating to any couple $(A, \beta)/S/B$ an element $f\big((A, \beta)/S/B\big)$ in $(\mathfrak{t}^*_{A/S})^{\otimes k}$. The rule f is compatible with isomorphisms and commutes with base change.

Using that the sheaf $(\mathfrak{t}^*_{A/S})^{\otimes k}$ over S has the property of being locally free (in general, for RM, we shall use that condition (R) holds) we arrive at the following reformulation:

DEFINITION 3.4. Let $N \geq 4$. A *modular form* over B of weight k and μ_N-level, is a rule f associating to any triple $(A, \beta, \omega)/R/B$, where R is a B-algebra, and where ω is an R-basis for $\mathfrak{t}^*_{A/R}$, an element

$$f(A, \beta, \omega) \in R \tag{3.3}$$

that depends only on the isomorphism class of $(A, \beta, \omega)/R$, commutes with base change, and satisfies

$$f(A, \beta, \alpha^{-1}\omega) = \alpha^k f(A, \beta, \omega), \quad \forall \alpha \in R^\times. \tag{3.4}$$

Unfortunately, it is quite hard to come up with examples of modular forms using this language. This leaves the analytic methods as the most powerful methods of generating modular forms. To the rescue comes the q-expansion principle. It allows one to study modular forms defined *analytically* as *arithmetic* objects.

First, note that since modular forms are sections of one scheme over another there is a Galois action. In particular, if B is a field then $\mathrm{Gal}(\overline{B}/B)$ acts on the modular forms defined over B.

THEOREM 3.5 (q-expansion principle). *Let $N \geq 4$ and let f be a complex modular form. Let $\sum_n a_n q^n$ be the q-expansion of f with respect to the cusp $i\infty$.*

(1) *The form f^σ, $\sigma \in \mathrm{Aut}(\mathbb{C})$, has q-expansion $\sum_n a_n^\sigma q^n$; moreover, f is defined over a ring $R \subset \mathbb{C}$ if and only if $a_n \in R$ for all n.*
(2) *Let $K \subset \mathbb{C}$ be a number field. Let $\mathcal{O}_K$ be its ring of integers, and let $\mathfrak{p} \lhd \mathcal{O}_K$ be a prime ideal. Let f be a modular form over K. Then there exists $a \in K$ such that $aa_n \in \mathcal{O}_K$ for all n. Assume that f is defined over $\mathcal{O}_K$, then f (mod $\mathfrak{p}$) (i.e., the section obtained after base change $(-) \times_{\mathcal{O}_K} \mathcal{O}_K/\mathfrak{p}$) is zero if and only if every a_n belongs to $\mathfrak{p}$.*

REMARK 3.6. Below (Chapter 5, §2) we introduce q-expansions in every characteristic. It would allow a sharper formulation of the q-expansion principle. Namely, that q-expansions commute with base change. For the q-expansion principle see [**79**, §§1.6, 1.9], [**39**, Chapter V, §1] and [**9**].

Finally, we define a modular form f over B of level 1 and weight k as a couple of modular forms g, h of levels μ_4, μ_5 respectively, that are defined over $B[\frac{1}{2}]$ and $B[\frac{1}{5}]$ (resp.), agree under pull back to level μ_{20} over $B[1/10]$, and are independent of the choice of μ_4 and μ_5 level structures (resp.) in the sense of Definition 3.4. Cf. [**79**, §1.9].

One should note that there could be modular forms in characteristic p that are not obtained from characteristic zero, e.g., there could be forms of weight 1 and level 1.

3.2. Operators. We discuss some operators acting on modular forms in characteristic p. We assume, for simplicity, that our modular forms are defined over a field $\mathbb{F}$.

DEFINITION 3.7 (Hecke Operators). For every prime ℓ, different from the characteristic of $\mathbb{F}$, we define the operator T_ℓ as follows: For every couple $(E,\omega)_{/R}$ consisting of an elliptic curve over an $\mathbb{F}$-algebra R and ω a non-vanishing differential (i.e., an R-basis to $\mathfrak{t}^*_{E/R}$),

$$(T_\ell f)(E,\omega) = \frac{1}{\ell}\sum_{H<E} f(E/H, \pi_{H*}\omega). \tag{3.5}$$

The sum extends over all subgroup schemes H of E of order ℓ; for every H we denote by $\pi_H\colon E \to E/H$ the natural morphism.

REMARK 3.8. A word of explanation is needed here. The sum really makes sense only étale locally on $\mathrm{Spec}(R)$. That is, when the $\ell+1$ distinct subgroup schemes of order ℓ are defined. One can verify that the local definitions "patch" to an element of R. Alternately, one may argue as follows: Let $\mathfrak{N}$ be the coarse moduli space over $\mathbb{Z}[\ell^{-1}]$ of elliptic curves with a $\Gamma_0(\ell)$-level structure. Namely, of pairs (E,H) consisting of an elliptic curve E and a finite étale subgroup scheme H of order ℓ of E. Let $\mathfrak{M}$ be the coarse moduli space of elliptic curves over $\mathbb{Z}[\ell^{-1}]$. There are two étale morphisms $\mathfrak{N} \to \mathfrak{M}$ of degree $\ell+1$ induced from the transformation of functors $(E,H) \mapsto E$ and $(E,H) \mapsto E/H$, denoted respectively p_1 and p_2. If $\underline{\omega}^{\otimes k}$ is the sheaf of modular forms of even weight k one checks that $p_1^*\underline{\omega}^{\otimes k} \cong p_2^*\underline{\omega}^{\otimes k}$. One deduces a homomorphism

$$\Gamma(\mathfrak{M}, \underline{\omega}^{\otimes k}) \to \Gamma(\mathfrak{M}, \underline{\omega}^{\otimes k}), \tag{3.6}$$

which is a composition of p_1^* with the trace map. This may be checked readily to be ℓT_ℓ (enough to check on couples (E,ω) defined over an algebraically closed field!).

One verifies immediately that $T_\ell f$ is a modular form over $\mathbb{F}$ of the same weight as f, and that the Hecke operators commute with each other. Thus the free algebra $\mathbb{Z}[T_\ell : \ell \neq \mathrm{char}(\mathbb{F})]$ acts on modular forms of weight k.

If $\mathbb{F} = \mathbb{C}$ the definition of the Hecke operators takes the usual form. Let $E = \mathbb{C}/(\mathbb{Z}+\tau\mathbb{Z})$. Then the subgroups H of order ℓ correspond to the $\ell+1$ points in the projective space obtained from $(\ell^{-1}\mathbb{Z}/\mathbb{Z}) + (\ell^{-1}\tau\mathbb{Z}/\tau\mathbb{Z})$ that we identify with $(\mathbb{Z}/\ell\mathbb{Z})^2$. The points $\{(a:1) : 0 \le a \le \ell-1\}$ and $(1:0)$ account for all the points and give us the subgroups H of E that are of order ℓ. Using the obvious notation, the lattices $\mathcal{L}_H = H + (\mathbb{Z} + \tau\mathbb{Z})$ all contain the lattice $\mathbb{Z} + \tau Z$ and the natural map

$$\mathbb{C}/(\mathbb{Z} + \tau Z) \to \mathbb{C}/\mathcal{L}_H, \tag{3.7}$$

is an isogeny with kernel H under which $dz \mapsto dz$.

Now, the point $(a:1)$ gives us the lattice spanned by $\{1, \tau, (\tau+a)/\ell\}$, or simply by $\{1, (\tau+a)/\ell\}$. Thus, the points $\{(a:1) : 0 \le a \le \ell - 1\}$ contribute to the sum in (3.5) $\ell^{-1}\sum_{a=0}^{\ell-1} f\big((\tau+a)/\ell\big)$. The other point $(1:0)$ defines the lattice spanned by $\{1/\ell, \tau\}$. The pair $(\mathbb{C}/(\ell^{-1}\mathbb{Z} + \tau\mathbb{Z}), dz)$ is isomorphic to $(\mathbb{C}/(\mathbb{Z} + \ell\tau\mathbb{Z}), \ell^{-1}dz)$ and contributes to the sum $\ell^{-1}f(\mathbb{C}/(\mathbb{Z} + \ell\tau\mathbb{Z}), \ell^{-1}dz) = \ell^{k-1}f(\ell\tau)$. Therefore, we find that for a modular form f of weight k,

$$T_\ell f(\tau) = \ell^{k-1}f(\ell\tau) + \frac{1}{\ell}\sum_{a=0}^{\ell-1} f\left(\frac{\tau+a}{\ell}\right). \tag{3.8}$$

This is the classical formula for Hecke operators.

One can calculate the effect of T_ℓ on q-expansions. See [**157**, Chapter I, Proposition 10.3]. We suggest it as an exercise:

EXERCISE 3.9. *If f has q-expansion $\sum_n a_n q^n$ then*

$$T_\ell f(q) = \sum_n a_{\ell n} q^n + \ell^{k-1}\sum_n a_n q^{\ell n}. \tag{3.9}$$

The last formula shows, by the q-expansion principle, that the action of T_p on modular forms in characteristic p that are reduction of modular forms from characteristic zero is well-defined and its effect on q-expansions is $\sum a_n q^n \mapsto \sum a_{pn} q^n$.

One may ask if there is a more intrinsic definition of this operator. For example, one may ask if there is an analogue of Definition 3.7 for the operator T_p. First, note that a problem immediately arises from the fact that if $\phi\colon E \to E'$ is an inseparable isogeny, as the one coming from dividing by the kernel of Frobenius, then ϕ^* acts as zero on the cotangent space and one can not push-forward differentials. However, the problem is *precisely* the one created by the Frobenius morphism and can be circumvented by thinking about "Frobenius the base change" and not "Frobenius the morphism." We therefore make the following:

DEFINITION 3.10. Let f be a modular form of level 1 defined over a field $\mathbb{F}$ of characteristic p. We define the modular form Vf of level 1 over $\mathbb{F}$, by

$$(Vf)(E, \omega) = f(E^{(p)}, \omega^{(p)}). \tag{3.10}$$

(See Appendix A for generalities on Frobenius and Verschiebung).

For example, if E is given by a Weierstrass equation $y^2 + a_1xy + a_3y = x^3 + a_2x^2 + a_4x + a_6$ with Néron differential $\omega = dx/(2y + a_1x + a_3)$ then $E^{(p)}$ is given by $y^2 + a_1^pxy + a_3^py = x^3 + a_2^px^2 + a_4^px + a_6^p$ and $\omega^{(p)}$ is again the Néron differential $dx/(2y + a_1^px + a_3^p)$.

Now, $Vf(E, \alpha^{-1}\omega) = f(E^{(p)}, (\alpha^{-1}\omega)^{(p)}) = f(E^{(p)}, \alpha^{-p}\omega^{(p)}) = \alpha^{pk}Vf(E, \omega)$. Therefore Vf is of weight pk. The effect of V on q-expansions can be computed easily once Tate objects are introduced. The q-expansion is the evaluation of the modular form at a particular Tate object, and, by definition, this evaluation commutes with base change.[5] If $f(q) = \sum_n a_n q^n$ then

$$(Vf)(q) = \sum_n a_n q^{pn}. \tag{3.11}$$

The definition of the U-operator is much more intricate. It turns out that a "naive" modular approach does not work. The problem seems to be, essentially, that of defining a trace map for non-separable extensions. For the time being we take the existence and properties of the U operator on faith. We return to the subject in Section 5.3.

FACT 3.11. There exists a linear operator U taking holomorphic modular forms f of weight k (and level μ_N, $(N,p)=1$) to meromorphic modular forms, with poles supported on the supersingular locus, of the same weight and level. If $k>1$ then U takes holomorphic modular forms to holomorphic modular forms. The effect of U on q-expansions is the following: If $f(q) = \sum_n a_n q^n$ then

$$Uf(q) = \sum_n a_{pn} q^n. \tag{3.12}$$

We notice that for $k>1$, if f is a reduction of a modular form from characteristic zero, Uf is the reduction of $T_p f$.

It follows immediately from the effect of these operators on q-expansions that U and V commute with the operators T_ℓ.

EXERCISE 3.12. *Given a modular form f (of level 1, say) of weight k over a field $\mathbb{F}$ of characteristic p we define two modular forms from it as follows. If E is an ordinary elliptic curve over an $\mathbb{F}$-algebra, there is a canonical separable isogeny*

$$\pi\colon E/E[\mathrm{Fr}] \to E, \tag{3.13}$$

whose kernel is $E[p]/E[\mathrm{Fr}]$. We define

$$Jf(E,\omega) = f(E/E[\mathrm{Fr}], \pi^*\omega). \tag{3.14}$$

Prove that Jf is a meromorphic modular form of weight k whose poles are supported on the supersingular locus.

For the second operator we consider the isogeny

$$\rho\colon E \to E/E[\mathrm{Ver}], \tag{3.15}$$

and define

$$Wf(E,\omega) = f(E/E[\mathrm{Ver}], \rho_*\omega). \tag{3.16}$$

This does not make sense in general, because E might not be ordinary, and even if it is, the subgroup $E[\mathrm{Ver}]$ need not be defined over the same base over which E is defined. We therefore consider the modular curve $X_0(p)^{\mathbb{Z}/p\mathbb{Z}}$ of Chapter 1, §5 and define Wf as the pull back f by the morphism $X_0(p)^{\mathbb{Z}/p\mathbb{Z}} \to X_0(1)$ that takes a pair (E,H) to E/H. Prove that Wf is a modular form of weight k on $X_0(p)^{\mathbb{Z}/p\mathbb{Z}}$.

Prove that for every modular form f of weight k we have

$$E_{p-1}^k \cdot Jf = Vf, \tag{3.17}$$

[5]The base change of the Tate object ($\mathbb{G}_m$/periods, dt/t) over $\mathbb{F}_p[\![q]\!]$ is the "same" object but considered over $\mathbb{F}_p[\![q^p]\!]$.

TABLE 3.1. Effect of operators on q-expansion.

operator	effect on weight	effect on q-expansion
T_ℓ $(\ell \neq p)$	$k \mapsto k$	$\sum_n a_{\ell n} q^n + \ell^{k-1} \sum_n a_n q^{\ell n}$
U	$k \mapsto k$	$\sum_n a_{pn} q^n$
V	$k \mapsto pk$	$\sum_n a_n q^{pn}$
θ	$k \mapsto k+p+1$	$\sum_n n a_n q^n$
$[h]$	$k \mapsto k+p-1$	$\sum_n a_n q^n$

and that

$$JWf = WJf = f, \tag{3.18}$$

where f stands also for the pull back of f to $X_0(p)^{\mathbb{Z}/p\mathbb{Z}}$ via the forgetful morphism $X_0(p)^{\mathbb{Z}/p\mathbb{Z}} \to X_0(1)$ that takes a pair (E, H) to E.

Note that Equation (3.17) bounds the poles introduced by J and shows that $Jf(q) = \sum_n a_n q^{pn}$, and that Equation (3.18) gives $Wf(q) = \sum_n a_n q^{n/p}$. Thus, morally speaking, $U = \operatorname{Tr} W$.

The next operator acting on modular forms in characteristic p that we mention is the operator θ discussed above. It takes modular forms of weight k to modular forms of weight $k+p+1$ and its effect on q-expansions is

$$\sum_n a_n q^n \xrightarrow{\theta} \sum_n n a_n q^n. \tag{3.19}$$

It therefore follows that θ^{p-1} raises weights by $p^2 - 1$ and acts on q-expansions by

$$\theta^{p-1}\left(\sum_n a_n q^n\right) = \sum_{(n,p)=1} a_n q^n. \tag{3.20}$$

The following identities on q-expansions are immediate:

$$UV = 1, \quad VU = 1 - \theta^{p-1}. \tag{3.21}$$

Let us denote the operator of multiplication by the modular form $\widetilde{E_{p-1}}$, whose q-expansion is 1, by $[h]$, i.e., $[h]f = \widetilde{E_{p-1}} f$. It raises the weight by $p-1$ and commutes with T_ℓ, U, V and θ.

We summarize all this. Let f be a modular form over a field $\mathbb{F}$ of characteristic p, of weight k and level 1. Let

$$f(q) = \sum_n a_n q^n \tag{3.22}$$

be its q-expansion. Table 3.1 summarizes the action of the operators in terms of q-expansions and weights. Table 3.2 gives some relations satisfied by these operators on q-expansions.

Let $\widetilde{\mathcal{N}}$ be the reduction modulo p of the q-expansions of all p-integral modular forms on $\mathrm{SL}_2(\mathbb{Z})$ as in §2. We note that there is a well-defined action of T_ℓ, $[h]$, θ, V and U on $\widetilde{\mathcal{N}}$.

PROPOSITION 3.13. *There is an exact sequence*

$$0 \to \widetilde{\mathcal{N}} \xrightarrow{V} \widetilde{\mathcal{N}} \xrightarrow{\theta} \widetilde{\mathcal{N}} \xrightarrow{U} \widetilde{\mathcal{N}} \to 0. \tag{3.23}$$

TABLE 3.2. Relations between operators.

$T_\ell U = UT_\ell$,	$T_\ell V = VT_\ell$;
$\theta T_\ell = \ell T_\ell \theta$;	
$UV = 1$,	$VU = 1 - \theta^{p-1}$;
$U\theta = 0$,	$\theta V = 0$.

PROOF. We already observed that $U\theta = 0$ and $\theta V = 0$. Clearly V is injective and since $UV = 1$, U is surjective. Let $f \in \operatorname{Ker}(\theta)$. Then $f = (I - \theta^{p-1})f = VUf$ and hence $f \in \operatorname{Im}(V)$. Finally, let $f \in \operatorname{Ker}(U)$. Then $f(q) = \sum_{(n,p)=1} a_n q^n$. Therefore, $f = \theta^{p-1} f$ and hence $f \in \operatorname{Im}(\theta)$. □

We note that the definition of the operators T_ℓ for $(\ell, N) = 1$, U and V extends verbatim to the case of modular forms of level μ_N, where N is prime to char($\mathbb{F}$). The operator θ extends as well: just repeat the discussion taking level μ_N into account. These operators are also defined on all modular forms in characteristic p. In addition there are the diamond operators $\langle d \rangle$ for every class $d \in (\mathbb{Z}/N\mathbb{Z})^\times$. Their action is given by

$$(\langle d \rangle f)(E, \omega, \beta_N) = f(E, \omega, [d] \circ \beta_N). \tag{3.24}$$

3.3. Filtration and systems of eigenvalues. We use the notations of §2. We consider the reduction modulo p of modular forms of level 1, or rather their q-expansions, modulo p. Let us fix $0 \le \alpha < p - 1$ and view $\widetilde{\mathcal{N}}^\alpha$ as an ascending union:

$$\widetilde{\mathcal{N}}_\alpha \xhookrightarrow{\times E_{p-1}} \widetilde{\mathcal{N}}_{\alpha+p-1} \xhookrightarrow{\times E_{p-1}} \widetilde{\mathcal{N}}_{\alpha+2p-2} \xhookrightarrow{\times E_{p-1}} \cdots. \tag{3.25}$$

DEFINITION 3.14. Let $f \in \widetilde{\mathcal{N}}^\alpha$. Define the *filtration* of f, $w(f)$, to be the least k such that $f \in \widetilde{\mathcal{N}}_k$.

In fact $w(f) \equiv \alpha \bmod p - 1$.

LEMMA 3.15. (1) *Let $f \in \mathcal{N}_k$, $f = \Phi(Q, R) \in \mathbb{Z}_{(p)}[Q, R]$, and suppose that $\tilde{f} \neq 0$. Then*

$$w(\tilde{f}) < k \iff \tilde{A} \mid \tilde{\Phi}. \tag{3.26}$$

(2) *Let $f \in \widetilde{\mathcal{N}}^\alpha$. Then*

$$w(\theta f) \le w(f) + p + 1, \tag{3.27}$$

with equality iff $w(f) \not\equiv 0 \bmod p$.

(3) *Let $f \in \widetilde{\mathcal{N}}^\alpha$. Then*

$$w(Vf) = pw(f). \tag{3.28}$$

(4) *Let $f \in \widetilde{\mathcal{N}}^\alpha$. Then*

$$w(Uf) \le \frac{1}{p}(w(f) + p^2 - 1), \tag{3.29}$$

with equality iff $w(f) \equiv 1 \bmod p$.

PROOF. (1) Say $\widetilde{\Phi} = \tilde{A}\widetilde{\Psi}$. Let $\Psi \in \mathbb{Z}_{(p)}[Q, R]$ be any isobaric lift of $\widetilde{\Psi}$ and $g = \Psi(Q, R)$. Then $\widetilde{g(q)} = \widetilde{f(q)}$, and $\mathrm{wt}(\widetilde{\Psi}) < \mathrm{wt}(\widetilde{\Phi})$, hence $w(\tilde{f}) < k$.

Conversely, if $w(\tilde{f}) < k$ let g be a modular form of weight $w(\tilde{f})$ such that $\widetilde{g(q)} = \widetilde{f(q)}$. Write $g = \Psi(Q, R)$. Then $\widetilde{g(q)} = \widetilde{f(q)}$ implies $(\tilde{A} - 1) \mid (\widetilde{\Phi} - \widetilde{\Psi})$. Since $\mathrm{wt}(\widetilde{\Psi}) < \mathrm{wt}(\widetilde{\Phi})$, we get $\tilde{A} \mid \widetilde{\Phi}$.

(2) Let $g \in \mathcal{N}_{w(f)}$ such that $\tilde{g} = f$. Say $g = \Phi(Q, R)$. Recall that by Theorem 2.16 (see Equation (2.48) in particular)

$$12\theta f = 12\theta\tilde{g} = (\omega(f)E_{p+1}g + E_{p-1}\delta g)^{\sim} \in \widetilde{\mathcal{N}}_{\omega(f)+p+1}. \tag{3.30}$$

Then

$$\begin{aligned} w(\theta f) < w(f) + p + 1 &\iff \tilde{A} \mid (\omega(f)B\Phi + A\delta\Phi)^{\sim} \\ &\iff \tilde{A} \mid (\omega(f)B\Phi)^{\sim} \\ &\underset{(\tilde{A},\widetilde{B})=1}{\iff} \tilde{A} \mid \widetilde{\omega(f)}\widetilde{\Phi} \\ &\iff \text{either } w(f) \equiv 0 \bmod p \text{ or } \tilde{A} \mid \widetilde{\Phi}, \end{aligned} \tag{3.31}$$

but the latter is impossible, because of Part 1.

(3) We write $f = \widetilde{\Phi}(\widetilde{Q}, \widetilde{R})$ with $\deg(\Phi) = w(f)$, and by Part 1 $\tilde{A} \nmid \widetilde{\Phi}$. Clearly, $Vf = \widetilde{\Phi^p}(\widetilde{Q}, \widetilde{R})$ and thus $w(Vf) \leq pw(f)$. But again, Part 1 implies that $w(Vf) < pw(f) \iff \tilde{A} \mid \widetilde{\Phi^p}$. This cannot happen, because $\tilde{A}$ has *simple* factors.

(4) Part 2 implies that $w(\theta^{p-1}f) \leq w(f) + p^2 - 1$. The inequality is strict iff there exists a j, $0 \leq j < p-1$ such that $p \mid \big(w(f) + j(p+1)\big)$; equivalently, $w(f) \not\equiv 1 \bmod p$.

Now, $VUf = f - \theta^{p-1}f$, and using Part 3 we get

$$pw(Uf) = w(VUf) = w(f - \theta^{p-1}f) \leq \max\{w(f), w(\theta^{p-1}f)\}. \tag{3.32}$$

This gives the inequality $w(Uf) \leq (w(f) + p^2 - 1)/p$. The equality occurs iff $w(f) \equiv 1 \bmod p$. □

COROLLARY 3.16. (1) *If* $w(f) \equiv 1 \bmod p$, $Uf \neq 0$.
(2) *If* $w(f) > p + 1$, *then* $w(Uf) < w(f)$.

DEFINITION 3.17. Let $W_k = \widetilde{\mathcal{N}}_k \otimes \overline{\mathbb{F}_p} / \widetilde{\mathcal{N}_{k-p+1}} \otimes \overline{\mathbb{F}_p}$.

We note that W_k is a Hecke module and that Corollary 3.16 gives the following: If $k > p + 1$, then U annihilates W_k.

We now focus our attention on the structure of $\widetilde{N}$ as a Hecke algebra. The first question is about how many different systems of eigenvalues exist modulo p? To that end we first make a

DEFINITION 3.18. A set $\{\lambda_\ell : \lambda_\ell \in \overline{\mathbb{F}_p}, \ell \neq p \text{ prime}\}$ is called *a system of eigenvalues* if there exists an integer k and $0 \neq f \in \widetilde{\mathcal{N}}_k \otimes \overline{\mathbb{F}_p}$ such that $T_\ell f = \lambda_\ell f$ for all $\ell \neq p$.

Even though we have infinitely many such systems in characteristic zero, in characteristic p, the situation is different:

THEOREM 3.19 (Jochnowitz [**77**]). *There exist only finitely many systems of eigenvalues.*[6]

[6] This theorem was superseded by the results of Ash–Stevens. See [**1, 2**].

PROOF. It is enough to prove the finiteness of systems for all the W_k's. Given $1 \leq a \leq p-1$, we define the twisted Hecke module

$$W_k[a] := W_k \otimes_{\overline{\mathbb{F}}} \overline{\mathbb{F}}, \tag{3.33}$$

where T_ℓ acts on $f \otimes r$ by $T_\ell(f) \otimes \ell^a r$.

Note that if $\{\lambda_\ell\}$ is a system in W_k then $\{\lambda_\ell \ell^a\}$ is a system in $W_k[a]$. Therefore, it is enough prove the following

CLAIM. *If $j > 2p$ then $W_j \cong W_m[a]$ as Hecke modules for some $1 \leq a \leq p-1$ and $m < j$.*

CASE 1. $j \not\equiv 1 \bmod p$.

Lemma 3.15 states that $w(\theta f) \leq w(f) + p + 1$ with equality if $w(f) \not\equiv 0 \bmod p$. Thus, $\theta \colon W_k \hookrightarrow W_{k+p+1}$ if $k \not\equiv 0 \bmod p$. If $k \geq p+1$, this is an isomorphism. This follows from calculating the dimensions of the two vector spaces—a verification we leave to the reader.

Let us apply these observations for $k = j - p - 1$ noting that $\theta \circ T_\ell = \ell T_\ell \circ \theta$. Since $k \geq p+1$ and $k \not\equiv 0 \bmod p$, we get $W_{j-p-1}[1] \cong W_j$.

CASE 2. $j \equiv 1 \bmod p$.

In this case, since $j > 2p > p+1$, by Corollary 3.16 and Lemma 3.15, U induces a homomorphism:

$$U \colon W_j \hookrightarrow W_{(j+p^2-1)/p}. \tag{3.34}$$

Composing with $V \colon W_{(j+p^2-1)/p} \hookrightarrow W_{j+p^2-1}$, we get

$$V \circ U \colon W_j \hookrightarrow W_{j+(p+1)(p-1)}. \tag{3.35}$$

This map must be an isomorphism by the same dimension count. Therefore, U is an isomorphism. □

3.4. Congruences mod p^m. We follow Serre [**146**]. Let $p \geq 5$. Let $f \in \mathbb{Q}[Q, R] = \mathcal{M}(\mathbb{Q}, \mathrm{SL}_2(\mathbb{Z}))$ be a modular form of level one with rational Fourier coefficients. Let $f(q) = \sum a_n q^n$ be the q-expansion of f. By assumption, $a_n \in \mathbb{Q}$. Define

$$\mathrm{val}_p(f) := \inf\{\mathrm{val}_p(a_n) : n = 0, 1, 2, \dots\}. \tag{3.36}$$

It follows from the q-expansion principle (Theorem 3.5) that $\mathrm{val}_p(f) > -\infty$.

THEOREM 3.20 (Serre). *Let $m \geq 1$. Let $f, f' \in \mathbb{Q}[Q, R]$ of weight k, k', respectively. Suppose $f \neq 0$. If $\mathrm{val}_p(f - f') \geq \mathrm{val}_p(f) + m$, then*

$$k' \equiv k \bmod (p-1)p^{m-1}. \tag{3.37}$$

PROOF. By multiplying by $p^{-\mathrm{val}_p(f)}$, we may assume that $\mathrm{val}_p(f) = 0$ and hence we are given that $\mathrm{val}_p(f - f') \geq m$. Namely, both f and f' are p-integral and

$$f \equiv f' \bmod p^m. \tag{3.38}$$

In particular, $f \equiv f' \bmod p$ and by Corollary 2.15

$$k' \equiv k \bmod p - 1. \tag{3.39}$$

Hence, the theorem is true for $m = 1$. Assume now that $m \geq 2$. We shall use the following

EXERCISE 3.21.

$$E_r \equiv 1 \bmod p^m \iff r \equiv 0 \bmod (p-1)p^{m-1}. \tag{3.40}$$

Let $h = k' - k$. Replacing f' by $f'E_{(p-1)p^n}$ ($n \gg 0$) we may assume that $h \geq 4$. We know that $h \equiv 0 \bmod p-1$. Let $r = \mathrm{val}_p(h) + 1$. We want to show that $r \geq m$.

Suppose that $r < m$. We claim that $p^{-r}(fE_h - f')$ is p-integral and

$$p^{-r}(fE_h - f') \equiv p^{-r}f(E_h - 1) \bmod p. \tag{3.41}$$

We have

$$fE_h - f' = f - f' + f(E_h - 1). \tag{3.42}$$

Now, $f - f' \equiv 0 \bmod p^m$ and $E_h - 1 \equiv 0 \bmod p^r$. Thus $p^{-r}(fE_h - f')$ is p-integral, and $r < m$ implies (3.41).

We have:

$$p^{-r}(E_h - 1) = \lambda\phi, \tag{3.43}$$

where $\lambda \in \mathbb{Z}_p^\times$ and $\phi = \sum_{n=1}^{\infty} \sigma_{h-1}(n)q^n$. Let

$$g := \lambda^{-1}p^{-r}(fE_h - f'). \tag{3.44}$$

Then (3.41) implies that

$$g = f\phi \bmod p. \tag{3.45}$$

Put:

$$\tilde{\phi} = \tilde{g}/\tilde{f}. \tag{3.46}$$

Note that $\tilde{\phi}$ is well-defined since $\tilde{f}$ is not zero. Since $\mathrm{wt}(g) = k'$ is congruent to $\mathrm{wt}(f) = k$ modulo $p-1$, $\tilde{\phi}$ belongs to the quotient field of $\widetilde{\mathcal{N}}^0$. We shall use the following fact.

FACT 3.22 (See Corollary 5.6). $\widetilde{\mathcal{N}}^0$ is a Dedekind domain.

We claim that $\tilde{\phi}$ is integral over $\widetilde{\mathcal{N}}^0$. To see this, first note that

$$\tilde{\phi} - \tilde{\phi}^p \equiv \sum_{(p,n)=1} \sigma_{h-1}(n)q^n \bmod p. \tag{3.47}$$

Define, for $r \in \mathbb{Z}$, $(n,p) = 1$,

$$\tilde{\sigma}_r(n) = \sum_{d|n} d^r \bmod p. \tag{3.48}$$

Then

$$\tilde{\sigma}_{-1}(n) = \tilde{\sigma}_1(n)/n, \quad \tilde{\sigma}_r(n) = \tilde{\sigma}_{pr}(n) = \tilde{\sigma}_{r+p-1}(n). \tag{3.49}$$

Consider the following identity:

$$\theta^{h-1}\Big(\sum_{n=1}^{\infty}\sigma_1(n)q^n\Big) \equiv \sum_{n=1}^{\infty} n^{h-1}\tilde{\sigma}_1(n)q^n \tag{3.50}$$

$$\equiv \sum_{(n,p)=1} \frac{\tilde{\sigma}_1(n)}{n}q^n \qquad (h\equiv 0 \bmod p-1)$$

$$\equiv \sum_{(n,p)=1} \tilde{\sigma}_{-1}(n)q^n$$

$$\equiv \sum_{(n,p)=1} \tilde{\sigma}_{h-1}(n)q^n \qquad (h\equiv 0 \bmod p-1)$$

$$\equiv \tilde{\phi}-\tilde{\phi}^p.$$

Therefore,

$$\tilde{\phi}-\tilde{\phi}^p = -\frac{1}{24}\theta^{h-1}(\widetilde{P}) = -\frac{1}{24}\theta^{h-1}(\widetilde{E_{p+1}}) = -\frac{1}{24}\theta^{p-2}(\widetilde{E_{p+1}}). \tag{3.51}$$

(θ^{p-1} is idempotent). The last expression shows that $\tilde{\phi}-\tilde{\phi}^p$ belongs to

$$\widetilde{\mathcal{N}}_{p+1+(p-2)(p+1)} = \widetilde{\mathcal{N}}_{p^2-1} \subset \widetilde{\mathcal{N}}^0. \tag{3.52}$$

This proves the claim of integrality, hence that $\tilde{\phi}-\tilde{\phi}^p$ belongs to $\widetilde{\mathcal{N}}^0$, plus the fact that

$$\tilde{\phi}-\tilde{\phi}^p = -\frac{1}{24}\theta^{p-2}(\widetilde{E_{p+1}}). \tag{3.53}$$

Now, the filtration of $\tilde{\phi}-\tilde{\phi}^p$ is $\max\{\omega(\tilde{\phi}),\omega(\tilde{\phi}^p)\} = \max\{\omega(\tilde{\phi}), p\cdot\omega(\tilde{\phi})\} = p\cdot\omega(\tilde{\phi})$. But Lemma 3.15 says that $\omega(\theta^{p-2}\widetilde{E_{p+1}}) = p^2-1$, and this is a contradiction. □

4. Serre's p-adic modular forms and p-adic zeta functions

Throughout this section $p\geq 5$.

Let m be a positive integer. Put

$$X_m = \mathbb{Z}/\big(p^{m-1}(p-1)\big)\mathbb{Z} = \mathbb{Z}/p^{m-1}\mathbb{Z}\times\mathbb{Z}/(p-1)\mathbb{Z}, \tag{4.1}$$

and

$$X = \varprojlim X_m = \mathbb{Z}_p\times\mathbb{Z}/(p-1)\mathbb{Z}. \tag{4.2}$$

Since the maps $\mathbb{Z}\to X_m$ are surjective, we get that the injection $\mathbb{Z}\hookrightarrow X$ has a dense image. Define for $v=v_1v_2\in\mathbb{Z}_p^\times = U_1\times\mu_{p-1}$ and $k=(k_1,k_2)\in\mathbb{Z}_p\times\mathbb{Z}/(p-1)\mathbb{Z}$

$$v^k := v_1^{k_1}v_2^{k_2}\in\mathbb{Z}_p^\times. \tag{4.3}$$

In this way

$$X = \mathrm{End}_{\mathrm{cont}}(\mathbb{Z}_p^\times). \tag{4.4}$$

DEFINITION 4.1. We say $k=k_1k_2$ in X is *even* if $k\in 2X$. Equivalently, k_2 is even or $(-1)^k=1$.

DEFINITION 4.2. A *p-adic modular form* (à la Serre) is a formal power series $f=\sum a_nq^n$, where:

(1) $a_n\in\mathbb{Q}_p$.

(2) There exists a sequence $(f_i)_{i\in\mathbb{N}}$ of modular forms of level 1 and weight k_i with rational Fourier coefficients, such that

$$\mathrm{val}_p(f - f_i) \xrightarrow{i\to\infty} \infty. \tag{4.5}$$

THEOREM 4.3 (Serre). *The weights k_i have a limit in X. This limit depends only on f and not on the particular sequence $(f_i)_{i\in\mathbb{N}}$. It is called the weight of f.*

PROOF. For $i \gg 0$, $\mathrm{val}_p(f_i) = \mathrm{val}_p(f)$ (uniform convergence), and given any m for $j \gg i \gg 0$, we have

$$\mathrm{val}_p(f_i - f_j) \geq \mathrm{val}_p(f) + m = \mathrm{val}_p(f_i) + m. \tag{4.6}$$

Since by Theorem 3.20 $k_i \equiv k_j \bmod (p-1)p^{m-1}$, there exists a limit to the k_i's.

If $f_1', f_2', \ldots$ is another sequence converging to f, consider the sequence

$$f_1, f_1', f_2, f_2', \ldots ! \tag{4.7}$$

□

The weight of f is always even, simply because it is a limit of even weights. Note that under val_p, the space of p-adic modular forms is a p-adic Banach space. That is, it is a vector space over $\mathbb{Q}_p$ with a norm $p^{-\mathrm{val}_p(f)}$ and the space is complete with respect to this norm. That means that if f_i are p-adic modular forms forming a Cauchy sequence with respect to the norm the limit of the f_i's exists.

EXERCISE 4.4. *Prove that* $1/E_{p-1} = \lim_{m\to\infty} E_{p-1}^{p^m-1}$.

THEOREM 4.5. *Let m be a positive integer. Let f, f' be two p-adic modular forms of weight $k, k' \in X$, respectively, and assume $f \neq 0$. Then $\mathrm{val}_p(f - f') \geq \mathrm{val}_p(f) + m$ implies that $k = k'$ in $X_m = \mathbb{Z}/p^{m-1}\mathbb{Z} \times \mathbb{Z}/(p-1)\mathbb{Z}$.*

EXERCISE 4.6. *Prove Theorem* 4.5.

The following corollary is one of the most amazing results of the theory of p-adic modular forms. It is an integral analog of Siegel's Theorem on the rationality of values of zeta functions at negative integers (Chapter 2, Remark 4.5). Recall that the idea behind the proof of that theorem was that if all the higher coefficients of the q-expansion of a modular form are rational then so is the constant coefficient. Thus there is a "rational" influence of the higher coefficients of the q-expansion on the zeroth coefficient. The following corollary says that this influence can be refined to an integral influence: If all the higher coefficients are p-adic integers then the constant coefficient has a valuation bounded from below in terms of the weight alone!

COROLLARY 4.7. *Let $f = a_0 + a_1 q + a_2 q^2 + \cdots$ be a p-adic modular form of weight $k \neq 0$. Let $m \geq 0$ be such that $k \neq 0$ in X_{m+1}. Then*

$$\mathrm{val}_p(a_0) + m \geq \inf_{n\geq 1}\{\mathrm{val}_p(a_n)\}. \tag{4.8}$$

PROOF. If $a_0 = 0$ then $\mathrm{val}_p(a_0) = \infty$ and the assertion holds. Else $a_0 \neq 0$ and we let $f' := a_0$ be a modular form of weight 0. We have $\mathrm{val}_p(f - f') = \inf_{n\geq 1}\{\mathrm{val}_p(a_n)\}$. If, on the contrary, $\inf_{n\geq 1}\{\mathrm{val}_p(a_n)\} > \mathrm{val}_p(a_0) + m$, i.e., if $\mathrm{val}_p(f - f') \geq \mathrm{val}_p(f') + (m+1)$, then Theorem 4.5 implies that $k \equiv 0$ in X_{m+1}, a contradiction. □

COROLLARY 4.8. *Let k be such that $(p-1) \nmid k$. Let $f = a_0 + a_1 q + a_2 q^2 + \cdots$ be a p-adic modular form of weight k. If the coefficients a_i are integral for every $i \geq 1$ then a_0 is also integral.*

PROOF. Take $m = 0$ in Corollary 4.7. □

COROLLARY 4.9. *Let L be a totally real field of degree g and $k \geq 2$ an integer. For p such that $kg \not\equiv 0 \bmod p-1$, we have that $\zeta_L(1-k)$ is p-integral. More generally,*

$$\operatorname{val}_p\big(\zeta_L(1-k)\big) \geq -1 - \operatorname{val}_p(kg). \tag{4.9}$$

PROOF. Recall the modular form of level 1 and weight kg defined in Chapter 2, §6:

$$\Phi^*\big(2^{-g}\zeta_L(1-k)E_{k,\mathcal{D}_L^{-1}}\big) = \frac{\zeta_L(1-k)}{2^g} + \sum_{n=1}^{\infty}\Bigg(\sum_{\substack{\nu\in\mathcal{D}_L^{-1+} \\ \mathrm{Tr}(\nu)=n}} \sigma_{k-1}\big((\nu)\mathcal{D}_L\big)\Bigg)q^n, \tag{4.10}$$

where $\Phi\colon \mathcal{H} \to \mathcal{H}^g$ is the diagonal map.

The leading coefficient of this modular form is $\zeta_L(1-k)/2^g$, and the higher coefficients are p-integral (in fact, integers). If $kg \not\equiv 0 \bmod p-1$, then Corollary 4.8 says that the leading coefficient is also p-integral.

To get the general inequality, we may assume $\tilde{k} = kg$ is congruent to 0 modulo $p-1$. Let $m = \operatorname{val}_p(\tilde{k}) + 1$; then $\tilde{k} \neq 0$ in X_{m+1}, hence

$$\operatorname{val}_p\big(\zeta_L(1-k)\big) + m \geq \inf\{\operatorname{val}_p(a_n)\} \geq 0. \tag{4.11}$$

□

COROLLARY 4.10. *Let*

$$f^{(i)} = \sum_{n=0}^{\infty} a_n^{(i)} q^n, \quad i = 1, 2, \ldots, \tag{4.12}$$

be a sequence of p-adic modular forms of weight $k^{(i)}$. Suppose:

(1) *For $n \geq 1$, $a_n^{(i)} \to a_n \in \mathbb{Q}_p$ uniformly in n. (That is, there exists a power series limit $\lim_i(\sum_{n=1}^{\infty} a_n^{(i)} q^n)$ in the val_p-topology.)*

(2) *$k^{(i)} \to k \in X$, and $k \neq 0$.*

Then $a_0^{(i)} \to a_0 \in \mathbb{Q}_p$ and

$$f = \sum_{n=0}^{\infty} a_n q^n \tag{4.13}$$

is a p-adic modular form of weight k.

PROOF. By deleting some $f^{(i)}$'s, we may assume that there exists an $m \geq 0$ such that all the $k^{(i)}$'s have non-zero image in X_{m+1}. Now, uniform convergence implies there is a $t \in \mathbb{Z}$ such that

$$\operatorname{val}_p(a_n^{(i)}) \geq t, \quad \forall n, \forall i \geq 1. \tag{4.14}$$

By Corollary 4.7 $\operatorname{val}_p(a_0^{(i)}) \geq t - m$. Hence, there exists a subsequence i_j such that $a_0^{(i_j)} \xrightarrow{j\to\infty} a_0 \in \mathbb{Q}_p$. Then clearly,

$$f = \lim f^{(i_j)} = a_0 + a_1 q + a_2 q^2 + \cdots \tag{4.15}$$

is a p-adic modular form of weight k.

Recall that in a compact metric space a sequence converges if and only if every converging subsequence converges to the same limit. If i_j is another converging subsequence $a_0^{(i_j)} \to a_0$, then $f = a_0' + \sum_{n=1}^{\infty} a_n q^n$ is a p-adic modular form of weight k, hence

$$f - f = a_0 - a_0' \tag{4.16}$$

of weight k, but also of weight zero! Since $k \neq 0$ we must have $a_0 - a_0' = 0$. □

EXAMPLE 4.11 (p-adic Eisenstein series). Let $k \in X$, and define

$$\sigma_{k-1}^*(n) = \sum_{\substack{d|n \\ (d,p)=1}} d^{k-1}, \quad n \geq 1, n \in \mathbb{Z}. \tag{4.17}$$

Assume k is even, choose $k_i \geq 4$ such that $k_i \to k$ in X, *and* $k_i \to \infty$ in $\mathbb{R}$. Then $\sigma_{k-1}(n) \to \sigma_{k-1}^*(n)$ in $\mathbb{Z}_p$ uniformly in n. Therefore, by Corollary 4.10, the modular forms $E_{k_i}^*$ converge p-adically to a p-adic modular form G_k^*. Moreover,

$$G_k^* = \frac{1}{2}\zeta_p^*(1-k) + \sum_{n=1}^{\infty} \sigma_{k-1}^*(n) q^n \tag{4.18}$$

is a p-adic modular form of weight k, where we define

$$\zeta_p^*(1-k) = \lim_i \zeta_{\mathbb{Q}}(1-k_i). \tag{4.19}$$

(The existence of the limit is a consequence of Corollary 4.10!) One calls G_k^* the *p-adic Eisenstein series of weight k*. Note that even if k is an integer $G_k^* \neq E_k^*$.

THEOREM 4.12 (Serre [**146**, Théorème 3]). *Let p be an odd prime. Let $(s, n) \in X = \mathbb{Z}_p \times \mathbb{Z}/(p-1)\mathbb{Z}$ be an odd element different from 1. Then*

$$\zeta_p^*\big((s,n)\big) = L_p(s, \omega^{1-n}), \tag{4.20}$$

where ω is the Teichmüller character.

Recall that $L_p(s, \chi)$ interpolates the special values of the classical L-function $L(s, \chi)$. Thus Serre's construction creates the "correct" p-adic zeta functions: they interpolate p-adically special values of classical L-functions. One can generalize Serre's construction to get a p-adic zeta function for L totally real (see Serre's paper in [**146**]).

5. A geometric approach to congruences

The methods of the previous sections rely very much, so it seems, on the specific structure of the ring of modular forms on $\mathrm{SL}_2(\mathbb{Z})$. To obtain a more general theory that works for level structures as well as for Hilbert modular forms, we reconsider the question of congruences, this time from a purely characteristic p approach. Thus, our initial object of study are modular forms in characteristic p and not the reduction of modular forms from characteristic zero. Already for elliptic modular forms there is a difference, but for Hilbert modular forms the difference becomes much more dramatic, with important characteristic p modular forms that cannot be lifted to characteristic zero.

We shall only treat the case of elliptic curves here, allowing ourselves a simpler picture than in the general Hilbert modular case that will be discussed in Chapter 5.

Let $N \geq 4$ be an integer prime to p. Let $\mathfrak{M}(\mathbb{F}_p, \mu_N)$ be the fine moduli space parameterizing elliptic curves E/R over $\mathbb{F}_p$-algebras R, endowed with a μ_N-level structure:

$$\beta_N \colon \mu_{N/R} \hookrightarrow E/R. \tag{5.1}$$

We may compactify $\mathfrak{M}(\mathbb{F}_p, \mu_N)$ to a proper scheme $\mathfrak{M}^*(\mathbb{F}_p, \mu_N)$ over $\mathbb{F}_p$ by adding finitely many cusps. In fact, for $N > 4$ the scheme $\mathfrak{M}^*(\mathbb{F}_p, \mu_N)$ represents a similar moduli problem involving generalized elliptic curves as in [**26**] and Appendix B. The scheme $\mathfrak{M}^*(\mathbb{F}_p, \mu_N)$ is an integral, geometrically irreducible, regular proper scheme over $\mathbb{F}_p$ of dimension 1.

Let $\mathfrak{M}(\mathbb{F}_p, \mu_{Np})$ be the fine moduli space representing elliptic curves E/R over $\mathbb{F}_p$-algebras R, endowed with a μ_{Np}-level structure:

$$\beta_N \times \beta_p \colon \mu_{N/R} \times \mu_{p/R} \hookrightarrow E/R. \tag{5.2}$$

Let $\mathfrak{M}^*(\mathbb{F}_p, \mu_{Np})$ be the scheme obtained from $\mathfrak{M}(\mathbb{F}_p, \mu_{Np})$ by adding the cusps. The scheme $\mathfrak{M}^*(\mathbb{F}_p, \mu_{Np})$ is not proper. In fact, if we let $\mathfrak{M}^*(\mathbb{F}_p, \mu_N)^{\mathrm{ord}}$ denote the open subscheme of $\mathfrak{M}^*(\mathbb{F}_p, \mu_N)$ obtained by deleting the points corresponding to the supersingular elliptic curves then we have a surjective étale Galois morphism

$$\mathfrak{M}^*(\mathbb{F}_p, \mu_{Np}) \to \mathfrak{M}^*(\mathbb{F}_p, \mu_N)^{\mathrm{ord}}, \tag{5.3}$$

with Galois group $(\mathbb{Z}/p\mathbb{Z})^\times$.

REMARK 5.1 (About q-expansions). We offer some remarks on q-expansions in the algebraic setting. We suggest that the reader takes them on faith at this point. A more systematic discussion is given in Chapter 5, §2. Let $\mathfrak{Y} = \mathfrak{M}(\mathbb{F}_p, \mu_{Np^r})$. As we have already remarked, one may add finitely many "cusps" to $\mathfrak{Y}$ to obtain an open immersion $\mathfrak{Y} \to \mathfrak{X}$, where $\mathfrak{X} = \mathfrak{M}^*(\mathbb{F}_p, \mu_{Np^r})$.

The scheme $\mathfrak{X}$ is a moduli space in its own right (though this insight is not necessary at this point). It parameterizes "generalized elliptic curves" (ordinary if $r > 0$) over characteristic p base schemes S. A generalized elliptic curve is essentially a flat family of curves over S such that every fiber is either an elliptic curve or a cyclic configuration of projective lines together with some extra data reflecting the level and group law. We refer to [**26**] and Appendix B for more details.

Over the formal neighborhood of every cusp of $\mathfrak{X}$ one can construct explicitly the fiber of the universal elliptic curve. One chooses a local parameter q at the cusp and constructs the restriction of the universal family to $\operatorname{Spec} \mathbb{F}_p((q))$. The construction in the elliptic curves case can be done completely explicitly and is precisely the Tate curve $\mathbf{T}$ (see [**26**, §VII], [**157**]) together with some μ_N or μ_{Np^r}-level structure on it. Because of the rigid analytic approach one denotes that Tate object by $\mathbb{G}_m/\underline{q}(\mathbb{Z})$. At this point the reader may just regard $\mathbb{G}_m/\underline{q}(\mathbb{Z})$ as a formal symbol for an elliptic curve defined over $\operatorname{Spec} \mathbb{F}_p((q))$ isomorphic to the restriction of the universal family over the image of $\operatorname{Spec} \mathbb{F}_p((q))$ under the classifying morphism $\operatorname{Spec} \mathbb{F}_p((q)) \to \mathfrak{Y} \to \mathfrak{X}$.

Let B be a ring of characteristic p. Let $\mathbf{T} \otimes B$ be the base change of $\mathbf{T}$ to $\operatorname{Spec} B((q))$. Since the Tate object $\mathbf{T} \otimes B$ is an elliptic scheme over $\operatorname{Spec} B((q))$ we can evaluate any modular form on it, obtaining an element of the base $B((q))$.

DEFINITION 5.2. A holomorphic modular form f over B of level μ_{Np^r} is a modular form in $\mathcal{F}(B, k, \mu_{Np^r})$ such that every q-expansion of f at any Tate object

$\mathbf{T} \otimes B$ lies in $B[\![q]\!]$. We denote the space of holomorphic modular forms defined over B, of weight k and level μ_{Np^r} by

$$\mathcal{M}(B, k, \mu_{Np^r}). \tag{5.4}$$

If the q-expansion of f at every cusp has no constant term, i.e., is of the form $q \cdot h(q)$, where $h(q) \in B[\![q]\!]$, then we say that f is a cusp form. We denote the space of cusp forms defined over B, of weight k and level μ_{Np^r} by

$$\mathcal{S}(B, k, \mu_{Np^r}). \tag{5.5}$$

5.1. The Hasse invariant. Let $(E, \beta_N, \omega)_{/R}$ be an elliptic curve over an $\mathbb{F}_p$-algebra R, together with a μ_N-level β_N and a non-vanishing differential ω. Let $\pi\colon E \to \mathrm{Spec}(R)$ be the structural morphism. We define a modular form H, called the Hasse invariant, as follows. We consider the exact sequence:

$$0 \to R^0\pi_*(\Omega^1_{E/R}) \to H^1_{dR}(E) \to R^1\pi_*(\mathcal{O}_E) \to 0. \tag{5.6}$$

(If R is a field this is the sequence

$$0 \to H^0(E, \Omega^1_{E/R}) \to H^1_{dR}(E) \to H^1(E, \mathcal{O}_E) \to 0.) \tag{5.7}$$

The modules $R^0\pi_*(\Omega^1_{E/R})$, $H^1_{dR}(E)$ and $R^1\pi_*(\mathcal{O}_E)$ are projective R-modules of rank 1, 2 and 1, respectively. (See also Chapter 5, §3.1). After localizing, we may assume that they are free R-modules. There is a perfect pairing

$$R^0\pi_*(\Omega^1_{E/R}) \times R^1\pi_*(\mathcal{O}_E) \xrightarrow{\cup} R^1\pi_*(\Omega^1_{E/R}) \cong R^0\pi_*(\mathcal{O}_E)^* = R \tag{5.8}$$

(using Serre's duality) and we let $\eta \in R^1\pi_*(\mathcal{O}_E)$ be the element dual to ω under this pairing.

The sheaf $\mathcal{O}_E$ is a sheaf of rings of characteristic p and therefore the Frobenius map $f \mapsto f^p$ induces a homomorphism of sheaves of abelian groups $\mathcal{O}_E \to \mathcal{O}_E$ and hence a map (called Frobenius) on cohomology

$$\mathrm{Fr}\colon R^1\pi_*(\mathcal{O}_E) \to R^1\pi_*(\mathcal{O}_E), \tag{5.9}$$

which is semi-linear: $\mathrm{Fr}(\alpha v) = \alpha^p \,\mathrm{Fr}(v)$ for $\alpha \in R$ and $v \in R^1\pi_*(\mathcal{O}_E)$. We let $H(E, \beta_N, \omega)$ be the unique element of R such that

$$\mathrm{Fr}(\eta) = H(E, \beta_N, \omega) \cdot \eta. \tag{5.10}$$

One easily verifies that the definition commutes with base change and depends only on the isomorphism class of (E, ω). Moreover, if $\alpha \in R^\times$ then the element dual to $\alpha^{-1}\omega$ is $\alpha\eta$ and $\mathrm{Fr}(\alpha\eta) = \alpha^p \,\mathrm{Fr}(\eta) = \alpha^{p-1}H(E, \omega)\alpha\eta$. That is,

$$H(E, \beta_N, \alpha^{-1}\omega) = \alpha^{p-1}H(E, \beta_N, \omega). \tag{5.11}$$

Thus, we proved the first half of the following proposition. The second part follows from the more general discussion of Tate objects given in Chapter 5, §2.

PROPOSITION 5.3. *The Hasse invariant H is a modular form over $\mathbb{F}_p$ of level one and weight $p-1$. Its q-expansion at every cusp is 1, hence H is equal to the reduction of E_{p-1} modulo p.*

5.2. The kernel of the q-expansion map. Fix a Tate object $\mathbf{T}$. Let $\mathcal{M}(\mathbb{F}_p, k, \mu_N)$ ($N > 4$ prime to p) be the vector space of holomorphic modular forms defined over $\mathbb{F}_p$, of weight k and level μ_N.

THEOREM 5.4 (Serre–Swinnerton-Dyer). *The kernel of the q-expansion map with respect to the cusp* $\mathbf{T}$

$$\bigoplus_{k\in\mathbb{Z}} \mathcal{M}(\mathbb{F}_p, k, \mu_N) \to \mathbb{F}_p[\![q]\!], \tag{5.12}$$

is generated by $H - 1$.

PROOF. Let R be the ring of regular functions on the curve $\mathfrak{M}^*(\mathbb{F}_p, \mu_{Np})$. Let $\mathbb{G}_m/\underline{q}(\mathbb{Z})$ be a Tate object represented by the cusp $\mathbf{T}$ of $\mathfrak{M}^*(\mathbb{F}_p, \mu_{Np})$.

PROPOSITION 5.5. *There is a surjective ring homomorphism*

$$r\colon \bigoplus_{k\in\mathbb{Z}} \mathcal{M}(\mathbb{F}_p, k, \mu_N) \twoheadrightarrow R, \tag{5.13}$$

such that the composition

$$\bigoplus_{k\in\mathbb{Z}} \mathcal{M}(\mathbb{F}_p, k, \mu_N) \twoheadrightarrow R \hookrightarrow \widehat{R_{\mathbf{T}}} \tag{5.14}$$

is the q-expansion map at the cusp $\mathbf{T}$.

Given the proposition, if we further prove that $\operatorname{Ker}(r) = (H - 1)$ then the theorem is proved.

To construct the map r we first construct modular forms $a(k)$ for $k \in \mathbb{Z}$ with the following properties (compare [**58**]):

- $a(k)$ is a modular form over $\mathbb{F}_p$ of μ_p-level and weight k.
- $a(k)a(k') = a(kk')$.
- $a(k)$ does not vanish on $\mathfrak{M}^*(\mathbb{F}_p, \mu_{Np})$.

Indeed, given a μ_p-level structure

$$\beta_p\colon \mu_p \hookrightarrow E, \tag{5.15}$$

we have an induced isomorphism

$$d\beta_p\colon \mathfrak{t}^*_{E/R} \to \mathfrak{t}^*_{\mu_p/R} = R \cdot \frac{dt}{t}. \tag{5.16}$$

Thus we get a canonical element $\omega_{\text{can}} = (d\beta_p)^{-1}(dt/t)$ in $\mathfrak{t}^*_{E/R}$. This gives the modular form $a(1)$. We note that since ω_{can} is non-vanishing $a(1)$ is non-vanishing. We simply put $a(k) = a(1)^k$, a non-vanishing modular form of weight k and level μ_p. Moreover, since for the Tate object ω_{can} *is* the canonical differential, the q-expansion of $a(1)$ (and hence of every $a(k)$) is 1.

We now define the map r by

$$r(f) = \frac{f}{a(k)}, \quad f \in \mathcal{M}(\mathbb{F}_p, k, \mu_N). \tag{5.17}$$

The properties of the forms $a(k)$ guarantee that this provides a well-defined ring homomorphism $\bigoplus_{k\in\mathbb{Z}} \mathcal{M}(\mathbb{F}_p, k, \mu_N) \to R$. Furthermore, since the q-expansion of a modular form is obtained by evaluating it at a Tate object defined over a base which is isomorphic to the completion of the moduli space at this point, the composition of r with the inclusion $R \hookrightarrow \widehat{R_{\mathbf{T}}}$ *is* the q-expansion.

We define an action of $(\mathbb{Z}/p\mathbb{Z})^\times$ on $\bigoplus_{k\in\mathbb{Z}} \mathcal{M}(\mathbb{F}_p, k, \mu_N)$ and on R. Given a modular form f on $\mathfrak{M}(\mathbb{F}_p, \mu_N)$ of weight k and $\alpha \in (\mathbb{Z}/p\mathbb{Z})^\times$ we define

$$[\alpha]f = \alpha^k f. \tag{5.18}$$

That is, the action of $(\mathbb{Z}/p\mathbb{Z})^\times$ is just the $\mathbb{Z}/(p-1)\mathbb{Z}$ grading on $\bigoplus_{k\in\mathbb{Z}} \mathcal{M}(\mathbb{F}_p, k, \mu_N)$. Given a function $g \in R$ we define

$$([\alpha]g)(E, \beta_N \times \beta_p) = g\big(E, \beta_N \times (\beta_p \circ \alpha)\big), \tag{5.19}$$

where we denote by $\alpha\colon \mu_p \to \mu_p$ the homomorphism of "raising to the α power." This action comes from identifying the group $(\mathbb{Z}/p\mathbb{Z})^\times$ with the Galois group of $\mathfrak{M}^*(\mathbb{F}_p, \mu_{Np}) \to \mathfrak{M}^*(\mathbb{F}_p, \mu_N)^{\mathrm{ord}}$.

We claim that the map r is equivariant for this action. Indeed, the definition given for functions $g \in R$ may well be given for modular forms on $\mathfrak{M}^*(\mathbb{F}_p, \mu_{Np})$, namely, by twisting the μ_p-level. Thus

$$[\alpha]r(f) = \frac{[\alpha]f}{[\alpha]a(k)} = \frac{f}{[\alpha]a(k)}. \tag{5.20}$$

(Note: as a modular form on $\mathfrak{M}^*(\mathbb{F}_p, \mu_{Np})$ the form f is invariant under the action of $(\mathbb{Z}/p\mathbb{Z})^\times$. Thus, $[\alpha]f = f$. But, as a modular form on $\mathfrak{M}^*(\mathbb{F}_p, \mu_N)^{\mathrm{ord}}$ we have by definition $[\alpha]f = \alpha^k f$). Now,

$$\begin{aligned}\big([\alpha]a(1)\big)(E, \beta_p) &= a(1)(E, \beta_p \circ \alpha)\\ &= d(\beta_p \circ \alpha)^{-1}(dt/t)\\ &= d\beta_p^{-1} d\alpha^{-1}(dt/t).\end{aligned} \tag{5.21}$$

Since $d\alpha(dt/t) = dt^\alpha/t^\alpha = \alpha t^{\alpha-1}dt/t^\alpha = \alpha dt/t$, we find that $d\alpha^{-1}(dt/t) = \alpha^{-1}dt/t$ and hence that $[\alpha]a(1) = \alpha^{-1}a(1)$. Therefore,

$$[\alpha]a(k) = \alpha^{-k}a(k), \tag{5.22}$$

and

$$[\alpha]r(f) = \alpha^k r(f) = r([\alpha]f). \tag{5.23}$$

We note that

$$R = \bigoplus_{k=0}^{p-2} R^k, \tag{5.24}$$

where $R^k = \{f \in R : [\alpha]f = \alpha^k f\}$. Then $r\big(\mathcal{M}(\mathbb{F}_p, k, \mu_N)\big) \subset R^k$.

We already know that $H - 1 \in \mathrm{Ker}(r)$. Let now $f_1 + \cdots + f_a$ be in the kernel of r. Say, f_i is of weight k_i. We assume that for $i \neq j$, $k_i \neq k_j$. By replacing f_i by $f_i + f_i(H-1)$ for suitable i's and sufficiently many times, we may assume that if $i \neq j$ then $k_i \neq k_j$ modulo $p-1$. But then every $r(f_i)$ lies in a different summand of the right-hand side of Equation (5.24). Hence, every $r(f_i) = 0$.

Note that on any fixed weight the q-expansion, equivalently the map r, is injective. This is evident from Equation (5.17). Thus, for every i, $f_i = 0$.

Finally, the map r is surjective. Let $g \in R^k$. Define $f = g \cdot a(k) \cdot H^n$. Then f is invariant under the Galois group of $\mathfrak{M}^*(\mathbb{F}_p, \mu_{Np}) \to \mathfrak{M}^*(\mathbb{F}_p, \mu_N)^{\mathrm{ord}}$. Hence, it defines a meromorphic modular form on $\mathfrak{M}^*(\mathbb{F}_p, \mu_N)$, with poles supported on the supersingular locus. Now, the Hasse invariant vanishes on the supersingular locus. This follows readily from interpreting $H^1_{dR}(E)$ as the Dieudonné module of $E[p]$.

See Appendix A. Thus, for $n \gg 0$ the modular form f would be holomorphic. Clearly, $r(f) = g$. □

COROLLARY 5.6. *We have*

$$\widetilde{\mathcal{N}}^0 = \bigoplus_{k \equiv 0 \bmod p-1} \mathcal{M}(\mathbb{F}, k, \mu_N)/(H-1) \cong R^0. \tag{5.25}$$

The ring R^0 is the ring of regular functions on the affine regular modular curve $\mathfrak{M}^*(\mathbb{F}, \mu_N)^{\mathrm{ord}}$. Hence, $\widetilde{\mathcal{N}}^0$ is a Dedekind domain.

5.3. Operators revisited. We use the method of §5.2 to interpret the operators $T_\ell, \langle d\rangle, U, V$ and θ via the isomorphism

$$r\colon \bigoplus_k \mathcal{M}(\mathbb{F}_p, k, \mu_N)/(H-1) \cong R, \tag{5.26}$$

where R is the ring of regular functions on the affine curve $\mathfrak{M}^*(\mathbb{F}_p, \mu_{Np})$. The a priori observation that the formulae for those operators are given solely in terms of their q-expansion and that for T_ℓ it depends on the weight only modulo $p-1$ indicates that "they are coming from R." To define those operators we shall think of the points of R as pairs (E, β_{Np}).

Let $g \in R$. Define

$$(T_\ell g)(E, \beta_{Np}) = \frac{1}{\ell} \sum_{H \subset E} g(E/H, \pi_H \circ \beta_{Np}); \tag{5.27}$$

the summation ranging over all subgroups of order ℓ of E and $\pi_H\colon E \to E/H$ is the projection. (Again, some care is needed with this definition. See Remark 3.8). We claim that

$$r(T_\ell f) = T_\ell r(f). \tag{5.28}$$

First note that

$$r(f)(E, \beta_{Np}) = f\big(E, \beta_N, \beta_{p*}(dt/t)\big), \tag{5.29}$$

where we put $\beta_{p*} = (d\beta_p)^{-1}$ for the induced map $\mathfrak{t}^*_{\mu_p/k} \to \mathfrak{t}^*_{E/k}$. This follows immediately from checking the q-expansions. Hence,

$$\begin{aligned} r(T_\ell f)(E, \beta_{Np}) &= (T_\ell f)\big(E, \beta_N, \beta_{p*}(dt/t)\big) \\ &= \frac{1}{\ell} \sum_{H \subset E} f(E/H, \pi_H \circ \beta_N, \omega_H), \end{aligned} \tag{5.30}$$

where ω_H is the differential induced from $\beta_{p*}(dt/t)$ via $E \to E/H$. That is, $\omega_H = (\pi_H \circ \beta_p)_*(dt/t)$. On the other hand,

$$\begin{aligned} \big(T_\ell r(f)\big)(E, \beta_{Np}) &= \frac{1}{\ell} \sum_{H \subset E} r(f)(E/H, \pi_H \circ \beta_{Np}) \\ &= \frac{1}{\ell} \sum_{H \subset E} f\big(E/H, \pi_H \circ \beta_N, (\pi_H \circ \beta_p)_*(dt/t)\big). \end{aligned} \tag{5.31}$$

Formula (5.28) follows.

Let $d \in (\mathbb{Z}/N\mathbb{Z})^\times$. We defined the action of the diamond operator $\langle d\rangle$ on a modular form f by

$$(\langle d\rangle f)(E, \beta_N, \omega) = f(E, \beta_N \circ d, \omega). \tag{5.32}$$

We define for $g \in R$

$$(\langle d \rangle g)(E, \beta_{Np}) = g(E, \beta_N \circ d, \beta_p). \tag{5.33}$$

Clearly,

$$r(\langle d \rangle f) = \langle d \rangle r(f). \tag{5.34}$$

We defined for a modular form f

$$(Vf)(E, \beta_N, \omega) = f(E^{(p)}, \beta_N^{(p)}, \omega^{(p)}). \tag{5.35}$$

The scheme $\mathfrak{M}^*(\mathbb{F}_p, \mu_{Np})$ is a scheme defined over $\mathbb{F}_p$. We thus have a Frobenius morphism

$$\mathrm{Fr}\colon \mathfrak{M}^*(\mathbb{F}_p, \mu_{Np}) \to \mathfrak{M}^*(\mathbb{F}_p, \mu_{Np}). \tag{5.36}$$

Given a function $g \in R$ we let

$$Vg := g \circ \mathrm{Fr}\,. \tag{5.37}$$

From a moduli perspective, since the point corresponding to $(E^{(p)}, \beta_{Np}^{(p)})$ is Fr of the point corresponding to (E, β_{Np}):

$$(Vg)(E, \beta_{Np}) = g(E^{(p)}, \beta_{Np}^{(p)}). \tag{5.38}$$

In particular,

$$\begin{aligned}\big(Vr(f)\big)(E, \beta_{Np}) &= f\big(E^{(p)}, \beta_N^{(p)}, (\beta_p^{(p)})_*(dt/t)\big) \\ &= f\big(E^{(p)}, \beta_N^{(p)}, \big(\beta_{p*}(dt/t)\big)^{(p)}\big) \\ &= (Vf)\big(E, \beta_N, \beta_{p*}(dt/t)\big) \\ &= r(Vf)(E, \beta_{Np}).\end{aligned} \tag{5.39}$$

That is,

$$r(Vf) = Vr(f). \tag{5.40}$$

Given a function $g \in R$ we define

$$\theta g = \frac{dg}{\mathrm{KS}(a(2))}. \tag{5.41}$$

We explain our notation. There is an isomorphism, called the Kodaira–Spencer isomorphism,

$$\mathrm{KS}\colon (\mathrm{t}^*_{\mathcal{E}^U/\mathfrak{M}^*(\mathbb{F}_p,\mu_N)})^{\otimes 2} \to \Omega^1_{\mathfrak{M}^*(\mathbb{F}_p,\mu_N)/\mathbb{F}}(\mathbf{cusps}), \tag{5.42}$$

for every field k of characteristic prime to N ($\mathcal{E}^U$ is the universal object). See [**79**, 1.5, A1.3], [**94**, I.4.3]. In particular, one can associate to a modular form f of weight 2 over $\mathfrak{M}^*(\mathbb{F}_p, \mu_N)$ a differential $\mathrm{KS}(f)$ on that curve. This differential is holomorphic iff the modular form is a cusp form.

Over the complex numbers this is very familiar: If f is a modular form of weight 2 (i.e., a section of $(\mathrm{t}^*_{\mathcal{E}^U/\mathfrak{M}^*(\mathbb{C},\mu_N)})^{\otimes 2}$) then $f(\tau)d\tau$ is a meromorphic differential with at most simple poles supported at the cusps. Indeed

$$\begin{aligned}f(\gamma\tau)d\gamma\tau &= (c\tau + d)^2 f(\tau) \cdot \frac{(c\tau + d)a - (a\tau + b)c}{(c\tau + d)^2} d\tau \\ &= f(\tau)d\tau.\end{aligned} \tag{5.43}$$

At $i\infty$, if $f(q) = \sum_n a_n q^n$ then $q = \exp(2\pi i \cdot \tau)$ implies $dq/q = 2\pi i \cdot d\tau$ and therefore $f(\tau) d\tau = 2\pi i \cdot f(q)(dq/q)$, which is holomorphic if and only if f is a cusp form.

The covering

$$\mathfrak{M}^*(\mathbb{F}_p, \mu_{Np}) \to \mathfrak{M}^*(\mathbb{F}_p, \mu_N)^{\mathrm{ord}}, \tag{5.44}$$

of non-singular affine curves, extends uniquely to a covering of non-singular complete curves

$$\mathfrak{M}^\dagger(\mathbb{F}_p, \mu_{Np}) \to \mathfrak{M}^*(\mathbb{F}_p, \mu_N). \tag{5.45}$$

Moreover, $\mathfrak{M}^\dagger(\mathbb{F}_p, \mu_{Np})$ may be defined intrinsically as $\mathfrak{M}^*(\mathbb{F}_p, \mu_N)[a(1)]$—the scheme obtained from $\mathfrak{M}^*(\mathbb{F}_p, \mu_N)$ by adjoining the $(p-1)$st root of the Hasse invariant H, which is a section of the line bundle $(\mathfrak{t}^*_{\mathcal{E}^U/\mathfrak{M}^*(\mathbb{F}_p,\mu_N)})^{\otimes p-1}$.

The morphism $\mathfrak{M}^\dagger(\mathbb{F}_p, \mu_{Np}) \to \mathfrak{M}^*(\mathbb{F}_p, \mu_N)$ is thus a finite separable morphism of degree $p-1$, commuting with the action of $(\mathbb{Z}/p\mathbb{Z})^\times$ (in fact, $\mathfrak{M}^*(\mathbb{F}_p, \mu_N)$ is the *quotient* by this action), and its ramification divisor is $(p-2)W_1$, where W_1 is the supersingular locus. In particular, it is totally ramified over the supersingular locus.

Now, when dealing with level μ_{Np}, $(N, p) = 1$, in characteristic p, the Kodaira–Spencer isomorphism needs to be modified. On the one hand, if we let f denote the morphism $\mathfrak{M}^\dagger(\mathbb{F}_p, \mu_{Np}) \xrightarrow{f} \mathfrak{M}^*(\mathbb{F}_p, \mu_N)$ then

$$f^*(\mathfrak{t}^*_{\mathcal{E}^U/\mathfrak{M}^*(\mathbb{F}_p,\mu_N)}) = \mathfrak{t}^*_{\mathcal{E}^U/\mathfrak{M}^\dagger(\mathbb{F}_p,\mu_{Np})}. \tag{5.46}$$

On the other, since the separable morphism $f\colon \mathfrak{M}^\dagger(\mathbb{F}_p, \mu_{Np}) \to \mathfrak{M}^*(\mathbb{F}_p, \mu_N)$ is completely (but tamely) ramified over the supersingular locus W_1, we get

$$f^*\Omega^1_{\mathfrak{M}^*(\mathbb{F}_p,\mu_N)} = \Omega^1_{\mathfrak{M}^\dagger(\mathbb{F}_p,\mu_{Np})}\big((2-p)W_1\big) \tag{5.47}$$

(see [**68**, IV.2]). Putting it all together, we get an isomorphism

$$\mathrm{KS}\colon (\mathfrak{t}^*_{\mathcal{E}^U/\mathfrak{M}^\dagger(\mathbb{F}_p,\mu_{Np})})^{\otimes 2} \to \Omega^1_{\mathfrak{M}^\dagger(\mathbb{F}_p,\mu_{Np})}(\mathbf{cusps} + (2-p)W_1). \tag{5.48}$$

It follows that $a(2)$ defines a differential whose order of vanishing along W_1 is $p = 2 + (p-2)$.

We claim that $r(\theta f) = \theta r(f)$. This is readily checked via the q-expansions. If the q-expansion of a function g is equal to $\sum a_n q^n$ then θg has q-expansion $(\sum n a_n q^{n-1} dq)/(dq/q)$ (one can prove that the differential on the Tate curve corresponding to $a(2)$ is dq/q). Thus,

$$r(\theta f) = \theta r(f). \tag{5.49}$$

Some of the delicate behaviour of the operator θ can be explained by this interpretation. First

LEMMA 5.7. *Let $g \in R^{k_0}$. Then the expansion of g around every supersingular point P has the form*

$$a_{k_P} x^{k_P} + a_{k_P+p-1} x^{k_P+p-1} + a_{k_P+2(p-1)} x^{k_P+2(p-1)} + \dots, \tag{5.50}$$

where $k_P \equiv k_0 \pmod{p-1}$. The filtration of the q-expansion $g(q)$ is

$$w\big(g(q)\big) = -\min\{k_P : P \in W_1\}. \tag{5.51}$$

PROOF. The shape of the Taylor expansion follows immediately from the fact that every supersingular point is a fixed point of the group action $(\mathbb{Z}/p\mathbb{Z})^\times$ and $g \in R^{k_0}$. The second fact is also evident, because $a(b)$, as a modular form, vanishes to order b at such point. □

COROLLARY 5.8. *Let f be a modular form of weight k and filtration $w(f)$. Then:*

(1) $w(f) < k$ *iff f vanishes along the supersingular locus W_1 or, equivalently,* $H \mid f$;
(2) $w(Vf) = p \cdot w(f)$;
(3) $w(\theta f) \leq w(f) + p + 1$ *with equality holding iff* $p \nmid w(f)$.

PROOF. The first assertion is immediate. Note that the Taylor expansion of $r(f)$ at a supersingular point P starts with x^{-k+t} $(t \geq 0)$, where t is not zero iff f vanishes at P.

To see the effect of V, we note that the Taylor expansion of $r(f)$ at some supersingular point starts with $a_{-w(f)}x^{-w(f)}$ $(a_{-w(f)} \neq 0)$ and at all other points with terms of not smaller degree. The Taylor expansion of $Vr(f)$ starts at some supersingular point with $cx^{-pw(f)}$, because $Vr(f) = r(f)\circ \mathrm{Fr}$ and at all other points with terms of not smaller degree.

To prove the last assertion, note that if at some supersingular point $r(f)$ has expansion $a_{k_P}x^{k_P} + a_{k_P+p-1}x^{k_P+p-1} + a_{k_P+2(p-1)}x^{k_P+2(p-1)} + \cdots$ then $\theta r(f)$ has expansion starting with $a_{k_P}k_P x^{k_P-1-p}$ (using that $dr(f)$ starts with $a_{k_P}k_P x^{k_P-1}$ and $a(2)$ with x^{-p}). Thus, $w(\theta f) \leq w(f) + p + 1$ with equality iff $p \nmid w(f)$. □

We now define the U-operator.[7] Given a function $g \in R$ we define

$$Ug = V^{-1}(I - \theta^{p-1})g. \tag{5.52}$$

We explain this definition. First, from the q-expansions we see that $\theta(I-\theta^{p-1})g = 0$. Since θ is a non-zero derivation of the quotient field K of R and K is a field of transcendence degree 1 we get that every derivation is trivial on $(I - \theta^{p-1})g$ and that implies that g is a p-power. It follows that also $g = Vg_1$ for a unique function g_1 that we call Ug.

The definition allows one to check easily that if g is of weight $k \pmod{p-1}$ then so is Ug. We therefore obtain an operator U taking holomorphic modular forms of weight k to modular forms of weight k with poles supported along the supersingular locus by defining

$$Uf = a(k) \cdot U\big(r(f)\big). \tag{5.53}$$

It follows immediately from the definition and the properties of θ that if $f(q) = \sum_n a_n q^n$ then

$$Uf(q) = \sum_n a_{pn} q^n. \tag{5.54}$$

PROPOSITION 5.9. *If $k > 1$ then U takes holomorphic modular forms to holomorphic modular forms.*

---

[7]This approach and its generalization to Hilbert modular forms is part of a joint work with F. Andreatta.

PROOF. Let f be a holomorphic modular form of (non-negative) weight k. We have an equality of meromorphic modular forms of weight $pk + (p-1)(p+1)$

$$H^{p+1}VUf = H^k(H^{p+1} - \theta^{p-1})f, \tag{5.55}$$

because the weights and the q-expansions of both sides agree. Suppose that Uf has a pole of order m at some supersingular point. Then VUf has a pole of order pm at some supersingular point. On the other hand $H^{k-p-1}(H^{p+1} - \theta^{p-1})f$ has a pole of order at most $p+1-k$ at any supersingular point because $(H^{p+1} - \theta^{p-1})f$ is a holomorphic modular form. Therefore, $k \leq 1$. □

Another advantage of our definition is that it explains why U tends to reduce filtration.

EXERCISE 5.10. *Prove, by analyzing Formulas* (5.52), (5.53), *that*

$$w(Uf) \leq \frac{1}{p}(w(f) + p^2 - 1). \tag{5.56}$$

In particular, if f is an ordinary eigenform, i.e., $Uf = \lambda f$ for some $\lambda \neq 0$, then

$$w(f) \leq p + 1. \tag{5.57}$$

6. p-adic elliptic modular forms

The main reference for this section is Katz's paper [**79**]. In fact, everything that we describe here can be found in that paper. The reader may also consult Gouvea's book [**51**]. A still very partial generalization for Hilbert modular forms is sketched in the next chapters. P. Kassaei [**78**] has dealt with some aspects of the theory for Shimura curves over $\mathbb{Q}$.

6.1. Test objects and overconvergent forms. Let B be a p-adic ring. By that we mean that $B \cong \varprojlim B/p^nB$, and so, in particular, B is a $\mathbb{Z}_p$-algebra. Note that examples are provided by the ring of integers R of a finite extension of $\mathbb{Q}_p$, and by any quotient $R/(\pi^n)$ of R, where π is a uniformizer. One may take $n = 1$, or more generally any field of characteristic p.

We fix:

- B—a p-adic ring ("The base ring").
- r—an element of B ("The growth gauge").

DEFINITION 6.1. Let A be a p-adic ring which is a B-algebra. A *test object* of:

- level μ_N,
- growth condition r,
- over B,

is a quadruple:

$$(E, \omega, \beta_N, Y)_{/A}, \tag{6.1}$$

such that:

- E/A is an elliptic curve, i.e., a proper, smooth group scheme over $\mathrm{Spec}(A)$, such that the geometric fibers are connected curves of genus one;
- $\omega \in \mathbf{t}^*_{E/A}$ is a relative non-vanishing differential;
- $\beta_N : \mu_{N/A} \hookrightarrow E$ is an embedding of group schemes over $\mathrm{Spec}(A)$;
- $Y \in A$ satisfies $Y \cdot E_{p-1}(E, \omega) = r$.

EXAMPLE 6.2. (1) Given $(E, \omega, \beta_N)_{/A}$ such Y need not exist. For example, if $r = 1$ (or any unit) the condition is that $E_{p-1}(E, \omega) \in A^\times$. This excludes every elliptic curve E/A such that $E \bmod \mathfrak{m}$ is supersingular for some maximal ideal $\mathfrak{m}$ of A.

EXERCISE 6.3. *The Jacobson radical J of A is the intersection of all maximal ideals. Prove that $J = \{a \in A : 1 + ab \in A^\times, \forall b \in A\}$.*

The ring A is p-adic, so $pA \neq A$, and we take $\mathfrak{m}$ to be any maximal ideal. Note that if $a \in pA$ then $(1-a)^{-1} = 1 + a + a^2 + \cdots$ converges in A. Therefore pA is contained in the Jacobson radical of A. That is, every maximal ideal contains pA. The reduction $(E \bmod \mathfrak{m}, \omega \bmod \mathfrak{m})$ is an elliptic curve together with a non-vanishing differential over the characteristic p field $A/\mathfrak{m}$. Since the reduction modulo p of E_{p-1} is the Hasse invariant, it vanishes at all supersingular elliptic curves over fields of characteristic p with any choice of differential. Thus

$$E_{p-1}(E, \omega) \bmod \mathfrak{m} \equiv E_{p-1}(E \bmod \mathfrak{m}, \omega \bmod \mathfrak{m}) = 0 \bmod \mathfrak{m} \quad \text{iff } E \bmod \mathfrak{m} \text{ is supersingular.} \tag{6.2}$$

(2) Examining the idea further (say $r = 1$), note that if $E_{p-1}(E, \omega) = p$, then indeed it may happen that $p^{1/(p-1)} \in A$ and then $E_{p-1}(E, p^{1/(p-1)}\omega) = 1$, but $p^{1/(p-1)}\omega$ is not a non-vanishing differential *over* A. That is, $(E, p^{1/(p-1)}\omega)$ is *not* part of a test object over A.

Say $r = p^n$ for some positive integer n. The $\overline{\mathbb{Q}_p}$ points of the moduli curve $\mathfrak{M}(\mathbb{Z}_p, \mu_N)$ over $\mathbb{Z}_p$ for elliptic curves with μ_N-level structure (E, β_N) that appear as part of test objects are exactly those elliptic curves (E, β_N) defined over some finite extension K of $\mathbb{Q}_p$ having good reduction and such that for some $\omega \in \Omega^1_{E/\mathcal{O}_K}$ we have $v(E_{p-1}(E, \omega)) \leq n$ (the existence of Y is really independent of the choice of ω). Note that if $\{x_i\}$ are the zeros of E_{p-1} on $\mathfrak{M}(\overline{\mathbb{Q}_p}, \mu_N)$ then we can define an equivalent collection of open sets as follows: choose an affine model for $\mathfrak{M}(\overline{\mathbb{Q}_p}, \mu_N)$. We may consider, for every n', points x on $\mathfrak{M}(\overline{\mathbb{Q}_p}, \mu_N)$ such that the distance of x from every x_i is at least n' (with respect to some norm on the ambient affine space).

(3) Given $(E, \omega, \beta_N, Y)_{/A}$, any other Y is of the form $Y + t$, with $t \cdot E_{p-1}(E, \omega) = 0$. Note that *if $E_{p-1}(E, \omega)$ is not a zero divisor Y is uniquely determined.*

(4) If $(E, \omega, \beta_N, Y)_{/A}$ is a test object, so is $(E, \lambda\omega, \beta_N, \lambda^{p-1}Y)_{/A}$ for $\lambda \in A^\times$. Indeed,

$$(\lambda^{p-1}Y) \cdot (E_{p-1}(E, \lambda\omega)) = \lambda^{p-1}Y\lambda^{-(p-1)}E_{p-1}(E, \omega) = r. \tag{6.3}$$

Equivalently, a test object could be thought of as data

$$(E, \beta_N, Y)_{/A}, \quad Y \in (\mathfrak{t}^*_{E/A})^{-(p-1)} \tag{6.4}$$

such that $Y \cdot E_{p-1}(E) = r$, where E_{p-1} is interpreted as a rule associating to E/A a section of $(\mathfrak{t}^*_{E/A})^{p-1}$.

DEFINITION 6.4. A *p-adic modular form* (à la Katz) of:

- weight $k \in \mathbb{Z}$,
- level μ_N,
- growth r,

- defined over B,

is a rule associating to a test object (of level μ_N and growth r) $(E, \omega, \beta_N, Y)_{/A}$ an element $f(E, \omega, \beta_N, Y) \in A$ such that:

- $f(E, \omega, \beta_N, Y)$ depends only on the isomorphism class of the test object $(E, \omega, \beta_N, Y)_{/A}$;
- the rule f commutes with base change;
- for every $\lambda \in A^\times$ and every test object $(E, \omega, \beta_N, Y)_{/A}$

$$f(E, \lambda\omega, \beta_N, \lambda^{p-1}Y)_{/A} = \lambda^{-k} f(E, \omega, \beta_N, Y)_{/A}. \tag{6.5}$$

DEFINITION 6.5. The space of modular forms over B, of μ_N-level, weight k and growth condition r is denoted by $\mathbb{F}(B, k, \mu_N; r)$. If $r \notin B^\times$, they are called *overconvergent* modular forms.

Given $s \in B$, we have a B-module homomorphism

$$\begin{aligned} &\mathbb{F}(B, k, \mu_N; sr) \to \mathbb{F}(B, k, \mu_N; r), \\ &\qquad f \mapsto f', \quad f'(E, \omega, \beta_N, Y)_{/A} = f(E, \omega, \beta_N, sY)_{/A}. \end{aligned} \tag{6.6}$$

If $s \in B^\times$, then this is an isomorphism. Thus, whenever $r \in B^\times$, we might as well take $r = 1$. We shall see in Corollary 6.22 that for $r = 1$ these are Serre's p-adic modular forms.

In the case $r = 1$, as we pointed out, we discard all elliptic curves with supersingular reduction; but if we consider a growth condition $r \notin B^\times$, the modular form in question is defined also on some portion of the supersingular disks of radius 1 in the moduli space of elliptic curves with level μ_N-structure, besides being defined on the complement. This is why such forms are called "overconvergent." To be precise, the very definition of the supersingular disks uses E_{p-1} ([**14**, §§1–2]). The disk of radius n may be defined as all the $\overline{\mathbb{Q}_p}$ points x of $\mathfrak{M}(\overline{\mathbb{Q}_p}, \mu_N)$ such that E_x has good reduction modulo p and for one (equivalently, any) differential ω_x on E_x such that (E_x, ω_x) has good reduction (i.e., ω_x extends to a non-vanishing differential over $\mathcal{O}_{\overline{\mathbb{Q}_p}}$) we have $|E_{p-1}(E_x, \omega_x)| \leq p^{-n}$. In fact the growth r actually controls the growth of the coefficients of the Laurent expansions of these forms around the supersingular disks.

Every classical modular form f of weight k, level N, over B defines a p-adic modular form (still denoted f) in $\mathbb{F}(B, k, \mu_N; r)$:

$$f(E, \omega, \beta_N, Y)_{/A} := f(E, \omega, \beta_N)_{/A}. \tag{6.7}$$

As an example of a truly p-adic modular form, consider the modular form in $\mathbb{F}(B, 1-p, \mu_N; r)$ given by:

$$f(E, \omega, \beta_N, Y)_{/A} = Y. \tag{6.8}$$

This is indeed a p-adic modular form. The requirement on base change and isomorphism class being obvious. By definition,

$$f(E, \lambda\omega, \beta_N, \lambda^{p-1}Y) = \lambda^{p-1}Y = \lambda^{-(1-p)} f(E, \omega, \beta_N, Y). \tag{6.9}$$

Therefore, the weight is $1 - p$.

If $r = 1$, then $Y \cdot E_{p-1}(E, \omega) = 1$, i.e.,

$$Y = \frac{1}{E_{p-1}}. \tag{6.10}$$

We remark that (in accordance with Corollary 6.22) this is also a modular form à la Serre (Exercise 4.4).

More interesting examples are the following (we follow the exposition of [**15**]):

(1) For every $n \geq 1$, classical modular forms over $\mathbb{Z}_p$ of level $\mu_N \cap \Gamma_0(p^n)$ are p-adic overconvergent modular forms of level μ_N for every growth parameter r satisfying $\mathrm{val}(r) < p^{2-n}/(p+1)$.
(2) If f is a p-adic modular form of level μ_N with q-expansion $\sum a_n q^n$ and $\chi\colon \mathbb{Z}_p^\times \to \mathbb{Z}_p^\times$ is a character, then there exists a p-adic modular form f^χ whose q-expansion is $\sum_{(p,n)=1} \chi(n) a_n q^n$.
(3) Since E_2 is a p-adic modular form (though loc. cit. shows it is transcendental over the ring of overconvergent modular forms), the theta operator is an operator on p-adic modular forms, given on q-expansions by $\Theta(\sum a_n q^n) = \sum n a_n q^n$. If f is a non-zero overconvergent modular form of weight $k \neq 0$, then $\Theta(f)$ is not overconvergent.

6.2. q-expansion for p-adic modular forms. If $E_{p-1}(E,\omega)$ is invertible, then in any test object (E,ω,β_N,Y) we must take $Y = r \cdot E_{p-1}(E,\omega)^{-1}$. This applies in particular to the Tate curve $\mathbf{T}(q) = \mathbb{G}_m/\underline{q}(\mathbb{Z})$ over B. The Tate curve carries a canonical μ_N-level:

$$\beta_{\mathrm{can}}\colon \mu_N \hookrightarrow \mathbb{G}_m \to \mathbf{T}(q), \tag{6.11}$$

and a canonical differential ω_{can} induced from the differential dt/t on $\mathbb{G}_m$. The cusp $(\mathbf{T}(q),\beta_{\mathrm{can}})$ is called the standard cusp. Every other cusp is obtained by changing the embedding of μ_N into $\mathbf{T}(q)$.

DEFINITION 6.6. Let f be a p-adic modular form, $f \in \mathbb{F}(B,k,\mu_N;r)$. The q-expansion of f is

$$f\left(\mathbf{T}(q),\omega_{\mathrm{can}},\beta_{\mathrm{can}},\frac{r}{E_{p-1}(\mathbf{T}(q),\omega_{\mathrm{can}})}\right) \in B(\!(q)\!). \tag{6.12}$$

Similarly, for any cusp $(\mathbf{T}(q),\beta_N)$ we define the q-expansion of f by

$$f\left(\mathbf{T}(q),\omega_{\mathrm{can}},\beta_N,\frac{r}{E_{p-1}(\mathbf{T}(q),\omega_{\mathrm{can}})}\right) \in B(\!(q)\!). \tag{6.13}$$

We call f *holomorphic* (respectively, *cusp*) form if all its q-expansions lie in $B[\![q]\!]$ (respectively, in $q \cdot B[\![q]\!]$) at every cusp. We denote the holomorphic (resp. cusp) forms by

$$\mathbb{M}(B,k,\mu_N;r) \quad (\text{resp. } \mathbb{S}(B,k,\mu_N;r)). \tag{6.14}$$

PROPOSITION 6.7. *Take any $\mathbb{X} \in \{\mathbb{F},\mathbb{M},\mathbb{S}\}$. Then*

$$\mathbb{X}(B,k,\mu_N;r) = \varprojlim \mathbb{X}(B/p^nB,k,\mu_N;r). \tag{6.15}$$

PROOF. All our objects are p-adic: $B = \varprojlim B/p^nB$, and to give an elliptic curve over A is equivalent, by Grothendieck's algebraization theorem [**60**, Theorem 5.4.5, p. 157] (see also [**64**]), to giving an inductive system

$$E_{1/(A/pA)} \hookrightarrow E_{2/(A/p^2A)} \hookrightarrow E_{3/(A/p^3A)} \hookrightarrow \cdots. \tag{6.16}$$

□

EXERCISE 6.8. *Let A be a p-adic dvr. Prove, without using Grothendieck's theorem, that to give an elliptic curve E/A is equivalent to giving an inductive system $E_{1/(A/pA)} \hookrightarrow E_{2/(A/p^2A)} \hookrightarrow E_{3/(A/p^3A)} \hookrightarrow \cdots$, in the following sense: Given*

E/A one obtains an inductive system $\{E \otimes_A (A/p^iA)\}$. Conversely, given an inductive system $E_{1_{/(A/pA)}} \hookrightarrow E_{2_{/(A/p^2A)}} \hookrightarrow E_{3_{/(A/p^3A)}} \hookrightarrow \cdots$ there exists an elliptic curve E/A, unique up to isomorphism over A, and an isomorphism of inductive systems $\{E_i\} \cong \{E \otimes_A (A/p^iA)\}$.

We offer some remarks:

(1) The q-expansion of a p-adic modular form f' induced from a classical modular form f as in (6.7) is the same q-expansion as of f.
(2) The q-expansion map is injective. See Corollary 6.20.

6.3. Standing assumptions and their raison d'être. *From this point until the end of the chapter, unless otherwise said, we assume that $p \geq 5$, $N > 4$ and $k \geq 2$.*

The assumptions are made to simplify technical issues and in many cases are unnecessary. The most important benefit is that under these assumptions we have base change properties for modular forms. Let $\mathfrak{M}(B, \mu_N) \to \operatorname{Spec}(B)$ be the moduli space parameterizing isomorphism classes of elliptic curves with μ_N-level structure over B-algebras, and let $\mathfrak{M}^*(B, \mu_N) \to \operatorname{Spec}(B)$ be the scheme obtained by adding the cusps. Assume that N is invertible in B. Then $\mathfrak{M}(B, \mu_N) \to \operatorname{Spec}(B)$ is a smooth morphism of relative dimension one and $\mathfrak{M}^*(B, \mu_N) \to \operatorname{Spec}(B)$ is a smooth proper morphism.

Before explaining that base change property, we remark on some of the assumptions. First, we point out that although $\mathfrak{M}(B, \mu_N) \to \operatorname{Spec}(B)$ is a fine moduli space also for $N = 4$, and although $\mathfrak{M}^*(B, \mu_N) \to \operatorname{Spec}(B)$ is well-defined, $\mathfrak{M}^*(B, \mu_N)$ is not a fine moduli space for $N = 4$. Two of the cusps of $X_1(4)$ have an automorphism of order 2 (In fact, they are the 4-gons $\mathbb{Z}/4\mathbb{Z} \times \mathbb{G}_m$ with the points of order four $(2, i)$, $(0, i)$). That is why we take $N > 4$.

The assumption that $p \geq 5$ is made to get a specific lift for the Hasse invariant. Namely, the Eisenstein series E_{p-1}. This, in fact, is not a serious point. The reader interested in the most general case should consult [**79**] and Chapter 5.

Let $\underline{\omega}$ be the relative cotangent sheaf at the origin $\mathrm{t}^*_{\mathcal{E}^B/\mathfrak{M}(B,\mu_N)}$ of the universal elliptic curve with μ_N-level structure $\mathcal{E}^B \to \mathfrak{M}(B, \mu_N)$, and similarly for $\mathfrak{M}^*(B, \mu_N) \to \operatorname{Spec}(B)$. Recall that modular forms of weight k defined over B are the global sections $H^0(\mathfrak{M}(B, \mu_N), \underline{\omega}^k)$, the holomorphic modular forms are $H^0(\mathfrak{M}^*(B, \mu_N), \underline{\omega}^k)$, and the cusp forms are $H^0\big(\mathfrak{M}^*(B, \mu_N), \underline{\omega}^k(-\mathbf{cusps})\big)$. We recall the Kodaira–Spencer isomorphism

$$\underline{\omega}^2 \cong \Omega^1_{\mathfrak{M}^*(B,\mu_N)/\operatorname{Spec}(B)}(\mathbf{cusps}). \tag{6.17}$$

(This uses the existence of the universal curve over the compactified curve! The reader may benefit from calculating the case $N = 4$, assuming a Kodaira–Spencer isomorphism and obtaining a contradiction by calculating the degree of $\underline{\omega}$).

LEMMA 6.9. *Let N be invertible in B. We have for every k*

$$H^0(\mathfrak{M}(B, \mu_N), \underline{\omega}^k) = H^0(\mathfrak{M}(\mathbb{Z}[N^{-1}], \mu_N), \underline{\omega}^k) \otimes B. \tag{6.18}$$

The same holds for holomorphic modular forms if $k \geq 2$.

PROOF. By base change theorems one needs to prove that $H^1(\mathfrak{M}(B, \mu_N), \underline{\omega}^k) = 0$ and $H^1(\mathfrak{M}^*(B, \mu_N), \underline{\omega}^k) = 0$ if $k \geq 2$, respectively.

For $H^1(\mathfrak{M}(B, \mu_N), \underline{\omega}^k)$ this is easy since the morphism $\mathfrak{M}(B, \mu_N) \to \operatorname{Spec}(B)$ is affine and $\underline{\omega}$ is a coherent sheaf. To establish the other vanishing result, we first note that $\deg \underline{\omega}^k = \frac{1}{2}k\big((2g-2)+C\big)$, where C is the number of cusps. Thus, for $k \geq 2$ we have $\deg \underline{\omega}^k > 2g-2$. The vanishing of H^1 follows from the Riemann–Roch Theorem. $\square$

EXERCISE 6.10. *The extent to which the conditions in Lemma 6.9 are necessary is a very subtle issue. See* [**79**, §§1.7, 1.8], [**94**, II.4]. *Just to illustrate the point, prove that for $N=1$ base change may fail for $k=1,2$.*

EXERCISE 6.11. *Prove that if $N>4$, $g>1$ and $k>2$ then*

$$H^0\big(\mathfrak{M}^*(B,\mu_N),\underline{\omega}^k(-\mathbf{cusps})\big) = H^0\big(\mathfrak{M}^*(\mathbb{Z}[N^{-1}],\mu_N),\underline{\omega}^k(-\mathbf{cusps})\big)\otimes B. \tag{6.19}$$

Prove that $g>1$ if $N>15$.

6.4. The case when p is nilpotent. We keep the assumptions described in §6.3. As before, B is a p-adic ring. We now assume further that p is nilpotent in B.

We use our usual notation $\mathcal{M}(B,k,\mu_N) = H^0(\mathfrak{M}^*(B,\mu_N),\underline{\omega}^k)$ for classical holomorphic modular forms over B, of μ_N-level structure and weight k. We denote by $\mathcal{F}(B,k,\mu_N) = H^0(\mathfrak{M}(B,\mu_N),\underline{\omega}^k)$ the classical modular forms over B, of μ_N-level structure and weight k with possible poles at infinity. As above, $\mathbb{F}(B,k,\mu_N;r)$ is the space of p-adic modular forms over B, of weight k, μ_N-level and growth r.

Given $j\geq 0$, define a map:

$$\mathcal{F}(B,k+j(p-1),\mu_N) \to \mathbb{F}(B,k,\mu_N;r) \tag{6.20}$$

by the formula

$$f\mapsto \tilde{f},\quad \tilde{f}(E,\omega,\beta_N,Y)_{/A} = Y^j f(E,\omega,\beta_N). \tag{6.21}$$

We claim that $\tilde{f}$ is a p-adic modular form of weight k. Indeed

$$\begin{aligned}\tilde{f}(E,\lambda\omega,\beta_N,\lambda^{p-1}Y) &= (\lambda^{p-1}Y)^j f(E,\lambda\omega,\beta_N)\\ &= \lambda^{j(p-1)}Y^j\lambda^{-(k+j(p-1))}f(E,\omega,\beta_N)\\ &= \lambda^{-k}\tilde{f}(E,\omega,\beta_N,Y).\end{aligned} \tag{6.22}$$

(Alternately, Y^j is a modular form of weight $-j(p-1)$. See (6.9).) Under this map, $E_{p-1}f$ is sent to $\widetilde{E_{p-1}f}$, and

$$\begin{aligned}\widetilde{E_{p-1}f}(E,\omega,\beta_N,Y) &= Y^{j+1}f(E,\omega,\beta_N)\cdot E_{p-1}(E,\omega,\beta_N)\\ &= r\cdot Y^j f(E,\omega,\beta_N)\\ &= r\cdot\tilde{f}.\end{aligned} \tag{6.23}$$

Therefore, we have obtained a well-defined homomorphism of B-modules

$$\Big(\bigoplus_{j\geq 0}\mathcal{F}(B,k+j(p-1),\mu_N)\Big)/(E_{p-1}-r)\to \mathbb{F}(B,k,\mu_N;r). \tag{6.24}$$

Here $(E_{p-1}-r)$ stands for the submodule generated by the elements $(E_{p-1}-r)f$, where $f\in\mathcal{F}(B,k+j(p-1),\mu_N)$ for some $j\geq 0$.

PROPOSITION 6.12. *The map*

$$\Big(\bigoplus_{j\geq 0}\mathcal{F}(B,k+j(p-1),\mu_N)\Big)/(E_{p-1}-r)\to \mathbb{F}(B,k,\mu_N;r) \tag{6.25}$$

is an isomorphism.

PROOF. Consider the functors:

F_1:

$$A \in B\text{-}\mathbf{alg} \mapsto F_1(A), \tag{6.26}$$

where $F_1(A)$ consists of isomorphism classes of the following data:

- $(E, \beta_N, Y)_{/A}$, such that
- $Y \in \underline{\omega}_E^{-(p-1)}$ and satisfies $Y \cdot E_{p-1} = r$,

where $\underline{\omega}_E = \mathfrak{t}^*_{E/A}$.

This functor is equivalent to the functor associating to $A \in B$-**alg** the following data:

- $g\colon \operatorname{Spec}(A) \to \mathfrak{M}(B, \mu_N)$,
- $Y \in g^*(\mathcal{L})$, such that $Y \cdot g^*E_{p-1} = r$,

where $\mathcal{L} = \underline{\omega}^{-(p-1)}$, $\underline{\omega} = \mathfrak{t}^*_{\mathcal{E}^U/\mathfrak{M}(B,\mu_N)}$.

The functor F_1 is a subfunctor of F_2:

F_2:

$$A \in B\text{-}\mathbf{alg} \mapsto F_2(A), \tag{6.27}$$

where $F_2(A)$ consists of isomorphism classes of the following data:

- $(E, \beta_N, Y)_{/A}$, where
- $Y \in \underline{\omega}_E^{-(p-1)}$.

The functor F_2 is by the same token equivalent to the functor associating to A the following:

- $g\colon \operatorname{Spec}(A) \to \mathfrak{M}(B, \mu_N)$,
- $Y \in g^*(\mathcal{L})$.

The last functor is representable by the $\mathfrak{M}(B, \mu_N)$-scheme $\mathbf{Spec}(S(\mathcal{L}^\vee))$.[8] The condition that a section Y satisfies $Y \cdot g^*(E_{p-1}) = r$ is exactly that locally, if the homomorphism $\mathbf{Spec}(S(\mathcal{L}^\vee)) \to \mathcal{M}(B, \mu_N)$ describing Y is given by $A_i[y] \xrightarrow{Y} A_i$, we have $Y(E_{p-1}) = Y \cdot E_{p-1} = r$.

Hence, F_1 is representable by the $\mathfrak{M}(B, \mu_N)$ scheme:

$$\alpha := \mathbf{Spec}\big(S(\underline{\omega}^{p-1})/(E_{p-1} - r)\big) \hookrightarrow \mathbf{Spec}\big(S(\underline{\omega}^{p-1})\big) \to \mathfrak{M}(B, \mu_N). \tag{6.28}$$

We write $f\colon \alpha \to \mathfrak{M}(B, \mu_N)$.

Note that since p is nilpotent in B, *any* B-algebra is p-adically complete. This gives the identification between $\mathbb{F}(B, k, \mu_N; r)$ and $H^0(\alpha, f^*\underline{\omega}^k)$. We may now calculate the space $\mathbb{F}(B, k, \mu_N; r)$ of p-adic modular forms over B of weight k and of growth r as follows:

$$\begin{aligned} \mathbb{F}(B, k, \mu_N; r) &= H^0(\alpha, f^*\underline{\omega}^k) \\ &= H^0(\mathfrak{M}(B, \mu_N), \underline{\omega}^k \otimes f_*\mathcal{O}_\alpha) \end{aligned} \tag{6.29}$$

---

[8] The notation is as follows: for a locally free sheaf $\mathcal{L}$ on X we let $\mathcal{L}^\vee$ denote the dual sheaf on X. We let $S(\mathcal{L}^\vee)$ denote the sheaf of symmetric algebras on X constructed from $\mathcal{L}^\vee$. We let $\mathbf{Spec}(S(\mathcal{L}^\vee)) \to X$ denote the associated affine morphism. See [**68**, Exercises II 5.16–5.18].

Locally, X is of the form $\operatorname{Spec}(A_i)$ and $\mathcal{L}$ is the free A_i-module $A_i \cdot x$. Then $\mathcal{L}^\vee$ is the module $A_i \cdot y$ with the pairing given by $a \cdot x \otimes b \cdot y \mapsto ab$. Note that giving the element $b \cdot y$ is equivalent to giving a homomorphism over A_i from $S(A_i \cdot x) = A_i[x] \to A_i$, which itself is a local section of $\mathbf{Spec}(S(\mathcal{L}))$. Finally, we replace the datum $b \cdot y$ by the homomorphism $Y\colon S(A_i \cdot y) = A_i[y] \to A_i$ over A_i (a local section of $\mathbf{Spec}(S(\mathcal{L}^\vee))$) that takes y to b.

(Leray's spectral seq., [**68**, Ex. III, 4.1])

$$= H^0\Big(\mathfrak{M}(B,\mu_N), \underline{\omega}^k \otimes \bigoplus_{j=0}^{\infty} \underline{\omega}^{j(p-1)}/(E_{p-1}-r)\Big)$$

$$= H^0\Big(\mathfrak{M}(B,\mu_N), \bigoplus_{j=0}^{\infty} \underline{\omega}^{k+j(p-1)}/(E_{p-1}-r)\Big)$$

$$\stackrel{*}{=} H^0\Big(\mathfrak{M}(B,\mu_N), \bigoplus_{j=0}^{\infty} \underline{\omega}^{k+j(p-1)}\Big)/(E_{p-1}-r)$$

$$= \bigoplus_{j=0}^{\infty} \mathcal{F}(B, k+j(p-1), N)/(E_{p-1}-r).$$

Let us explain the equality marked by $*$. Note that $b = \bigoplus_{j=0}^{\infty} \underline{\omega}^{k+j(p-1)}$ is a quasi-coherent sheaf on the *affine* scheme $\mathfrak{M}(B,\mu_N)$. We have an exact sequence:

$$0 \to b \xrightarrow{\times(E_{p-1}-r)} b \to b/(E_{p-1}-r) \to 0. \tag{6.30}$$

Since all higher cohomology groups of quasi-coherent sheaves on affine schemes vanish we get an exact sequence:

$$\begin{aligned}0 \to H^0(\mathfrak{M}(B,\mu_N), b) &\xrightarrow{\times(E_{p-1}-r)} H^0(\mathfrak{M}(B,\mu_N), b)\\ &\to H^0\big(\mathfrak{M}(B,\mu_N), b/(E_{p-1}-r)\big) \to 0.\end{aligned} \tag{6.31}$$

□

REMARK 6.13. A similar argument, using the compactified curve $\mathfrak{M}^*(B,\mu_N)$ gives

$$\mathbb{M}(B,k,\mu_N;r) = H^0\Big(\mathfrak{M}^*(B,\mu_N), \Big(\bigoplus_{j=0}^{\infty} \underline{\omega}^{k+j(p-1)}\Big)/(E_{p-1}-r)\Big). \tag{6.32}$$

Though $\big(\bigoplus_{j=0}^{\infty} \mathcal{M}(B, k+j(p-1), \mu_N)\big)/(E_{p-1}-r)$ is contained in this space, generally they are not equal. However, under the assumptions we have made (notably $k \geq 2$), if r is not a zero divisor in B, they are. See [**79**, Theorem 2.5.1]. Another instance where one may prove equality (under more general conditions) is when r is a p-adic unit.

6.5. The case of r a unit.

PROPOSITION 6.14. *Let p be nilpotent and let r be a (p-adic) unit. Let $N > 4$. Then for every k*

$$\mathbb{M}(B,k,\mu_N;r) \cong \Big(\bigoplus_{j=0}^{\infty} \mathcal{M}(B, k+j(p-1), \mu_N)\Big)/(E_{p-1}-r), \tag{6.33}$$

and

$$\mathbb{S}(B,k,\mu_N;r) \cong \Big(\bigoplus_{j=0}^{\infty} \mathcal{S}(B, k+j(p-1), \mu_N)\Big)/(E_{p-1}-r). \tag{6.34}$$

PROOF. The idea is to use the *affine* curve

$$\mathfrak{M}^*(B,\mu_N)^{\mathrm{ord}} := \mathfrak{M}^*(B,\mu_N) - [\text{supersingular locus}] \tag{6.35}$$

instead of the affine curve $\mathfrak{M}(B,\mu_N)$ used before. We may assume without loss of generality that $r=1$. Consider:

F_1: The functor associating to a B-algebra A the data:
- $g\colon \operatorname{Spec}(A)\to\mathfrak{M}^*(B,\mu_N)$,
- $Y\in g^*(\mathcal{L})$ such that $Y\cdot g^*E_{p-1}=1$, where $\mathcal{L}=\underline{\omega}^{-(p-1)}$.

F_2: The functor associating to a B-algebra A the data:
- $g\colon \operatorname{Spec}(A)\to\mathfrak{M}^*(B,\mu_N)$,
- $Y\in g^*(\mathcal{L})$.

The functor F_1 is a subfunctor of the functor F_2, which is represented by the scheme

$$\mathbf{Spec}\big(S(\mathcal{L}^\vee)\big)\to\mathfrak{M}^*(B,\mu_N). \tag{6.36}$$

A similar reasoning gives that F_1 is represented by the scheme

$$\beta=\mathbf{Spec}\big(S(\mathcal{L}^\vee)/(E_{p-1}-1)\big)\to\mathfrak{M}^*(B,\mu_N). \tag{6.37}$$

Still better, the structural morphism of β factors through $\mathfrak{M}^*(B,\mu_N)^{\mathrm{ord}}$. Thus F_1 is represented by

$$f\colon\beta\to\mathfrak{M}^*(B,\mu_N)^{\mathrm{ord}}. \tag{6.38}$$

Let $\underline{\omega}$ denote the sheaf of modular forms on $\mathfrak{M}^*(B,\mu_N)^{\mathrm{ord}}$. Let $\mathbb{X}$ be one of the symbols $\mathbb{M}$, $\mathbb{S}$. We define a sheaf b by

$$\mathbb{X}=\begin{cases}\underline{\omega}^k, & \mathbb{X}=\mathbb{M},\\ \underline{\omega}^k(-\mathbf{cusps}), & \mathbb{X}=\mathbb{S}.\end{cases} \tag{6.39}$$

Then

$$\begin{aligned}\mathbb{X}(B,k,\mu_N;1)&=H^0(\beta,b)\\ &=H^0(\mathfrak{M}(B,\mu_N)^{\mathrm{ord}},b\otimes f_*\mathcal{O}_\beta)\\ &=H^0\Big(\mathfrak{M}(B,\mu_N)^{\mathrm{ord}},b\otimes\Big(\bigoplus_{j=0}^{\infty}\underline{\omega}^{j(p-1)}\Big)/(E_{p-1}-1)\Big)\\ &=H^0\Big(\mathfrak{M}(B,\mu_N)^{\mathrm{ord}},b\otimes\bigoplus_{j=0}^{\infty}\underline{\omega}^{j(p-1)}\Big)/(E_{p-1}-1)\\ &=\bigoplus_{j=0}^{\infty}X(B,k+j(p-1),\mu_N)/(E_{p-1}-1).\end{aligned} \tag{6.40}$$

Here $X=\mathcal{M}$ (resp. $\mathcal{S}$) if $\mathbb{X}=\mathbb{M}$ (resp. $\mathbb{S}$), and the equalities follows from the same considerations as in the proof of Proposition 6.12. In particular, the first equality uses that p is nilpotent in B and the fourth equality uses that $\mathfrak{M}^*(B,\mu_N)^{\mathrm{ord}}$ is affine. □

An immediate consequence of Propositions 6.14, 6.7 and base change properties as in §6.3 is the following

THEOREM 6.15. *Let $N > 4$ be an integer. Let $p \geq 5$ be prime and let B be a p-adic ring. Let $r \in B^\times$. Then*

$$\mathbb{M}(B, k, \mu_N; r) = \varprojlim_n \Big(\bigoplus_{j=0}^{\infty} \mathcal{M}(B/p^n B, k + j(p-1), \mu_N)\Big)/(E_{p-1} - r) \tag{6.41}$$
$$= \varprojlim_n \Big(\bigoplus_{j=0}^{\infty} \mathcal{M}(\mathbb{Z}, k + j(p-1), \mu_N) \otimes B/p^n B\Big)/(E_{p-1} - r),$$

and

$$\mathbb{S}(B, k, \mu_N; r) = \varprojlim_n \Big(\bigoplus_{j=0}^{\infty} \mathcal{S}(B/p^n B, k + j(p-1), \mu_N)\Big)/(E_{p-1} - r) \tag{6.42}$$
$$= \varprojlim_n \Big(\bigoplus_{j=0}^{\infty} \mathcal{S}(\mathbb{Z}, k + j(p-1), \mu_N) \otimes B/p^n B\Big)/(E_{p-1} - r),$$

where the last equality is valid for $k > 2, g > 1$.

Note that the sums appearing on the right-hand sides are not finite anymore. To gain some firm grip on the space of p-adic modular form we introduce a kind of "basis" for that space. This is the topic of the next section.

6.6. Katz's expansion. In this section we assume, apart from the standing assumptions of §6.3, that r is not a zero divisor in the p-adic ring B. Consider the map:

$$\mathcal{M}(\mathbb{Z}_p, k + j(p-1), \mu_N) \xrightarrow{\times E_{p-1}} \mathcal{M}(\mathbb{Z}_p, k + (j+1)(p-1), \mu_N). \tag{6.43}$$

We note that upon reduction modulo p this map is *injective*, as is obvious from looking at q-expansions. Furthermore, it has a finite free cokernel ([**79**, Lemma 2.6.1]). This implies:

- The map in (6.43) is injective.
- The map in (6.43) splits.

We choose complements:

$$\mathcal{M}(\mathbb{Z}_p, k + (j+1)(p-1), \mu_N) = \tag{6.44}$$
$$E_{p-1} \cdot \mathcal{M}(\mathbb{Z}_p, k + j(p-1), \mu_N) \oplus A(\mathbb{Z}_p, k + (j+1)(p-1), \mu_N)$$

(taking $A(\mathbb{Z}_p, k, \mu_N) = \mathcal{M}(\mathbb{Z}_p, k, \mu_N)$). We may tensor with B and we get the same equality with B-coefficients. Then

$$\bigoplus_{a=0}^{j} A(B, k + a(p-1), \mu_N) \cong \mathcal{M}(B, k + j(p-1), \mu_N); \tag{6.45}$$

the map given by

$$(f_0, \ldots, f_j) \mapsto \sum_{a=0}^{j} f_a \cdot E_{p-1}^{j-a}. \tag{6.46}$$

Consider the p-adically complete B-module:

$$(6.47)\quad A^{\text{rigid}}(B,k,\mu_N) = \left\{\sum_{a=0}^{\infty} b_a : b_a \in A(B,k+a(p-1),\mu_N),\ b_a \to 0 \ p\text{-adically uniformly}\right\}.$$

(i.e. $\forall n$, $\exists c(n)$ such that $a > c(n)$ implies $p^n \mid b_a$).

EXERCISE 6.16. *Let B is a dvr with quotient field K. Taking $A^{\text{rigid}}(B,k,\mu_N)$ to be the unit ball in $A^{\text{rigid}}(K,k,\mu_N)$ we get that $A^{\text{rigid}}(K,k,\mu_N)$ is a p-adic Banach space.*

PROPOSITION 6.17 (Katz's expansion). *Under the assumptions above, there exists an isomorphism:*

$$(6.48)\quad A^{\text{rigid}}(B,k,\mu_N) \underset{\psi}{\xrightarrow{\sim}} \mathbb{M}(B,k,\mu_N;r)$$

given by

$$(6.49)\quad \sum_{a=0}^{\infty} b_a \xrightarrow{\psi} \left\langle\!\!\left\langle \sum_{a=0}^{\infty} r^a b_a / E_{p-1}^a \right\rangle\!\!\right\rangle,$$

where the right-hand side stands for the p-adic modular form whose value on a test object $(E,\omega,\beta_N,Y)_{/A}$ (where $Y \cdot E_{p-1}(\underline{A},\omega) = r$) is:

$$(6.50)\quad \left\langle\!\!\left\langle \sum_{a=0}^{\infty} r^a b_a / E_{p-1}^a \right\rangle\!\!\right\rangle (E,\omega,\beta_N,Y) = \sum_{a=0}^{\infty} Y^a \cdot b_a(E,\omega,\beta_N).$$

PROOF. One easily sees that ψ is a well-defined continuous homomorphism of B-modules (in particular, the sum on the right-hand side converges). Note that by Remark 6.13 the space of p-adic holomorphic modular forms $\mathbb{M}(B,k,\mu_N;r)$ is the p-adic completion of $(\bigoplus_{j=0}^{\infty} \mathcal{M}(B,k+j(p-1),\mu_N))/(E_{p-1}-r)$. Thus, the Proposition is in essence comparison of two p-adic completions of two related spaces.

We first show that ψ is surjective. Every element x of $\mathbb{M}(B,k,\mu_N;r)$ can be written as

$$(6.51)\quad \sum_{j=0}^{\infty} s_j, \quad s_j \in \mathcal{M}(B,k+j(p-1),\mu_N),\ s_j \to 0 \ p\text{-adically}.$$

This follows from Remark 6.13. Note that when we say $\sum_{j=0}^{\infty} s_j \in \mathbb{M}(B,k,\mu_N;r)$, we really mean that as a function whose values on a test object are given by $(\sum s_j)(E,\omega,\beta_N,Y) := \sum Y^j s_j(E,\omega,\beta_N)$. Hence, this element really is $\psi(\sum s_j)$ *if* the sum belongs to $A^{\text{rigid}}(B,k,\mu_N)$!

Improving on Equation (6.51), one can write x as:

$$(6.52)\quad \sum_{j=0}^{\infty} s_j, \quad s_j \in A(B,k+j(p-1),\mu_N).$$

Moreover, still $s_j \to 0$ p-adically. Indeed, if p^N divides $s_j = a_j + b_j$, where $a_j \in A(B,k+j(p-1),\mu_N)$ and $b_j \in E_{p-1} \cdot \mathcal{M}(B,k+(j-1)(p-1),\mu_N)$ then $p^N \mid b_j$. Thus, collecting terms backwards does not destroy convergence! $s_j \to 0$ p-adically, and thus $\sum_{j=0}^{\infty} s_j \in A^{\text{rigid}}(B,k,\mu_N)$. This shows that ψ is surjective.

We now show ψ is injective. Using that r is not a zero divisor and base change for modular forms, we may assume (by localizing at r if necessary) that r is a unit in B. An easy reduction then allows one to assume that $r = 1$. Assume that henceforth.

Consider the following diagram:

$$\begin{array}{ccc} A^{\text{rigid}}(B,k,\mu_N) & \xrightarrow{\psi} & \mathbb{M}(B,k,\mu_N;1) \\ \uparrow{\scriptstyle\xi} & & \uparrow{\scriptstyle\phi} \\ \bigoplus_{a=0}^{j} A(B,k+a(p-1),\mu_N) & \xrightarrow{\eta} & \mathcal{M}(B,k+j(p-1),\mu_N). \end{array} \tag{6.53}$$

In this diagram:

$$\eta\Big(\sum_{a=0}^{j} b_a\Big) = \sum_{a=0}^{j} b_a \cdot E_{p-1}^{j-a} \qquad \text{(an isomorphism);}$$

$$\xi\Big(\sum_{a=0}^{j} b_a\Big) = \sum_{a=0}^{j} b_a \qquad \text{(the natural inclusion);}$$

$$\psi\Big(\sum_{a=0}^{\infty} b_a\Big)(E,\omega,\beta_N,Y) = \sum_{a=0}^{\infty} Y^a \cdot b_a(E,\omega,\beta_N);$$

$$\phi(f)(E,\omega,\beta_N,Y) = Y^j \cdot f(E,\omega,\beta_N).$$

The diagram commutes:

$$\begin{aligned} \phi\bigg(\eta\Big(\sum_{a=0}^{j} b_a\Big)\bigg)(E,\omega,\beta_N,Y) &= \phi\Big(\sum_{a=0}^{j} b_a \cdot E_{p-1}^{j-a}\Big)(E,\omega,\beta_N,Y) \\ &= Y^j\Big(\sum_{a=0}^{j} b_a E_{p-1}^{j-a}\Big)(E,\omega,\beta_N) \\ &= \sum_{a=0}^{j} Y^a \cdot b_a(E,\omega,\beta_N) \\ &= \psi\bigg(\xi\Big(\sum_{a=0}^{j} b_a\Big)\bigg)(E,\omega,\beta_N,Y). \end{aligned} \tag{6.54}$$

Assume that $\psi(\sum_{a=0}^{\infty} b_a) = 0$. Fix an integer N_0. Then for some N_1 we have $a > N_1 \implies p^{N_0} \mid b_a$. Thus

$$\psi\Big(\sum_{a=0}^{N_1} b_a\Big) \equiv 0 \bmod p^{N_0}. \tag{6.55}$$

However, $\psi(\sum_{a=0}^{N_1} b_a) = \phi\big(\eta(\sum_{a=0}^{N_1} b_a)\big) = \phi(\sum_{a=0}^{N_1} b_a E_{p-1}^{N_1-a})$. Therefore,

$$\phi\Big(\sum_{a=0}^{N_1} b_a E_{p-1}^{N_1-a}\Big) \equiv 0 \bmod p^{N_0}. \tag{6.56}$$

But,

$$\begin{aligned} \mathcal{M}(B,k+N_1(p-1),\mu_N)/(p^{N_0}) &= \mathcal{M}(B/(p^{N_0}),k+N_1(p-1),\mu_N) \\ &\hookrightarrow \mathbb{M}(B/(p^{N_0}),k,\mu_N;r) = \mathbb{M}(B,k,\mu_N;r)/(p^{N_0}). \end{aligned} \tag{6.57}$$

Thus $\sum_{a=0}^{N_1} b_a \equiv 0 \bmod p^{N_0}$, and hence $\sum_{a=0}^{\infty} b_a \equiv 0 \bmod p^{N_0}$ for every N_0. This implies $\sum_{a=0}^{\infty} b_a = 0$. □

COROLLARY 6.18. *Let $r_2 = rr_1$. Then the natural map*

$$\mathbb{M}(B,k,N;r_2) \to \mathbb{M}(B,k,N;r_1), \tag{6.58}$$

(given by $f \mapsto f'$ and $f'(E,\omega,\beta_N,Y) = f(E,\omega,\beta_N,rY)$) is given by

$$\sum_{a=0}^{\infty} b_a \mapsto \sum_{a=0}^{\infty} r^a b_a \tag{6.59}$$

and is thus injective.

6.7. Properties of q-expansions of p-adic modular forms.

PROPOSITION 6.19. *Let $b \in B$ be an element dividing a positive power of p. Let $f \in \mathbb{M}(B,k,\mu_N;1)$. The following assertions are equivalent:*

(1) *$f \in b \cdot \mathbb{M}(B,k,\mu_N;1)$.*
(2) *The q-expansion of f lies in $b \cdot B[\![q]\!]$.*

PROOF. Since the q-expansion is B-linear, the implication (1) $\Longrightarrow$ (2) is clear. Now, let us prove (2) $\Longrightarrow$ (1). Note that

$$\begin{aligned}\mathbb{M}(B/bB,k,\mu_N;1) &= A^{\text{rigid}}(B/bB,k,\mu_N) \\ &= A^{\text{rigid}}(B,k,\mu_N)/bA^{\text{rigid}}(B,k,\mu_N).\end{aligned} \tag{6.60}$$

Thus, replacing B by B/bB, we may assume that $b=0$ and that p is nilpotent in B. We aim to prove that the q-expansion map is injective.

For p nilpotent, f is a *finite* sum $\sum_{a=0}^{d} b_a$, $b_a \in A(B,k+a(p-1),\mu_N)$, and its q-expansion is

$$\begin{aligned} f(\mathbf{T}(q),\omega_{\text{can}},\beta_{\text{can}},E_{p-1}^{-1}) &= \sum_{a=0}^{d} E_{p-1}^{-a}\cdot(q\text{-expansion of } b_a) \\ &= \sum_{a=0}^{d} \frac{q\text{-expansion of } (E_{p-1}^{d-a}\cdot b_a)}{E_{p-1}^{d}}, \end{aligned} \tag{6.61}$$

so the q-expansion of $\sum_{a=0}^{d} E_{p-1}^{d-a}\cdot b_a$ is zero. By the q-expansion principle for classical modular forms $\sum_{a=0}^{d} E_{p-1}^{d-a}\cdot b_a$ is zero. This implies that each b_a is zero, because $(b_0,\ldots,b_d) \in \sum_{a=0}^{d} A(B,k+a(p-1),\mu_N) \cong \mathcal{M}(B,k+d(p-1),\mu_N)$. □

COROLLARY 6.20. *The q-expansion map on p-adic modular forms is injective.*

THEOREM 6.21. *Let $f(q) \in B[\![q]\!]$ be a power series. The following assertions are equivalent:*

(1) *$f(q)$ is the q-expansion of an element $f \in \mathbb{M}(B,k,\mu_N;1)$.*
(2) *For all n, there exists a positive integer $M(n)$, $M(n) \equiv 0 \bmod p^{n-1}$, and a classical modular form $g_n \in \mathcal{M}(B,k+M(n)(p-1),\mu_N)$ such that the q-expansion $g_n(q) \equiv f(q) \bmod p^n$.*

PROOF. First, let us show that (2) $\Longrightarrow$ (1). Writing $E_{p-1}(q) = 1+px$, we see that $E_{p-1}^{p^{n-1}} \equiv 1 \bmod p^n$. Now, multiplication of g_n by $E_{p-1}^{p^{n-1}}$ changes the weight by

$(p-1)p^{n-1}$, so we can assume $M(n)$ is increasing. Let $\Delta(n) = M(n+1) - M(n)$, so

$$g_{n+1} - g_n \cdot E_{p-1}^{\Delta(n)} \in p^n \cdot \mathcal{M}(B, k + (p-1)M(n+1), \mu_N), \tag{6.62}$$

(since $\Delta(n) \equiv 0 \bmod p^{n-1}$).

Hence, $f := g_0 + \sum_{a=0}^{\infty}(g_{a+1} - g_a \cdot E_{p-1}^{\Delta(a)}) \in \mathbb{M}(B, k, \mu_N; 1)$. Modulo p^n this sum is

$$g_0 + (g_1 - g_0 E_{p-1}^{\Delta(0)}) + \cdots + (g_n - g_{n-1} \cdot E_{p-1}^{\Delta(n-1)}). \tag{6.63}$$

But $E_{p-1} = 1$ in $\mathbb{M}(B, k, \mu_N; 1)$, so the telescopic sum is equal to g_n. Hence, the q-expansion of f is $\lim g_n(q)$—the power series $f(q)$.

The implication (1) $\implies$ (2) can be proved as follows: Let $f \in \mathbb{M}(B, k, \mu_N; 1)$. Then

$$f = \psi\Big(\sum_{a=0}^{\infty} b_a\Big), \quad b_a \in A(B, k + a(p-1), \mu_N). \tag{6.64}$$

Consider $c_n = \psi(\sum_{a=0}^{n} b_a) = \phi\Big(\eta(\sum_{a=0}^{n} b_a)\Big) \in \mathcal{M}(B, k + n(p-1), \mu_N)$ (we use the notation of the proof of Proposition 6.17). Take $M(n)$ to be suitably increasing powers of p and $g_n = c_{M(n)}$. □

We may now link Serre's approach and Katz's approach to p-adic modular forms:

COROLLARY 6.22. *Serre's p-adic modular forms of weight $k \in \mathbb{Z}$ and level μ_N are the same as Katz's p-adic modular forms $\mathbb{M}(B, k, \mu_N; 1)$ of growth condition* 1.

7. The ring of divided congruences

In this section we follow Katz [**80**]. We fix the following notation

p	a rational prime ≥ 5;
k	a perfect field of characteristic p;
$W(k)$	the ring of infinite Witt vectors over k;
$W_m(k)$	the ring of Witt vectors of length m over k (equal to $W(k)/p^m W(k)$);
N	an integer ≥ 3 and prime to p (auxiliary level);
ζ	a fixed root of unity of order N in k (or in $W(k)$ via the Teichmüller lift);
$\mathfrak{M}(W(k), N)$	the fine moduli scheme of elliptic curves over $W(k)$ algebras, together with a symplectic level N structure: $(\mathbb{Z}/N\mathbb{Z})^2 \cong E[N]$, the symplectic pairing given by $\langle(1,0),(0,1)\rangle = \zeta$ corresponds to the Weil pairing;
$\mathfrak{M}^*(W(k), N)$	the canonical compactification of $\mathfrak{M}(W(k), N)$ obtained by adding cusps;
M_m^0	is $\mathfrak{M}(W_m(k), N) = \mathfrak{M}(W(k), N) \otimes_{W(k)} W_m(k)$;
S_m^0	is the open subscheme of M_m^0 where the Hasse invariant is invertible;
M_m	is $\mathfrak{M}^*(W_m(k), N) = \mathfrak{M}^*(W(k), N) \otimes_{W(k)} W_m(k)$;
S_m	is the open subscheme of M_m where the Hasse invariant is invertible.

We remark that the schemes S_m and S_m^0 are affine and that the structural morphisms $S_m \to W_m(k)$ and $S_m^0 \to W_m(k)$ are smooth with irreducible special fibers. We have the compatibilities:

$$S_m = S_{m+1} \otimes_{W_{m+1}(k)} W_m(k), \quad S_m^0 = S_{m+1}^0 \otimes_{W_{m+1}(k)} W_m(k). \tag{7.1}$$

We let

$$T_{m,n} \to S_m \tag{7.2}$$

be the étale covering of μ_{p^n}-level: $\beta_{p^n} : \mu_{p^n} \hookrightarrow E$. That is, $T_{m,n}$ represents the moduli functor of elliptic curves over $W_m(k)$-algebras with symplectic level N structure and μ_{p^n} structure. The covering $T_{m,n} \to S_m$ is thus Galois with Galois group $(\mathbb{Z}/p^n\mathbb{Z})^\times$. It follows that $T_{m,n}$ is an affine scheme and that the morphism $T_{m,n} \to W_m(k)$ is smooth. We have the compatibility

$$T_{m,n} = T_{m+1,n} \otimes_{W_{m+1}(k)} W_m(k). \tag{7.3}$$

The schemes $T_{1,n}$ appeared before in Chapter 3, §6. We offer a panoramic view:

(7.4)

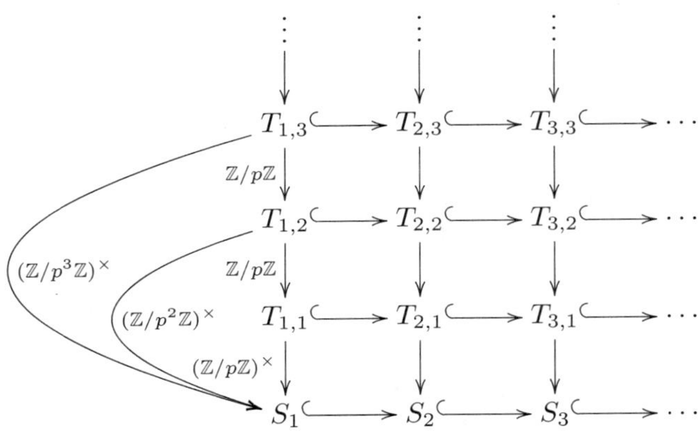

The horizontal arrows are all closed immersions. The vertical arrows are all étale Galois coverings. Let $V_{m,n}$ be the ring of regular functions on $T_{m,n}$. We get the dual diagram, with horizontal arrows being surjective and vertical arrows being inclusions.

$$\begin{array}{ccccccc}
\vdots & & \vdots & & \vdots & & \\
\uparrow & & \uparrow & & \uparrow & & \\
V_{1,3} & \longleftarrow & V_{2,3} & \longleftarrow & V_{3,3} & \longleftarrow & \cdots \\
\uparrow & & \uparrow & & \uparrow & & \\
V_{1,2} & \longleftarrow & V_{2,2} & \longleftarrow & V_{3,2} & \longleftarrow & \cdots \\
\uparrow & & \uparrow & & \uparrow & & \\
V_{1,1} & \longleftarrow & V_{2,1} & \longleftarrow & V_{3,1} & \longleftarrow & \cdots \\
\uparrow & & \uparrow & & \uparrow & & \\
V_{1,0} & \longleftarrow & V_{2,0} & \longleftarrow & V_{3,0} & \longleftarrow & \cdots
\end{array} \tag{7.5}$$

Note that

$$V_{m+1,n}/p^m V_{m+1,n} \cong V_{m,n}. \tag{7.6}$$

We let

$$T_{m,\infty} = \varprojlim_n T_{m,n}, \quad T_{\infty,\infty} = \varinjlim_m T_{m,\infty}. \tag{7.7}$$

We remark that the scheme $T_{\infty,\infty}$ represents the functor of $W(k)$- algebras

$$R \mapsto \text{Iso. classes of } (E, \beta_{p^\infty} : \widehat{\mathbb{G}}_m \xrightarrow{\cong} \widehat{E}). \tag{7.8}$$

Let

$$V_{m,\infty} = \varinjlim_n V_{m,n}, \quad V_{\infty,\infty} = \varprojlim_m V_{m,\infty}. \tag{7.9}$$

Note that

$$V_{m+1,\infty}/p^m V_{m+1,\infty} \cong V_{m,\infty}. \tag{7.10}$$

The rings $V_{m,n}$ are smooth W_m algebras and via the standard cusp we have

$$V_{m,n} \subset W_m[\![q]\!]. \tag{7.11}$$

LEMMA 7.1. *The cokernel $W_m[\![q]\!]/V_{m,n}$ is a flat W_m-module.*

PROOF. Since W_m is a local ring, the assertion amounts to

$$p \cdot f(q) \in V_{m,n} \implies f(q) \in V_{m,n} + p^{m-1} W_m[\![q]\!]. \tag{7.12}$$

Let $g \in V_{m,n}$ with a q-expansion $p \cdot f(q)$. Now

$$V_{m,n} \otimes_{W_m} W_1 = V_{m,n} \otimes_{W_m} k = V_{1,n} \hookrightarrow k[\![q]\!]. \tag{7.13}$$

Thus g is zero modulo $pV_{m,n}$. Hence, for some $f' \in V_{m,n}$ we have $p\big(f(q) - f'(q)\big)$ and it follows that $f(q) - f'(q) \in p^{m-1} W_m[\![q]\!]$. □

COROLLARY 7.2. *For $m \leq \infty$ we have an injection*

$$V_{m,\infty} \hookrightarrow W_m[\![q]\!] \tag{7.14}$$

with a W_m flat cokernel.

Consider the group scheme $\mu_{p^n/A} = \operatorname{Spec} A[x]/(x^{p^n} - 1)$. Let B denote the ring $A[x]/(x^{p^n} - 1)$. Then

$$\Omega_{\mu_{p^n}/\operatorname{Spec}(A)} = \Omega_{B/A} \tag{7.15}$$

is the B-module generated by dx with the relation $d(x^{p^n} - 1) = 0$. That is, $\Omega_{B/A}$ is the B-module

$$A[x]/(x^{p^n} - 1, p^n x^{p^n - 1}) = A[x]/(x^{p^n} - 1, p^n). \tag{7.16}$$

It is free over B iff $p^n = 0$ in A.

We conclude that the schemes $T_{m,n}$ with $m \leq n$ are special in that the level structure

$$\beta_{p^n} : \mu_{p^n/R} \to A_{/R}, \qquad R \in W_m\text{-}\mathbf{Alg}, \tag{7.17}$$

induces an isomorphism

$$\mathfrak{t}^*_{\mu_{p^n}/R} \cong \mathfrak{t}^*_{A/R}. \tag{7.18}$$

In particular, *there exists a canonical modular form of weight* 1, $a(1)$ *on $T_{m,n}$ for $m \leq n$ corresponding to the image of the canonical generator dx/x of $\mathfrak{t}^*_{\mu_{p^n}/R}$.*

EXERCISE 7.3. *Prove that under the maps*

$$T_{m,n+1} \to T_{m,n} \quad m \leq n \tag{7.19}$$

and

$$T_{m,n} \to T_{m+1,n} \quad m+1 \leq n \tag{7.20}$$

the modular forms $a(1)$ *agree. Conclude that there exists a canonical modular form of weight* 1 *on the schemes* $T_{m,\infty}$ *and* $T_{\infty,\infty}$.

We let

$$a(i) := a(1)^i, \tag{7.21}$$

denote the modular form of weight i on the scheme $T_{m,n}$ with $m \leq n$ (including $m = \infty$).

REMARK 7.4. Using the modular forms $a(j)$ one may realize any space of classical (resp. p-adic) modular forms $\mathcal{M}(W_m(k), j, \mu_{p^n}) = \Gamma(M_m, \underline{\omega}^k)$ (respectively, $\mathbb{M}(W_m(k), j, \mu_{p^n}; r)$, any r) inside $V_{m,\infty}$ for $m \leq \infty$.

Indeed, given a modular form f in one of the spaces above, pull it back via the morphism $T_{m,\infty} \to T_{m,n}$. Denote the pull back still by f and take the function $f/a(j)$. In terms of closed points, say, we may also describe that process as follows. Given $x \in T_{m,\infty}(\bar{k})$ let E_x be the associated elliptic curve and $\{\beta_{p^t}\}_{t\geq 0}$ the compatible level structures. Taking any $t \geq m$ we obtain a canonical non-vanishing differential ω_x. If f is classical, the value $\big(f/a(j)\big)(x)$ *is* $f(E_x, \omega_x, \beta_{p^n,x})$. If f is p-adic, since E_x has ordinary reduction the value $E_{p-1}(E_x, \omega_x)$ is invertible and there is a unique Y_x such that $(E_x, \omega_x, \beta_{p^n,x}, Y_x)$ is a test object. The value $\big(f/a(j)\big)(x)$ *is* $f(E_x, \omega_x, \beta_{p^n,x}, Y_x)$.

Finally, note that for every fixed weight j the map $\mathcal{M}(W_m(k), j, \mu_{p^n}) \to T_{m,\infty}$ (resp. $\mathbb{M}(W_m(k), j, \mu_{p^n}) \to T_{m,\infty}$) is injective. Thus, the function spaces $V_{m,\infty}$ appear as natural enveloping shells for the spaces of modular forms.

For every m let

$$R_m = \bigoplus_{k\geq 0} \Gamma(M_m, \underline{\omega}^k) \tag{7.22}$$

be the graded ring of classical modular forms on the scheme M_m ($\underline{\omega} = \pi_* \mathfrak{T}^*_{\mathcal{E}^U/M_m}$). Let

$$R_\infty = \bigoplus_{k\geq 0} \Gamma(M, \underline{\omega}^k), \tag{7.23}$$

be the graded ring of classical modular forms on the scheme $M = \mathfrak{M}(W(k), N)$.

Now, fix m and define a homomorphism

$$r_m \colon R_m \to V_{m,m} \hookrightarrow V_{m,\infty}, \tag{7.24}$$

as described in the remark above, by

$$r_m\Big(\sum f_i\Big) = \sum f_i/a(i), \quad f_i \in \Gamma(M_m, \underline{\omega}^i). \tag{7.25}$$

We define

$$r_\infty \colon R_\infty \to V_{\infty,\infty}, \tag{7.26}$$

by the composition

$$R_\infty \to \varprojlim_m R_m \xrightarrow{\varprojlim r_m} \varprojlim_m V_{m,m} = V_{\infty,\infty}. \tag{7.27}$$

Since the q-expansion of $a(1)$ is one at the standard cusp, the q expansion of $r_m(\sum f_i)$ is just $\sum f_i(q)$, where $f_i(q)$ is the q-expansion of f_i and it belongs to $W_m[\![q]\!]$.

Let

$$I_{m,m_1} \lhd R_m \tag{7.28}$$

be the ideal of R_m consisting of all sums $\sum f_i$ such that $\sum f_i(q) \equiv 0 \bmod p^{m_1}$ at the standard cusp.

LEMMA 7.5. *Let $\sum f_i \in I_{m,m_1}$, then the q expansion $\sum f_i(q)$ with respect to* any *cusp satisfies $\sum f_i(q) \equiv 0 \bmod p^{m_1}$.*

PROOF. We are given that $r_m(p^{m-m_1} \sum f_i)$, which is a *function* on the irreducible scheme $T_{m,m}$, is expressed in the local ring a certain point (i.e., the one belonging to the standard cusp) by zero. Therefore, the function $r_m(p^{m-m_1} \sum f_i)$ *is* zero, and hence is zero in every local ring on $T_{m,m}$. In particular, it is zero in the local ring of any other cusp. That is, all its q-expansions are zero. □

COROLLARY 7.6. *Let $\sum f_i \in I_{m,m_1}$. Then for all $a \in (\mathbb{Z}/p^m\mathbb{Z})^\times$*

$$\sum a^i f_i(q) \equiv 0 \mod p^{m_1}. \tag{7.29}$$

PROOF. The Galois action of $(\mathbb{Z}/p^m\mathbb{Z})^\times$ on $V_{m,m}$ is best described on points: The effect of $a \in (\mathbb{Z}/p^m/Z)^\times$ is $(E, \beta_{p^n}) \mapsto (E, \beta_{p^n} \circ a)$. As in the proof of Theorem 5.4 one verifies that r_m is equivariant with respect to this action when we let a act on a modular form of weight i by a^i. □

Let $\sum f_i \in I_{m,m_1}$ $(m_1 \leq m)$. Then $p^{m-m_1} r_m(\sum f_i)$ is zero in V_{m,m_1}. Since $W_m[\![q]\!]/V_{m,m_1}$ is a flat W_m-module, there exists an $h \in V_{m,m_1}$ such that

$$r_m\left(\sum f_i\right) = p^{m_1} h. \tag{7.30}$$

The function h is unique modulo p^{m-m_1} and we obtain a well-defined homomorphism

$$\gamma_{m_1} : I_{m,m_1} \to V_{m-m_1,m} \subset V_{m-m_1,\infty}. \tag{7.31}$$

We write symbolically,

$$\gamma_{m_1} = \frac{1}{p^{m_1}} r_m. \tag{7.32}$$

Passing to the limit, we get a homomorphism

$$\gamma_{m_1} : I_{\infty,m_1} \to V_{\infty,\infty}. \tag{7.33}$$

If $\sum f_i \in I_{\infty,m_1}$ then $\sum f_i \in R_\infty = \bigoplus_k \Gamma(M, \underline{\omega}^k)$ and $\sum f_i(q) \equiv 0 \pmod{p^{m_1}}$; furthermore,

$$\left(\gamma_{m_1}\left(\sum f_i\right)\right)(q) = \frac{1}{p^{m_1}} \sum f_i(q). \tag{7.34}$$

We have the following diagram

$$\begin{array}{ccccccccc} R_\infty & \longrightarrow & \frac{1}{p}I_{\infty,1} & \longrightarrow & \frac{1}{p^2}I_{\infty,2} & \longrightarrow & \cdots & \longrightarrow & D \subset R_\infty[1/p] \\ {\scriptstyle r_\infty}\downarrow & & \downarrow{\scriptstyle p\cdot\gamma_1} & & \downarrow{\scriptstyle p^2\cdot\gamma_2} & & \cdots & & \downarrow{\scriptstyle \beta} \\ V_{\infty,\infty} & \longrightarrow & V_{\infty,\infty} & \longrightarrow & V_{\infty,\infty} & \longrightarrow & \cdots & \longrightarrow & V_{\infty,\infty}. \end{array} \tag{7.35}$$

(Horizontal arrows being the natural inclusions). The $\mathbb{Z}_p$-algebra D is precisely the elements in $R_\infty[1/p]$ with p-integral q-expansion. It is called *the ring of divided congruences.* Its significance can be seen from the next beautiful theorem (the reader should bear in mind Remark 7.4).

THEOREM 7.7 (Katz). *For every m, the map*

$$\beta(m)\colon D/p^m D \to V_{\infty,\infty}/p^m V_{\infty,\infty} = V_{m,\infty}, \tag{7.36}$$

induced from β is an isomorphism.

PROOF. • $\beta(m)$ is injective.

Let $\sum f_i \in p^{-n} I_{\infty,n}$ be in the kernel of $\beta(m)$. Then $\sum f_i(q) \equiv 0 \pmod{p^m}$. Therefore, $\sum f_i \in p^{-n} I_{\infty,n+m} = p^m(p^{-n-m} I_{\infty,n+m}) \subset p^m D$.

• $\beta(m)$ is surjective.

We claim that if $\beta(1)\colon D/pD \to V_{1,\infty}$ is surjective then $\beta(m)\colon D/p^m D \to V_{m,\infty}$ is surjective. Indeed, if $\beta(1)\colon D/pD \to V_{1,\infty}$ is surjective and $x \in V_{m,\infty}$, there exists an element $g \in D$ such that $\beta(m)(g) \equiv x \bmod pV_{m,\infty}$. Therefore $\beta(m)(g) - x = px_1$ for some $x_1 \in V_{m,\infty}$, etc. The process stops because p is nilpotent in $V_{m,\infty}$. It remains to show that

• $\beta(1)$ is surjective.

We know that the composition

$$R_\infty \to V_{1,\infty} \to V_{1,1} \tag{7.37}$$

is surjective (see the proof of Theorem 5.4). We shall abuse notation and denote this map by $\beta(1)$ as well.

Artin–Schreier theory says the following: Let A be a ring of characteristic p (e.g., $A = V_{1,n}$) and $B \supset A$ a finite étale A algebra with Galois group $\mathbb{Z}/p\mathbb{Z}$ (e.g., $B = V_{1,n+1}$), then there exists $b \in B$ such that: (i) $b^p - b \in A$; (ii) $\ell \in \mathbb{Z}/p\mathbb{Z}$ acts by $b \mapsto b + \ell$. Such an element b is unique up to addition of elements from A. Note that any element $b \in B$ such that $b \mapsto b+1$ under some element of $\mathbb{Z}/p\mathbb{Z}$ generates B as an A-module. We call such an element an *Artin–Schreier generator* for B over A. See [**99**, III, §4].

We next note that the group $\mathbb{Z}_p^\times$ acts on D. First, $\mathbb{Z}_p^\times$ acts on $R_\infty[1/p]$ by

$$[a]\Big(\sum f_i\Big) = \sum a^i f_i. \tag{7.38}$$

By Corollary 7.6 the action preserves D.

LEMMA 7.8 (Key Lemma). *Let $n \geq 1$. There exists an element $d_n \in D$ such that for every $k \geq 0$*

$$[1 + p^{n+k}](d_n) \equiv d_n + p^k E_{p-1} \bmod p^{k+1} D. \tag{7.39}$$

We first explain how the theorem follows from the Key Lemma. We know that

$$\beta(1)\colon R_\infty \twoheadrightarrow V_{1,1}. \tag{7.40}$$

It follows that for $n \geq 1$ we have

$$V_{1,n+1} = V_{1,n}[\beta(1)(d_n)], \tag{7.41}$$

where d_n is the element provided by the Key Lemma. Indeed,

$$[1+p^{n+1}](d_n) \equiv d_n + pE_{p-1} \bmod p^2 D \tag{7.42}$$

and

$$d_n + pE_{p-1} \equiv d_n \bmod pD, \tag{7.43}$$

whence $\beta(1)(d_n) \in V_{1,n+1}$ (because $\mathrm{Aut}(V_{1,n+1}/V_{1,n}) = (1+p^n\mathbb{Z}_p)/(1+p^{n+1}\mathbb{Z}_p)$). Similarly,

$$[1+p^n](d_n) \equiv d_n + E_{p-1} \equiv d_n + 1 \bmod pD. \tag{7.44}$$

Thus, d_n is an Artin–Schreier generator for $V_{1,n+1}$ over $V_{1,n}$ and (7.41) follows. Therefore, the map

$$\beta(1)\colon R_\infty \to V_{1,n} \tag{7.45}$$

is surjective for every n and that implies that

$$\beta(1)\colon R_\infty \to V_{1,\infty} \tag{7.46}$$

is surjective because R_∞ is p-adically complete. □

PROOF OF KEY LEMMA. The proof is by induction on n. For $n = 1$ take

$$d_1 = \frac{1-E_{p-1}}{p}. \tag{7.47}$$

We calculate that for a suitable integer c

$$\begin{aligned} [1+p^{1+k}](d_1) &= [1+p^{1+k}]\left(\frac{1-E_{p-1}}{p}\right) \\ &= \frac{1-(1+p^{1+k})^{p-1}E_{p-1}}{p} \\ &= \frac{1-E_{p-1}}{p} - \frac{(p-1)p^{1+k}E_{p-1}}{p} - \frac{p^{2+2k}}{p}\cdot c\cdot E_{p-1} \\ &= d_1 + p^k E_{p-1} \bmod p^{k+1}D. \end{aligned} \tag{7.48}$$

Suppose that $d_1, \dots, d_n$ were constructed. As noted, for $\ell \leq n$

$$V_{1,\ell+1} = V_{1,\ell}[\beta(1)d_\ell], \quad (\beta(1)d_\ell) - (\beta(1)d_\ell)^p \in V_{1,\ell}. \tag{7.49}$$

Thus,

$$V_{1,n} = \beta(1)(R_\infty[d_1, \dots, d_{n-1}]), \tag{7.50}$$

and

$$(\beta(1)d_n) - (\beta(1)d_n)^p = \beta(1)c_n \tag{7.51}$$

for some $c_n \in R_\infty[d_1, \dots, d_{n-1}]$. Hence, $d_n - d_n^p - c_n \in \mathrm{Ker}\big(\beta(1)\big) = pD$. We let

$$d_{n+1} = \frac{d_n - d_n^p - c_n}{p}. \tag{7.52}$$

It is an element of D. We check the Galois action. First:

$$[1+p^{(n+1)+k}](d_{n+1}) = \frac{[1+p^{(n+1)+k}](d_n) - \left([1+p^{(n+1)+k}](d_n)\right)^p - [1+p^{(n+1)+k}](c_n)}{p}. \tag{7.53}$$

On the other hand, by induction,

- $[1+p^{n+(1+k)}](d_n) \equiv d_n + p^{k+1}E_{p-1} \bmod p^{k+2}D$.
- $\left([1+p^{n+(1+k)}](d_n)\right)^p \equiv d_n^p \bmod p^{k+2}D$.
- $[1+p^{n+(1+k)}](c_n) \equiv c_n \bmod p^{k+2}D$ (true for any $c \in R_\infty[d_1, \dots, d_n]$).

Put together this yields

$$[1+p^{(n+1)+k}](d_{n+1}) \equiv \frac{d_n - d_n^p - c_n + p^{k+1}E_{p-1}}{p} \tag{7.54}$$

$$\equiv d_{n+1} + p^k E_{p-1} \bmod p^{k+1}D. \tag{7.55}$$

□

COROLLARY 7.9. *The ring of divided congruences D is p-adically dense in the space of Katz's functions $V_{\infty,\infty}$.*

CHAPTER 5

p-adic Hilbert Modular Forms

As in Chapter 4 our interest is in a p-adic theory of modular forms and we follow the same conceptual frame work presented there. However, we shall not be able to present a theory as complete as for elliptic curves simply because the theory is not yet worked out!

The motivation for studying p-adic Hilbert modular forms includes that given in Chapter 4, i.e., the p-adic interpolation of L-functions associated to totally real fields (see Cassou-Noguès [**8**] and Deligne–Ribet [**27**]) and the theory of deformation of Galois representations of a totally real field.

When developing the theory of p-adic Hilbert modular forms one must abandon the hope of using specific knowledge of the ring of modular forms over $\mathbb{C}$ and using "down to earth arguments" as, say, in Chapter 4, §2. That distinguishes the case of $g > 1$ from the elliptic case. On a conceptual level this is expected: while for $g = 1$ there is "one modular curve" (i.e., up to correspondences; or $\varprojlim \Gamma(n)\backslash\mathfrak{H}$), for $g > 1$ there is "one modular variety" (in the same sense) for every totally real field of degree g over $\mathbb{Q}$.

For simplicity of exposition many statements are given under the condition of μ_N-level structure for $N \geq 4$, for schemes over $\mathcal{O}_{\tilde{L}}[d_L^{-1}]$ (where $\tilde{L}$ is a fixed Galois closure of L), and for modular forms of parallel weight.

We begin our study of p-adic modular forms using Katz's approach of defining them as functions on test objects that transform in a certain way that depends on the weight. Such modular forms have q-expansions and we discuss this concept in greater depth than before. The theory of q-expansions is perhaps the most technically demanding concept in this subject and we have tried to discuss this topic in sufficient detail to allow the reader an understanding of the ideas and difficulties involved, without overly immersing in the deep technical issues of this theory.

One basically thinks of q-expansions as special values of modular forms. One may complete the fine moduli scheme $\mathfrak{M}$ over $\mathbb{Z}$ parameterizing abelian varieties with real multiplication and sufficient level structure to a proper scheme $\mathfrak{M}^*$ over $\mathbb{Z}$ by "adding" finitely many sections called cusps. Over the complex numbers we get the compactified quotients of $\mathcal{H}^g$ by suitable discrete subgroups of $\mathrm{PGL}_2(L)^+$ that were already mentioned in Chapter 2, Section 2.2. The cusps correspond to orbits of these discrete subgroups in $\mathbb{P}^1(L)$.

Alternately, one may obtain this compactification by the method of Bailly–Borel, namely, by the taking $\mathbf{Proj}\big(\bigoplus_{k\geq 0}\Gamma(\mathfrak{M}, (\det\underline{\omega})^{\otimes k})\big)$ as the compactification of $\mathfrak{M}$, where $\det\underline{\omega}$ is the determinant of the relative cotangent bundle at the origin of the universal object over $\mathfrak{M}$, known to be an ample line bundle.

Ideally, there could have been a universal object over the compactified moduli scheme $\mathfrak{M}^*$, perhaps a universal family of semi-abelian varieties with real multiplication and appropriate level structure. This is so in the one-dimensional case where we deal with elliptic curves. However, this doesn't seem to exist in the higher-dimensional cases. This makes the theory more complicated. In the ideal situation, we would have had a semi-abelian variety with real multiplication over the formal neighborhood $U = \mathrm{Spf}\, R$ of every cusp, and we could have evaluated every modular form at this semi-abelian variety, obtaining an expression lying in R. Instead, the reality is that we must find many semi-abelian varieties over bases U' that lie over the formal neighborhood U of a cusp and evaluate modular forms at these objects. It turns out that one still gets a well-defined value in R.

The Bailly–Borel (or Satake, or minimal) compactification has other deficiencies besides the lack of universal object. For example, though one still has a normal scheme (at least over $\mathbb{Z}[d_L^{-1}]$) the singularities at the cusps are quite complicated. One may thus prefer to consider the toroidal compactifications of the moduli scheme as in [**9, 39**]. These are not canonical in any way. Every such compactification maps surjectively onto the minimal compactification. It has the advantage of being smooth over the base (if the polyhedral decomposition used to define it is fine enough), and there exists a semi-abelian variety over every such smooth toroidal compactification. In fact, the toroidal compactifications are used to prove the properties of the q-expansion map.

Sections 3 and 4 deal with the definition and properties of partial Hasse invariants and the kernel of the q-expansion map in characteristic p. The partial Hasse invariants $h_1, \dots, h_g$ are Hilbert modular in characteristic p that have non-parallel weight and q-expansion equal to one at every cusp. Therefore, they cannot be lifted to characteristic zero (Chapter 2, Proposition 3.6). The kernel of the q-expansion map is generated by $h_1 - 1, \dots, h_g - 1$. The product $H = h_1 \cdots h_g$ is the total Hasse invariant—the determinant, up to a sign, of the Hasse–Witt matrix. In Section 5 we give some applications of the theory q-expansion to special values of zeta function of totally real fields and to filtration on Hilbert modular forms.

Section 6 deals with the p-adic theory of Hilbert modular forms. We restrict our attention to modular forms of parallel weight. One satisfactory result is Corollary 6.14 that asserts that p-adic Hilbert modular forms of growth coefficient 1 and parallel weight (Katz's approach) are the same as taking p-adic uniform limit of classical Hilbert modular forms of parallel weight (Serre's approach).

1. Algebraic Hilbert modular forms

Let $N \geq 4$ be an integer. Let

$$\mathfrak{M}(\mu_N) \to \mathrm{Spec}(\mathcal{O}_L[d_L^{-1}]) \tag{1.1}$$

be the fine moduli scheme parameterizing polarized abelian varieties with real multiplication and μ_N-level structure $\underline{A} = (A, \iota, \lambda, \beta_N)$ satisfying condition (**R**) (it exists in fact over $\mathbb{Z}$). See Chapter 3, §5. We let $\mathfrak{M}^*(\mu_N)$ be its minimal compactification "at infinity" as in the introduction to this chapter.

We recall that one fixes representatives $(\mathfrak{a}, \mathfrak{a}^+)$ for $\mathrm{Cl}(L)^+$. The scheme $\mathfrak{M}(\mu_N)$ represents the functor of $\mathcal{O}_L[d_L^{-1}]$-algebras taking a scheme S to isomorphism classes of quadruples $(A, \iota, \lambda, \beta_N)/S$ consisting of:

- an abelian scheme $A \to S$ of relative dimension g;
- a ring embedding $\iota\colon \mathcal{O}_L \hookrightarrow \mathrm{End}_S(A)$;

- an isomorphism $\lambda\colon (\mathcal{M}_A, \mathcal{M}_A^+) \to (\mathfrak{a}, \mathfrak{a}^+)$ of sheaves in the étale topology (where $\mathcal{M}_A = \mathrm{Hom}_{\mathcal{O}_L}(A, A^t)^{\mathrm{sym}}$);
- a μ_N-level structure $\beta_N\colon \mu_{N/S} \otimes_{\mathbb{Z}} \mathcal{D}_{L/\mathbb{Q}}^{-1} \hookrightarrow A$ ($\mathcal{O}_L$-equivariant embedding of group schemes over S).

One imposes the condition that $\mathfrak{t}_{A/S}^*$ is a locally free $\mathcal{O}_L \otimes \mathcal{O}_S$-module of rank 1.

We remark that the morphism $\mathfrak{M}(\mu_N) \to \mathrm{Spec}(\mathcal{O}_L[d_L^{-1}])$ is smooth of relative dimension g and the components of the geometric fibers are parameterized by $\mathrm{Cl}(L)^+$. One may also consider the moduli of triples $(A, \iota, \beta_N)/S$ with locally free cotangent space, such that there exists an isomorphism $(\mathcal{M}_A, \mathcal{M}_A^+) \to (\mathfrak{a}, \mathfrak{a}^+)$ locally étale, but without actually fixing it. This functor is coarsely representable by a scheme over $\mathcal{O}_L[d_L^{-1}]$, which may not be smooth over $\mathcal{O}_L[d_L^{-1}]$.

For any scheme $f\colon B \to \mathrm{Spec}(\mathcal{O}_L[d_L^{-1}])$ let

$$\mathfrak{M}(B, \mu_N) = \mathfrak{M}(\mu_N) \times_{\mathrm{Spec}(\mathcal{O}_L[d_L^{-1}])} B \to B, \tag{1.2}$$

and similarly for $\mathfrak{M}^*(\mu_N)$. The scheme $\mathfrak{M}(B, \mu_N)$ is a fine moduli scheme for abelian varieties with real multiplication and μ_N-level structure over base schemes $S \in \mathbf{Sch}_{/B}$. If $B = \mathrm{Spec}(R)$ we just write $\mathfrak{M}(R, \mu_N)$. If we need to make the dependence on L explicit, we shall write $\mathfrak{M}_L(\mu_N)$, $\mathfrak{M}_L(B, \mu_N)$, etc.

Let

$$\pi\colon \underline{A}^U \to \mathfrak{M}(\mu_N) \tag{1.3}$$

be the universal object and let $\underline{A}^B$ be the induced universal object over $\mathfrak{M}(B, \mu_N)$. We let

$$\underline{\omega} = \mathfrak{t}^*_{\underline{A}^U/\mathfrak{M}(\mu_N)} \tag{1.4}$$

be the sheaf over $\mathfrak{M}(\mu_N)$ of relative cotangent spaces at the origin. Similarly, for $f\colon B \to \mathrm{Spec}(\mathcal{O}_L[d_L^{-1}])$ we let

$$\underline{\omega}_B = \mathfrak{t}^*_{\underline{A}^B/B} = f^*\underline{\omega}. \tag{1.5}$$

When there is no danger of confusion we write $\underline{\omega}$ for $\underline{\omega}_B$.

The sheaf $\underline{\omega}_B$ is a locally free $\mathcal{O}_B$-module of rank g and is a locally free $\mathcal{O}_L \otimes \mathcal{O}_B$-module of rank one.

Let B be an $\mathcal{O}_L[d_L^{-1}]$-algebra. In order to define weights for Hilbert modular forms we introduce the following functor $\mathbb{T}_B$. It is a functor from B-algebras to abelian groups given by

$$R \mapsto \mathbb{T}_B(R) = (\mathcal{O}_L \otimes_{\mathbb{Z}} R)^\times, \qquad R \in B\text{-}\mathbf{Alg}. \tag{1.6}$$

Then $\mathbb{T}_B$ is a group scheme over $\mathrm{Spec}(B)$, which is in fact a torus.[1] That is, for every geometric point $\mathrm{Spec}(k) \to \mathrm{Spec}(B)$ the scheme $\mathbb{T}_B \times_{\mathrm{Spec}(B)} \mathrm{Spec}(k)$, which represents the functor

$$R \mapsto (\mathcal{O}_L \otimes R)^\times \cong \bigoplus_{i=1}^{g} R^\times, \qquad R \in k\text{-}\mathbf{Alg}, \tag{1.7}$$

is isomorphic to $\mathbb{G}_{m/k}^g$.

[1] In fact $\mathbb{T}_B = (\mathbf{Res}_{\mathcal{O}_L/\mathbb{Z}}\, \mathbb{G}_m) \times_{\mathrm{Spec}(\mathbb{Z})} \mathrm{Spec}(B)$.

Let $\tilde{L}$ be a Galois closure of L. If we take B to be an $\mathcal{O}_{\tilde{L}}[d_L^{-1}]$-algebra, where $\mathcal{O}_{\tilde{L}}$ is the ring of integers of $\tilde{L}$, then $\mathbb{T}_B$ is a *split* torus: if we enumerate the embeddings of L into $\tilde{L}$ by

$$\sigma_1, \dots, \sigma_g \colon L \hookrightarrow \tilde{L}, \tag{1.8}$$

then

$$\mathcal{O}_L \otimes B = \bigoplus_{i=1}^{g} B, \tag{1.9}$$

under the map induced from

$$a \otimes 1 \mapsto \big(\sigma_1(a), \dots, \sigma_g(a)\big), \tag{1.10}$$

and we get a canonical isomorphism

$$\mathbb{T}_B = \mathbb{G}_{m/B}^g. \tag{1.11}$$

Assume henceforth that B is an $\mathcal{O}_{\tilde{L}}[d_L^{-1}]$-algebra. The character group $X(\mathbb{T}_B)$ is given by the free abelian group (written multiplicatively)

$$X(\mathbb{T}_B) = \langle \chi_1, \dots, \chi_g \rangle \tag{1.12}$$

on the characters $\chi_1, \dots, \chi_g$, where χ_i is the projection of $\mathbb{T}_B$ on the i-th component

$$\mathbb{T}_B = \mathbb{G}_{m/B}^g \xrightarrow{\chi_i} \mathbb{G}_{m/B}. \tag{1.13}$$

We are ready to define modular forms. Over $\mathfrak{M}(B, \mu_N)$ the vector bundle $\underline{\omega}$ decomposes as a direct sum of line bundles

$$\underline{\omega} = \underline{\omega}(\chi_1) \oplus \cdots \oplus \underline{\omega}(\chi_g). \tag{1.14}$$

This decomposition comes from Equation (1.9), $\mathcal{O}_L \otimes B = \bigoplus_{i=1}^g B$, where we also denote the i-th projection from $\mathcal{O}_L \otimes B$ to B by

$$\chi_i \colon \mathcal{O}_L \otimes B \to B. \tag{1.15}$$

Note that $\chi_i(a \otimes 1) = \sigma_i(a)$ as an element of B. Thus, $\mathcal{O}_L$ acts on $\underline{\omega}(\chi_i)$ via the homomorphism $\chi_i \colon \mathcal{O}_L \otimes B \to B$. We define for $\chi = \chi_1^{a_1} \cdots \chi_g^{a_g}$

$$\underline{\omega}(\chi) = \underline{\omega}(\chi_1)^{\otimes a_1} \otimes \cdots \otimes \underline{\omega}(\chi_g)^{\otimes a_g}. \tag{1.16}$$

It is a line bundle over $\mathfrak{M}(B, \mu_N)$. We remark that if $\underline{\omega}$ is defined by a cocycle $\xi \in H^1(\mathfrak{M}(B, \mu_N), (\mathcal{O}_L \otimes \mathcal{O}_{\mathfrak{M}(B,\mu_N)})^\times)$ then $\underline{\omega}(\chi)$ is defined by the image of ξ under the map

$$H^1(\mathfrak{M}(B, \mu_N), (\mathcal{O}_L \otimes \mathcal{O}_{\mathfrak{M}(B,\mu_N)})^\times) \xrightarrow{\chi} H^1(\mathfrak{M}(B, \mu_N), \mathcal{O}_{\mathfrak{M}(B,\mu_N)}^\times). \tag{1.17}$$

This process is called "change of structural group". Note that the same construction applies for every abelian scheme $\underline{A} \to S$ (where $S \to \operatorname{Spec}(\mathcal{O}_{\tilde{L}}[d_L^{-1}])$) with real multiplication and μ_N-level structure. Indeed, via the classifying morphism $S \to \mathfrak{M}(\mu_N)$, $\underline{A}$ *is* $\underline{A}^S$. If we wish to make the dependence on $\underline{A}$ clear, we shall write $\underline{\omega}_{\underline{A}}(\chi)$.

DEFINITION 1.1. Let B be an $\mathcal{O}_{\tilde{L}}[d_L^{-1}]$-algebra. A *Hilbert modular form* f over B, of μ_N-level and weight $\chi \in X(\mathbb{T}_B)$ is a section of $\underline{\omega}_B(\chi)$. We denote by

$$\mathcal{M}(B, \chi, \mu_N) \tag{1.18}$$

the space of Hilbert modular forms over B of weight χ and level μ_N.

Equivalently,

DEFINITION 1.2. Let B be an $\mathcal{O}_{\tilde{L}}[d_L^{-1}]$-algebra. A *Hilbert modular form* f over B, of μ_N-level and weight $\chi \in X(\mathbb{T}_B)$ is a rule

$$\underline{A} \mapsto f(\underline{A}) \in \underline{\omega}_{\underline{A}}(\chi), \tag{1.19}$$

commuting with base change and depending only on the isomorphism class of $\underline{A}$.

Let R be an $\mathcal{O}_{\tilde{L}}[d_L^{-1}]$-algebra and assume that $\underline{\omega}_{\underline{A}}$ is a free $\mathcal{O}_L \otimes R$ module of rank 1. A generator $\omega \in \underline{\omega}_{\underline{A}}$ is called a *non-vanishing differential.* We may equivalently define a Hilbert modular form as follows:

DEFINITION 1.3. Let B be an $\mathcal{O}_{\tilde{L}}[d_L^{-1}]$-algebra. A *Hilbert modular form* f over B, of μ_N-level and weight $\chi \in X(\mathbb{T}_B)$ is a rule

$$(\underline{A}, \omega)_{/R} \mapsto f(\underline{A}, \omega) \in R \tag{1.20}$$

(R a B-algebra, ω a non-vanishing differential), commuting with base change and depending only on the isomorphism class of $(\underline{A}, \omega)_{/R}$, such that

$$f(\underline{A}, \alpha^{-1}\omega) = \chi(\alpha) f(\underline{A}, \omega), \quad \alpha \in (\mathcal{O}_L \otimes R)^\times = \mathbb{T}_B(R). \tag{1.21}$$

REMARK 1.4. If $B = \mathbb{C}$, a Hilbert modular form of weight $\chi = \chi_1^{a_1} \cdots \chi_g^{a_g}$ is a complex Hilbert modular form of weight $(a_1, \ldots, a_g)$. See Chapter 3, p. 65. As a matter of notation we use

$$\mathrm{Norm} = \chi_1 \cdots \chi_g. \tag{1.22}$$

Thus, a Hilbert modular form of parallel weight k is of weight Norm^k in this terminology.

DEFINITION 1.5. Let $\mathfrak{c}$ be a fractional ideal of $\mathcal{O}_L$ endowed with its natural notion of positivity. A $\mathfrak{c}$-*polarized Hilbert modular form* f over B of level μ_N and weight χ is equivalently: (i) a section of $\underline{\omega}(\chi)$ over the component of the moduli space $\mathfrak{M}(B, \mu_N)$ corresponding to the polarization module $(\mathfrak{c}, \mathfrak{c}^+)$; (ii) a rule associating to a $\mathfrak{c}$-polarized abelian variety $\underline{A}/R$ and a non-vanishing differential ω on A a value $f(\underline{A}, \omega) \in R$ satisfying the same conditions as in Definition 1.3 above.

By abuse of notation, we shall denote also the B-module of $\mathfrak{c}$-polarized Hilbert modular forms over B of level μ_N and weight χ by $\mathcal{M}(B, \chi, \mu_N)$.

2. Tate objects and the q-expansion

The purpose of this section is to define and state the properties of the q-expansions of a Hilbert modular form over arbitrary base rings. These q-expansions are obtained as "special values" of the Hilbert modular form at particular abelian varieties with real multiplication—the *Tate objects.* For simplicity of exposition we shall treat only the case of parallel weight. The analogous assertions for non-parallel weight are correct with the necessary evident modifications, e.g., the base rings should be over the ring of integers of a Galois closure of L (so that the different weights are defined, etc.).

Given a closed point $x \in \mathfrak{M}(B, \mu_N)$ and a choice of local parameters $t_1, \ldots, t_g$ at x such that $\widehat{\mathcal{O}}_x \cong B[\![t_1, \ldots, t_g]\!]$ and a trivialization of $\mathfrak{t}^*_{\underline{A}^U/\mathfrak{M}(B,\mu_N)}$ around x, every modular form can be expressed uniquely as an element of $\widehat{\mathcal{O}}_x$ by a Taylor series expansion. Moreover, this expansion is the *value* of the modular form at $\underline{A}^U|_{\widehat{\mathcal{O}}_x}$. The q-expansion is obtained by performing a similar process around a cusp. That is, in its essence it is nothing more than a Taylor expansion. Alas, several problems arise:

- The cusps are not regular points of $\mathfrak{M}^*(B, \mu_N)$.
- The universal family $\underline{A}^U$ over $\mathfrak{M}(B, \mu_N)$ does not extend to a family of *abelian varieties* over $\mathfrak{M}^*(B, \mu_N)$ and, in particular, there does not exist an abelian scheme with real multiplication over the completion of the local ring of a cusp that extrapolates the existing family over the "punctured" local ring.

Now, while for elliptic curves there is a totally satisfactory solution for this problem obtained by considering *generalized elliptic curves* (a family of generalized elliptic curves is allowed to have fibers that are a cyclic configuration of $\mathbb{P}^1$'s—an "N-gon". See [**26**] and Appendix B)—a concept flexible enough to yield a universal object $\mathcal{E}^U/\mathfrak{M}^*_{\mathbb{Q}}(B, \mu_N)$ that extends the universal family of elliptic curves $\mathcal{E}^U/\mathfrak{M}_{\mathbb{Q}}(B, \mu_N)$, when $g > 1$ no such solution seems to be available at this time. Instead we must consider a whole class of polarized semi-abelian schemes with real multiplication $\mathcal{A} \to \mathrm{Spec}(S)$ over base schemes S, such that S is a normal local Noetherian ring; the family is abelian outside the unique closed point s_0; s_0 maps to a particular cusp of $\mathfrak{M}^*_L(B, \mu_N)$. We then *evaluate* any modular form (of parallel weight) at $\mathcal{A}|_{\mathrm{Spec}(S)-\{s_0\}}$ and obtain a value lying in a suitable localization of S. This value turns out to depend only on the cusp to which s_0 maps and "descends" to an element of the completion of the local ring of that cusp on $\mathfrak{M}^*_L(B, \mu_N)$.

We fixed a totally real field L of degree g over $\mathbb{Q}$, with ring of integers $\mathcal{O}_L$ and inverse different $\mathcal{D}^{-1}_{L/\mathbb{Q}}$. Let $\mathfrak{c}$ be a fixed fractional ideal of L. We think of $\mathfrak{c}$, or rather on its class in $\mathrm{Cl}(L)^+$, as defining a module with a notion of positivity and therefore a component of $\mathfrak{M}(B, \mu_N)$.

Let $\mathfrak{a}, \mathfrak{b}$ be two fractional ideals of L such that

$$\mathfrak{c} = \mathfrak{a}\mathfrak{b}^{-1}. \tag{2.1}$$

Recall (Chapter 2, §2.2) that over $\mathbb{C}$, $\mathfrak{c}$-polarized abelian varieties with real multiplication come from the lattice $\mathfrak{b} \oplus (\mathcal{D}_L\mathfrak{a})^{-1}$ via various embeddings

$$\mathfrak{b} \oplus (\mathcal{D}_L\mathfrak{a})^{-1} \to \mathfrak{b} \cdot \tau \oplus (\mathcal{D}_L\mathfrak{a})^{-1} \cdot 1 \subseteq \mathbb{C}^g, \quad \tau \in \mathcal{H}^g. \tag{2.2}$$

Indeed, we computed that the polarization module is $(\mathcal{D}_L \cdot \mathfrak{b} \cdot (\mathcal{D}_L\mathfrak{a})^{-1})^{-1} = \mathfrak{c}$. The moduli space over $\mathbb{C}$ of $\mathfrak{c}$-polarized abelian varieties with real multiplication is isomorphic to[2]

$$\mathrm{SL}(\mathfrak{b} \oplus (\mathcal{D}_L\mathfrak{a})^{-1}) \backslash \mathcal{H}^g. \tag{2.3}$$

The group $\mathrm{SL}(\mathfrak{b} \oplus (\mathcal{D}_L\mathfrak{a})^{-1})$ contains the translations

$$\tau \mapsto \tau + \alpha, \quad \alpha \in (\mathcal{D}_L\mathfrak{a}\mathfrak{b})^{-1}. \tag{2.4}$$

Therefore, every modular form over $\mathbb{C}$ has a q-expansion with respect to the totally positive elements of $\mathfrak{a}\mathfrak{b} = \left((\mathcal{D}_L\mathfrak{a}\mathfrak{b})^{-1}\right)^\vee$:

$$f_{(\mathfrak{a},\mathfrak{b})_{\mathrm{an}}} = a_0 + \sum_{\alpha \in (\mathfrak{a}\mathfrak{b})^+} a_\alpha q^\alpha, \quad q^\alpha = e^{2\pi i \cdot \mathrm{Tr}(\alpha \cdot \tau)}. \tag{2.5}$$

[2]For any two ideals $\mathfrak{d}$, $\mathfrak{e}$ such that $(\mathcal{D}_L\mathfrak{d}\mathfrak{e})^{-1} = \mathfrak{c}$ we may take $\mathrm{SL}(\mathfrak{d} \oplus \mathfrak{e}) \backslash \mathcal{H}^g$. The advantage in choosing different groups is that the stabilizer of $i\infty$ in $\mathrm{SL}(\mathfrak{d} \oplus \mathfrak{e})$ contains $\mathfrak{d}^{-1}\mathfrak{e}$. Thus, by choosing properly the group we mod by we get rid of the need to work with other cusps beside $i\infty$.

The ring $\mathbb{Z}[\![q^\alpha : \alpha = 0 \text{ or } \alpha \in (\mathfrak{ab})^+]\!]$ (where $q^0 = 1$ and $q^\alpha q^\beta = q^{\alpha+\beta}$) is not Noetherian—a somewhat awkward situation for algebraic geometry. We thus consider "approximations" of this ring. Let

$$\Delta = \{\ell_1, \cdots, \ell_g\} \tag{2.6}$$

be a set of $\mathbb{Q}$-linear functionals on the g-dimensional rational vector space L that are independent and have the property

$$\ell_i(L^+) \subseteq \mathbb{Q}^+. \tag{2.7}$$

We say that an element $m \in L$ is Δ-*positive* if

$$\ell_i(m) \geq 0, \quad \forall i. \tag{2.8}$$

Given a lattice $M \subset L$ we let

$$M_{\Delta\text{-pos}} = \{m \in M : m \text{ is } \Delta\text{-positive}\}. \tag{2.9}$$

One observes that $M_{\Delta\text{-pos}}$ is a finitely generated monoid, whilst $M^+ = \{m \in M : m \in L^+\}$ is not in general. We also note that

$$M^+ = \bigcap_{\Delta} M_{\Delta\text{-pos}}. \tag{2.10}$$

EXERCISE 2.1. *Prove the assertions made on $M_{\Delta\text{-pos}}$ and M^+. Prove that a discrete monoid $M \subset L$ is finitely generated iff $M \otimes \mathbb{Q}^+$ is finitely generated over $\mathbb{Q}^+$.*

EXAMPLE 2.2. Let $L = \mathbb{Q}(\sqrt{2}) = \{a + b\sqrt{2} : a, b \in \mathbb{Q}\}$. Let $M = \mathcal{O}_L$. Let f and g be the following functionals

$$f(a + b\sqrt{2}) = a, \quad g(a + b\sqrt{2}) = a - b. \tag{2.11}$$

Let $\Delta = \{f, g\}$. Note that $(\mathcal{O}_L)_{\Delta\text{-pos}}$ is generated by $1 + \sqrt{2}$ and $-\sqrt{2}$, but $\mathcal{O}_L^+$ is not finitely generated. See Figure 2.1.

We define

$$\mathbb{Z}[\![M; \Delta]\!] = \left\{ \sum a_\alpha q^\alpha : a_\alpha \in \mathbb{Z}, \alpha \in M_{\Delta\text{-pos}} \right\}. \tag{2.12}$$

It is a ring under the rules

$$q^0 = 1, \quad q^\alpha q^\beta = q^{\alpha+\beta}. \tag{2.13}$$

Note that if $x_1, \dots, x_n$ are generators of $M_{\Delta\text{-pos}}$ then $\mathbb{Z}[\![M; \Delta]\!]$ is a quotient of $\mathbb{Z}[\![x_1, \dots, x_n]\!]$. Therefore, $\mathbb{Z}[\![M; \Delta]\!]$ is a local Noetherian ring.

EXERCISE 2.3. *Prove that $\mathbb{Z}[\![M; \Delta]\!]$ is complete with respect to the ideal I generated by $\{q^\alpha : \alpha \in M^+\}$. Prove that it is a normal domain.*

We let

$$\mathbb{Z}(\!(M; \Delta)\!) = \mathbb{Z}[\![M; \Delta]\!][U^{-1}], \quad U = \{q^\alpha : \alpha \in M^+\}. \tag{2.14}$$

EXERCISE 2.4. *Let α be an element of M such that $\ell_i(\alpha) > 0$ for all i. Prove that*

$$\mathbb{Z}(\!(M; \Delta)\!) = \mathbb{Z}[\![M; \Delta]\!][(q^\alpha)^{-1}]. \tag{2.15}$$

Prove also that

$$\mathbb{Z}((M;\Delta)) = \Big\{\sum_\alpha a_\alpha q^\alpha : a_\alpha \in \mathbb{Z}, \alpha \in M, \quad \exists n \text{ s.t. } a_\alpha = 0 \text{ unless } \ell_i(\alpha) \geq n, \forall i\Big\}. \tag{2.16}$$

EXERCISE 2.5. *Prove that the natural map $\alpha \mapsto \epsilon\alpha$ defines an action of $\mathcal{O}_L^{\times +}$ on the ring $\mathbb{Z}((M;\Delta))$. Note that it doesn't act on $\mathbb{Z}[\![M;\Delta]\!]$.*[3]

Consider the torus $\mathbb{T} = \mathbb{G}_m \otimes_{\mathbb{Z}} \mathcal{D}_{L/\mathbb{Q}}^{-1}\mathfrak{a}^{-1}$ over the ring $S = \mathbb{Z}((\mathfrak{ab};\Delta))$. Recall that for any $R \in S$-**alg** we have

$$\mathbb{T}(R) = R^\times \otimes_{\mathbb{Z}} \mathcal{D}_{L/\mathbb{Q}}^{-1}\mathfrak{a}^{-1} \tag{2.17}$$

and that characterizes $\mathbb{T}$. The character group of $\mathbb{T}$ is $\mathfrak{a} = (\mathcal{D}_{L/\mathbb{Q}}^{-1}\mathfrak{a}^{-1})^\vee$.

We claim that there exists a canonical $\mathcal{O}_L$-linear homomorphism

$$\underline{q}\colon \mathfrak{b} \to \mathbb{G}_m \otimes_{\mathbb{Z}} \mathcal{D}_{L/\mathbb{Q}}^{-1}\mathfrak{a}^{-1}. \tag{2.18}$$

To begin with, there is a given $\mathbb{Z}$-linear map

$$q\colon \mathfrak{ab} \to \mathbb{G}_m(S), \tag{2.19}$$

given by

$$q(\alpha) = q^\alpha. \tag{2.20}$$

[3]When dealing with non-parallel weight one needs to modify the action by introducing the $r_k(\epsilon)$ factors.

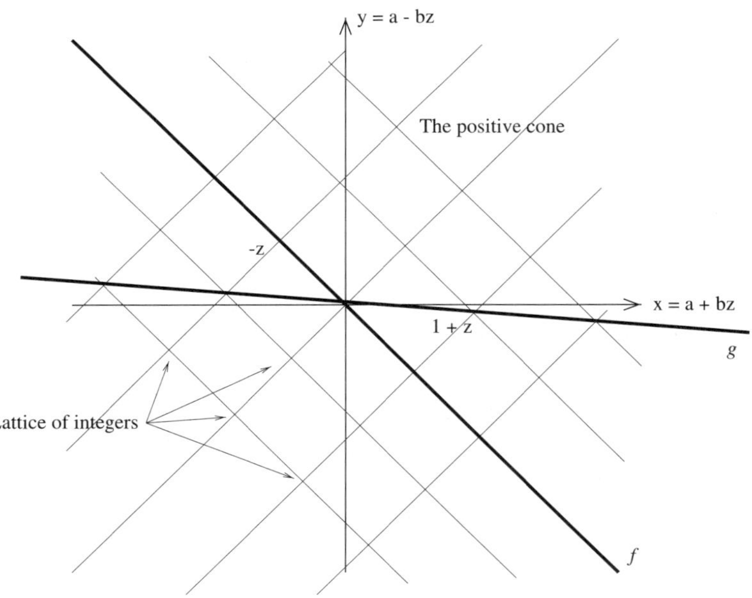

FIGURE 2.1. Δ-positive elements for $\mathbb{Q}(\sqrt{2})$ (z stands for $\sqrt{2}$).

Note that for every S-algebra R we get a (functorial) map $\mathfrak{ab} \to \mathbb{G}_m(R)$. We can define a unique $\mathcal{O}_L$-linear map,

$$\underline{q}\colon \mathfrak{ab} \to \mathbb{G}_m \otimes_{\mathbb{Z}} \mathcal{D}_{L/\mathbb{Q}}^{-1}, \tag{2.21}$$

having the property

$$(1 \otimes \operatorname{Tr}_{L/\mathbb{Q}})(\underline{q}(\alpha)) = q(\alpha) = q^{\alpha}, \tag{2.22}$$

where of course

$$1 \otimes \operatorname{Tr}_{L/\mathbb{Q}}\colon \mathbb{G}_m \otimes_{\mathbb{Z}} \mathcal{D}_{L/\mathbb{Q}}^{-1} \to \mathbb{G}_m \otimes_{\mathbb{Z}} \mathbb{Z} = \mathbb{G}_m. \tag{2.23}$$

Tensoring Equation (2.21) by $\otimes_{\mathcal{O}_L} \mathfrak{a}^{-1}$, we get the map

$$\underline{q}\colon \mathfrak{b} \to \mathbb{G}_m \otimes_{\mathbb{Z}} \mathcal{D}_{L/\mathbb{Q}}^{-1}\mathfrak{a}^{-1}. \tag{2.24}$$

FACT 2.6. The quotient $\mathbb{G}_m \otimes \mathcal{D}_{L/\mathbb{Q}}^{-1}\mathfrak{a}^{-1}/\underline{q}(\mathfrak{b})$, associating to a complete dvr $\phi\colon S \to R$ the abelian group $R^{\times} \otimes_{\mathbb{Z}} \mathcal{D}_{L/\mathbb{Q}}^{-1}\mathfrak{a}^{-1}/\phi \circ \underline{q}(\mathfrak{b})$, can be algebraized to an abelian scheme over the ring $S = \mathbb{Z}((\mathfrak{ab}; \Delta))$ with real multiplication by $\mathcal{O}_L$ that carries a canonical $\mathfrak{c} = \mathfrak{ab}^{-1}$-polarization. This abelian scheme is called a *Tate object* and we denote it by $\mathbf{T}_{\mathfrak{a},\mathfrak{b}}(\underline{q})$. The relative cotangent space is given by

$$\mathfrak{t}^*_{\mathbf{T}_{\mathfrak{a},\mathfrak{b}}(\underline{q})/S} \cong S\frac{dt}{t} \otimes_{\mathbb{Z}} \mathfrak{a}. \tag{2.25}$$

If $\mathfrak{a} = \mathcal{O}_L$ (and hence $\mathfrak{b} = \mathfrak{c}^{-1}$) the abelian scheme $\mathbf{T}_{\mathfrak{a},\mathfrak{b}}(\underline{q})$ carries a canonical differential

$$\omega_{\text{can}} = \frac{dt}{t} \otimes 1, \tag{2.26}$$

and a canonical μ_N-level for every N

$$\beta_{N,\text{can}}\colon \mu_N \otimes \mathcal{D}_{L/\mathbb{Q}}^{-1} \hookrightarrow \mathbb{G}_m \otimes \mathcal{D}_{L/\mathbb{Q}}^{-1}/\underline{q}(\mathfrak{c}^{-1}). \tag{2.27}$$

DEFINITION 2.7. The Tate objects $\mathbf{T}_{\mathcal{O}_L,\mathfrak{c}^{-1}}(\underline{q})$ over $S = \mathbb{Z}((\mathfrak{c}^{-1}; \Delta))$ are called *standard Tate objects* (or *standard cusps*). They come endowed with a canonical $\mathfrak{c}$-polarization, a canonical non-vanishing differential ω_{can}, and a canonical μ_N-level structure $\beta_{N,\text{can}}$ for every N. We shall denote such an object by $\mathbf{T}_{\mathfrak{c}}(\underline{q})$.

We note that

$$\begin{aligned} \mathbf{T}_{\mathfrak{a},\mathfrak{b}}(\underline{q}) \otimes_{\mathcal{O}_L} \mathfrak{c} &\cong \mathbb{G}_m \otimes_{\mathbb{Z}} \mathcal{D}_{L/\mathbb{Q}}^{-1}\mathfrak{a}^{-1}\mathfrak{ab}^{-1}/\underline{q}(\mathfrak{bab}^{-1}) \\ &\cong \mathbb{G}_m \otimes \mathcal{D}_{L/\mathbb{Q}}^{-1}\mathfrak{b}^{-1}/\underline{q}(\mathfrak{a}) \\ &= \mathbf{T}_{\mathfrak{b},\mathfrak{a}}(\underline{q}). \end{aligned} \tag{2.28}$$

That is, $\mathbf{T}_{\mathfrak{b},\mathfrak{a}}(\underline{q})$ is the dual abelian variety to $\mathbf{T}_{\mathfrak{a},\mathfrak{b}}(\underline{q})$.

Given a Hilbert modular form (of parallel weight) over a base ring B, we define its *q-expansion* at the standard cusp $\mathbf{T}_{\mathfrak{c}}(\underline{q})$ by

$$f_{\mathfrak{c}}(q) = f(\mathbf{T}_{\mathfrak{c}}(\underline{q}) \otimes_S S_B, \iota_{\text{can}}, \lambda_{\text{can}}, \omega_{\text{can}}, \beta_{N,\text{can}}) \in S_B := \mathbb{Z}((\mathfrak{c}^{-1}; \Delta)) \otimes_{\mathbb{Z}} B. \tag{2.29}$$

In general, to obtain q-expansions at every cusp we need to *specify* a differential and level structure. Thus, given B, one chooses an isomorphism

$$j\colon \mathfrak{a} \otimes_{\mathbb{Z}} B \cong \mathcal{O}_L \otimes_{\mathbb{Z}} B. \tag{2.30}$$

This allows one to identify $\mathbb{G}_m \otimes \mathcal{D}_{L/\mathbb{Q}}^{-1}\mathfrak{a}^{-1}$ with $\mathbb{G}_m \otimes \mathcal{D}_{L/\mathbb{Q}}^{-1}$ over the base $S_B = \mathbb{Z}((\mathfrak{c}^{-1}; \Delta)) \otimes_{\mathbb{Z}} B$. We let $\omega(j)$ and $\beta_{N,j}$ be the resulting non-vanishing differential and μ_N-level structure obtained via this identification.

DEFINITION 2.8. The q-expansion of a Hilbert modular form f defined over B (of parallel weight) at the cusp $\mathbf{T}_{\mathfrak{a},\mathfrak{b},j}(\underline{q})$ is

$$f_{\mathfrak{a},\mathfrak{b},j}(\underline{q}) = f(\mathbf{T}_{\mathfrak{a},\mathfrak{b},j}(\underline{q})_{/S_B}, \iota_{\text{can}}, \lambda_{\text{can}}, \omega(j), \beta_{N,j}) \in S_B := \mathbb{Z}((\mathfrak{ab}; \Delta)) \otimes_{\mathbb{Z}} B. \tag{2.31}$$

Note that immediately from the definition we obtain that the q-expansion commutes with base change and has bounded denominators.

THEOREM 2.9 (q-expansion principle). *Let $g > 1$.*[4]

(1) *The q-expansion $f_{\mathfrak{a},\mathfrak{b},j}(q)$ of a Hilbert modular form f defined over B, of polarization module $\mathfrak{c} = \mathfrak{ab}^{-1}$ and of level μ_N, at the cusp $\mathbf{T}_{\mathfrak{a},\mathfrak{b},j}(\underline{q})$, is independent of Δ and lies in fact in $\mathbb{Z}[\![(\mathfrak{ab})^+]\!]^{U_N^2} \otimes B$, where U_N is the group of units of $\mathcal{O}_L$ that are congruent to 1 modulo N. The ring $\mathbb{Z}[\![(\mathfrak{ab})^+]\!]^{U_N^2} \otimes B$ is isomorphic to the completion of the local ring of the cusp $(\mathfrak{a}, \mathfrak{b}, j)$ on $\mathfrak{M}^*(B, \mu_N)$.*

(2) *The q-expansion map is an injective Galois equivariant homomorphism*

$$\mathcal{M}(B, \mathrm{Norm}^k, \mu_N) \to \mathbb{Z}[\![\ (\mathfrak{ab})^+\ ]\!]^{U_N^2} \otimes B \tag{2.32}$$

that commutes with base change.

(3) *Let f be a complex Hilbert modular form of polarization module $\mathfrak{c}$. Let $j_{\mathbb{C}}\colon \mathfrak{a} \otimes_{\mathbb{Z}} \mathbb{C} \to \mathcal{O}_L \otimes_{\mathbb{Z}} \mathbb{C}$ be the natural identification. Then the algebraic q-expansion*

$$f_{\mathfrak{a},\mathfrak{b},j}(q) = f(\mathbf{T}_{\mathfrak{a},\mathfrak{b},j}(\underline{q})_{/S_B}, \iota_{\text{can}}, \lambda_{\text{can}}, \omega(j), \beta_{N,j}) \tag{2.33}$$

is equal to the analytic q-expansion

$$f_{(\mathfrak{a},\mathfrak{b})_{\text{an}}} = a_0 + \sum_{\alpha \in (\mathfrak{ab})^+} a_\alpha q^\alpha. \tag{2.34}$$

We refer the reader to [**9, 38, 39, 81, 135**] for a complete exposition of the theory of q-expansions.

3. Hasse invariants

3.1. Definition and main properties of partial Hasse invariants. Let R be a ring of characteristic p. Let $f\colon A \to \mathrm{Spec}(R)$ be an abelian scheme. The exact sequence for R a field

$$0 \to H^0(A, \Omega^1_{A/R}) \to H^1_{\text{dR}}(A/R) \to H^1(A, \mathcal{O}_A) \to 0 \tag{3.1}$$

can be interpolated to a sequence of sheaves on $\mathrm{Spec}(R)$[5]

$$0 \to R^0 f_* \Omega^1_{A/R} \to H^1_{\text{dR}}(A/R) \to R^1 f_* \mathcal{O}_A \to 0. \tag{3.11}$$

---

[4]For $g = 1$ see Chapter 4, Theorem 3.5. The only difference is the Köcher principle.

[5]Let $f\colon X \to S$ be a morphism of schemes and let $\Omega^*_{X/S}$ be the complex of relative Kahler differentials:

$$0 \to \mathcal{O}_X \xrightarrow{d} \Omega^1_{X/S} \xrightarrow{d} \Omega^2_{X/S} \xrightarrow{d} \Omega^3_{X/S} \xrightarrow{d} \cdots, \tag{3.2}$$

where $\Omega^i_{X/S} = \bigwedge^i \Omega^1_{X/S}$. The relative de Rham cohomology $H^*_{\text{dR}}(X/S)$ is by definition the hypercohomology of the complex $\Omega^*_{X/S}$, or more precisely, the hypercohomology of the global sections functor applied to $\Omega^*_{X/S}$:

$$H^*_{\text{dR}}(X/S) = \mathbb{H}^*(X, \Omega^*_{X/S}). \tag{3.3}$$

The identification $H^1(A,\mathcal{O}_A)\cong \mathfrak{t}_{A^t/R}$ for R a field, where $\mathfrak{t}_{A^t/R}$ is the tangent space at zero $\mathfrak{T}_{A^t,0}$ or equivalently $\mathrm{Lie}(A^t/R)$, can be extended to an identification

$$R^1 f_*\mathcal{O}_A \cong \mathfrak{t}_{A^t/R}. \tag{3.12}$$

There is a spectral sequence

$$H^q(X,\Omega^p_{X/S}) \implies H^*_{\mathrm{dR}}(X/S). \tag{3.4}$$

(See [**63**, §1], [**62**], [**119**, §5]). Recall the notion of Hypercohomology (See [**60**, §11.4], [**67**, Chapters I, II], [**46**, Chapter 4]). If X is a topological space and

$$F^* = \cdots \to 0 \to F^1 \to F^2 \to \cdots \tag{3.5}$$

is a complex of sheaves of abelian groups on X bounded from below, the hypercohomology of F^*, $\mathbb{H}^*(X,\mathbb{F}^*)$, is the derived functor of the global sections functor Γ, in the sense of derived categories.

Recall that the category of abelian sheaves on X has a derived category. The construction of derived category is applicable to any abelian category A ([**67**, Chapter I], [**46**, Chapter 4]): First, two complexes F^* and G^* of objects of A are called quasi-isomorphic if there exists a map $F^*\to G^*$ of complexes that induces isomorphisms on homology $h^i(F^*)\cong h^i(G^*)$, where $h^i(F^*) = \mathrm{Ker}(F^i\to F^{i+1})/\mathrm{Im}(F^{i-1}\to F^i)$ as usual (we refrain from using H^i as these stand for the usual right derived functors of the global sections functor!). The derived category $D(A)$ is obtained from the category of complexes of objects of A (bounded from below) by localizing at all quasi-isomorphisms.

Given a left exact functor $K\colon A\to B$ between two abelian categories A and B, one can define, under very general conditions, a derived functor $\mathbb{R}(K)\colon D(A)\to D(B)$ as follows. Assume that A has enough injectives so that every complex F^* of objects of A (bounded from below) is quasi-isomorphic to a complex I^* of injective objects of A, $F^*\cong I^*$. One defines then

$$\mathbb{R}(K)(F^*) = \{\mathbb{R}^i(K)(F^*)\} = \{h^i(KI^*)\}. \tag{3.6}$$

Note that we may think of an object F of A as a complex F^* concentrated in degree 0:

$$\cdots\to 0\to F\to 0\to 0\to\cdots. \tag{3.7}$$

If $0\to F\to I^0\to I^1\to\cdots$ is an injective resolution of F then we get a quasi isomorphism of complexes

$$\begin{array}{ccccccccc} \cdots 0 & \longrightarrow & F & \longrightarrow & 0 & \longrightarrow & 0 & \longrightarrow & \cdots \\ \downarrow & & \downarrow & & \downarrow & & \downarrow & & \\ \cdots 0 & \longrightarrow & I^0 & \longrightarrow & I^1 & \longrightarrow & I^2 & \longrightarrow & \cdots. \end{array} \tag{3.8}$$

One finds that $\mathbb{R}^i(K)(F^*) = R^iK(F)$.

Coming back to de Rham cohomology, we take the global section functors Γ, from abelian sheaves on X to abelian groups, and apply it to the object $\Omega^*_{X/S}$ considered in the derived category of abelian sheaves on X. We remark that there is a way to compute this effectively by creating a double complex, in which the i-th column is an injective resolution of $\Omega^i_{X/S}$, and calculating the cohomology of the associated total complex ([**60**, §11.4]). This gives the spectral sequence (3.4).

One may also form the de Rham cohomology as a sheaf on the base S ([**63**, §1], [**62**]). One lets

$$\mathcal{H}^i_{\mathrm{dR}}(X/S) = \mathbb{R}^i f_*(\Omega^*_{X/S}). \tag{3.9}$$

In this case we take the hypercohomology of the left exact functor of push forward from the category of abelian sheaves on X to abelian sheaves on S (i.e., the right hyper-derived functors from the derived category of abelian sheaves on X to abelian sheaves on S). Again, it comes with a spectral sequence

$$\mathbb{R}^q f_*(\Omega^p_{X/S}) \implies \mathcal{H}^*_{\mathrm{dR}}(X/S). \tag{3.10}$$

In our situation of abelian variety over an affine base $S=\mathrm{Spec}(R)$ we have $\mathcal{H}^*_{\mathrm{dR}}(X/S)(S) = \mathbb{H}^*_{\mathrm{dR}}(X/S)$.

REMARK 3.1. Note that over $\mathbb{C}$, the exact sequence of sheaves $0 \to \mathbb{Z} \to \mathcal{O}_A \to \mathcal{O}_A^\times \to 0$ gives $H^1(A, \mathbb{Z}) \to H^1(A, \mathcal{O}_A) \to H^1(A, \mathcal{O}_A^\times)^0 = A^t$, making the identification of $H^1(A, \mathcal{O}_A)$ with the tangent space of A^t clear in this case.

Given a polarization $\lambda\colon A \to A^t$ of degree prime to p we get an isomorphism

$$\lambda_*\colon \mathfrak{t}_{A/R} \to \mathfrak{t}_{A^t/R}, \tag{3.13}$$

and hence, a perfect pairing

$$\langle \cdot, \cdot \rangle_\lambda\colon R^0 f_* \Omega^1_A \times R^1 f_* \mathcal{O}_A \to R, \quad (a, b) \mapsto \langle a, \lambda_*^{-1}(b) \rangle. \tag{3.14}$$

Here $\langle a, \lambda_*^{-1}(b) \rangle$ stands for the natural pairing

$$\mathfrak{t}^*_{A/R} \times \mathfrak{t}_{A/R} \to R. \tag{3.15}$$

Assume that p is not ramified in L. Let $\mathbb{F}$ be a fixed finite field isomorphic to the residue field of any of the residue fields of the prime factors of p in $\tilde{L}$—the fixed Galois closure of L. We now define modular forms $h_1, \dots, h_g$ in characteristic p (more precisely, over $\mathbb{F}$) that we call *partial Hasse invariants.*

Assume that $\underline{A} = (A, \iota, \beta_N)_{/R}$ is an abelian variety with real multiplication and μ_N-level over an $\mathbb{F}$-algebra R. Assume further that $\mathfrak{t}^*_{A/R}$ is a free $\mathcal{O}_L \otimes R$ module. Let ω be a non-vanishing differential. Then ω gives an R-basis to $\mathfrak{t}^*_{A/R}$

$$\omega_1, \dots, \omega_g. \tag{3.16}$$

Letting e_i be the i-th idempotent of the ring $\mathcal{O}_L \otimes R = \bigoplus_{i=1}^g R$, we have $\omega_i = e_i \omega$. The Frobenius automorphism $\sigma \in \mathrm{Aut}(W(\mathbb{F}))$ acts on the idempotents e_i. Let $i \mapsto \phi(i)$ be the permutation giving the action. Note that $\sigma \circ \sigma_i = \sigma_{\phi(i)}$, where $\sigma_1, \dots, \sigma_g$ are the embedding of $\mathcal{O}_L$ into $W(\mathbb{F})$ corresponding to the idempotents $e_1, \dots, e_g$. Let λ be an $\mathcal{O}_L$-linear polarization of degree prime to p.

LEMMA 3.2. *Such λ exists.*

PROOF. This follows from the proof of Lemma 5.5 in Chapter 3. □

Let

$$\langle \cdot, \cdot \rangle_\lambda\colon \mathfrak{t}^*_{A/R} \times \mathfrak{t}_{A^t/R} \to R \tag{3.17}$$

be the associated pairing, and let

$$\eta_1, \dots, \eta_g \tag{3.18}$$

be the dual basis over R to $\mathfrak{t}_{A^t/R} = R^1 f_* \mathcal{O}_A$.

LEMMA 3.3. *The natural action of* $\mathrm{Fr}\colon R^1 f_* \mathcal{O}_A \to R^1 f_* \mathcal{O}_A$, *induced from $\mathcal{O}_A \to \mathcal{O}_A$ by $x \mapsto x^p$, permutes the $\mathcal{O}_L$-eigenspaces $R\eta_i$.*

PROOF. It is enough to prove that $[a] \cdot \mathrm{Fr}(\eta_{\phi^{-1}(i)}) = \sigma_i(a)\,\mathrm{Fr}(\eta_{\phi^{-1}(i)})$ for $a \in \mathcal{O}_L$. But the action of Fr commutes with endomorphisms.[6] Thus,

$$\begin{aligned} [a] \cdot \mathrm{Fr}(\eta_{\phi^{-1}(i)}) &= \mathrm{Fr}([a] \cdot \eta_{\phi^{-1}(i)}) \\ &= \mathrm{Fr}(\sigma_{\phi^{-1}(i)}(a)\eta_{\phi^{-1}(i)}) \\ &= \sigma^p_{\phi^{-1}(i)}(a)\,\mathrm{Fr}(\eta_{\phi^{-1}(i)}) \\ &= \sigma_i(a)\,\mathrm{Fr}(\eta_{\phi^{-1}(i)}). \end{aligned} \tag{3.19}$$

□

[6]Either because of the fact that endomorphisms induce homomorphisms of Dieudonné modules, or because we are using two *independent* functorialities of $H^1(A, \mathcal{O}_A)$: one with respect to the space A (endomorphisms), and one with respect to the coefficients $\mathcal{O}_A$ (Frobenius).

DEFINITION 3.4. The i-th partial Hasse invariant $h_i(\underline{A},\omega)$ is the unique element of R such that

$$\mathrm{Fr}(\eta_{\phi^{-1}(i)}) = h_i(\underline{A},\omega)\cdot\eta_i. \tag{3.20}$$

We shall write symbolically

$$h_i(\underline{A},\omega) = \frac{\mathrm{Fr}(\eta_{\phi^{-1}(i)})}{\eta_i}. \tag{3.21}$$

THEOREM 3.5. *The rule*

$$(\underline{A},\omega)\mapsto h_i(\underline{A},\omega) \tag{3.22}$$

is a Hilbert modular form over $\mathbb{F}$, *of level* 1 *and weight* $\chi^p_{\phi^{-1}(i)}\chi_i^{-1}$ *and is independent of* λ. *Its* q*-expansion at every standard cusp is* 1.

PROOF. To simplify notation we shall assume henceforth that p is inert in $\mathcal{O}_{\tilde{L}}$. In this case we may work over $\mathcal{O}_{L,p}$-algebras. We choose $\mathbb{F}=\mathcal{O}_L/(p)$. We write as before $\sigma_1,\dots,\sigma_g$ for the embeddings $L\to L_p$. We assume that

$$\sigma\circ\sigma_i=\sigma_{i+1}, \tag{3.23}$$

where σ is the unique lift of the Frobenius map $\mathrm{Fr}\colon\mathbb{F}\to\mathbb{F}$ given by $x\mapsto x^p$.

The verification that the rule h_i commutes with base change and depends only on the isomorphism class is straightforward. It is also evidently independent of the μ_N-level structure β_N. We calculate the weight:

Let $\alpha\in(\mathcal{O}_L\otimes R)^\times$. The basis for $\mathfrak{t}^*_{A/R}$ obtained from $\alpha^{-1}\omega$ is

$$e_i([\alpha^{-1}]\omega)=[e_i\cdot\alpha^{-1}]\cdot[e_i]\cdot\omega=\chi_i(\alpha)^{-1}[e_i]\omega,\quad i=1,\dots,g. \tag{3.24}$$

The dual basis is thus

$$\chi_1(\alpha)\eta_1,\cdots,\chi_g(\alpha)\eta_g, \tag{3.25}$$

and

$$\frac{\mathrm{Fr}(\chi_{i-1}(\alpha)\eta_{i-1})}{\chi_i(\alpha)\eta_i}=\chi^p_{i-1}(\alpha)\chi_i^{-1}(\alpha)\cdot\frac{\mathrm{Fr}(\eta_{i-1})}{\eta_i}. \tag{3.26}$$

That is,

$$h_i(\underline{A},\alpha^{-1}\omega)=(\chi^p_{i-1}\chi_i^{-1})(\alpha)h_i(\underline{A},\omega). \tag{3.27}$$

Hence, h_i is of weight $\chi^p_{i-1}\chi_i^{-1}$.

Let λ' be another $\mathcal{O}_L$-linear polarization of degree prime to p. Then $\lambda=m\lambda'$ for some $m\in L$ such that $(m,p)=1$. We have

$$\langle x,y\rangle_\lambda=\langle x,my\rangle_{\lambda'}. \tag{3.28}$$

Hence, if $\eta_1,\dots,\eta_g$ are λ-dual to $\omega_1,\dots,\omega_g$ then $m\eta_1,\dots,m\eta_g$ are λ'-dual to $\omega_1,\dots,\omega_g$. But,

$$\frac{\mathrm{Fr}(m\eta_{i-1})}{m\eta_{i-1}}=\frac{m\,\mathrm{Fr}(\eta_{i-1})}{m\eta_{i-1}}=\frac{\mathrm{Fr}(\eta_{i-1})}{\eta_{i-1}}. \tag{3.29}$$

Therefore, h_i is independent of λ.

We now address the issue of q-expansions. We first remark that it would follow from the map

$$r\colon\bigoplus_{\chi\in X(\mathbb{T})}\mathcal{M}(\mathbb{F},\chi,\mu_N)\to R, \tag{3.30}$$

(which is the analogue of the map (5.13) in Theorem 4.1) that it is enough to prove that the q-expansion of h_i is 1 at *one* cusp of *every* component of the moduli space $\mathfrak{M}(\mathbb{F}, \mu_N)$. Let $\mathbf{T} = \mathbb{G}_m \otimes \mathcal{D}_L^{-1}/\underline{q}(\mathfrak{b})$ be a standard Tate object with its canonical $\mathcal{O}_L$-structure ι_{can}, canonical μ_N-level structure $\beta_{N,can}$, and canonical non-vanishing differential ω_{can}. It is an abelian scheme over S where $S = \mathbb{F}((\mathfrak{b}; \Delta))$. Then

$$\mathfrak{t}^*_{\mathbf{T}/S} \cong S \otimes_{\mathbb{Z}} \mathcal{O}_L \cdot \frac{dt}{t}. \tag{3.31}$$

The canonical isomorphism is provided by the invariant differential dt/t of $\mathbb{G}_m$ with coordinate t.

We have, over a perfect ring of characteristic p:

$$\langle \mathrm{Fr}(\eta_{i-1}), \omega_i\rangle_\lambda = \langle \eta_{i-1}, \mathrm{Ver}(\omega_i)\rangle_\lambda^{(1/p)}. \tag{3.32}$$

Therefore, it is enough to show that

$$\mathrm{Ver}(w_{\mathrm{can},i}) = w_{\mathrm{can},i-1}, \quad \forall i. \tag{3.33}$$

This amounts to the identity

$$e_i \,\mathrm{Ver} = \mathrm{Ver}\, e_{i-1} \tag{3.34}$$

as operators on $\mathfrak{t}^*_{\mathbf{T}/S}$. Now,

$$(a \otimes 1)\,\mathrm{Ver} = \mathrm{Ver}(a \otimes 1), \quad (a \otimes r^p)\,\mathrm{Ver} = \mathrm{Ver}(1 \otimes r). \tag{3.35}$$

Hence, we just assert that the automorphism $v\colon \mathcal{O}_L \otimes R \to \mathcal{O}_L \otimes R$ given by $v(a \otimes r) = a \otimes r^p$ takes e_{i-1} to e_i. Since clearly $v(e_{i-1})$ is a multiple of e_i (consider the action of $\mathcal{O}_L$) the assertion follows from e_i being orthogonal idempotents

$$1 = e_1 + \cdots + e_g, \quad e_i^2 = e_i, \ \forall i. \tag{3.36}$$ □

For further study of the partial Hasse invariants see [**48**]. In Chapter 6, Corollary 11.8 we prove the following

THEOREM 3.6. *The divisor of h_i is reduced.*

Let

$$H = h_1 \cdots h_g \tag{3.37}$$

be the total Hasse invariant. It is, up to a sign, the determinant of the Hasse–Witt matrix and is a Hilbert modular form over $\mathbb{F}$ (in fact over $\mathbb{Z}/p\mathbb{Z}$) of level 1 and of weight Norm^{p-1}. It has q-expansion 1 at every standard cusp. In Chapter 6, Section 11.4 we relate the divisors of the modular forms h_i to a certain stratification of $\mathfrak{M}(\mathbb{F}, \mu_N)$ studied in [**50**]. In particular it follows that the divisors

$$W_i = (h_i), \tag{3.38}$$

are reduced and "as transversal to each other as possible."

3.2. Further properties. Following [**166**] we prove the following

LEMMA 3.7. *Let $\underline{A} = (A, \iota, \beta_N)_{/k}$ be an abelian variety with real multiplication (and μ_N-level) over a perfect field $k \supseteq \mathbb{F}$. Then $A[p]^0 \cong \mu_p \otimes \mathcal{D}_L^{-1}$ iff there exists a non-vanishing differential $\omega \in \Omega^1_{A/k} = \Omega^1_{A[p]^0/k}$ such that $h_i(\underline{A}, \omega) = 1$ for all i.*

PROOF. If $A[p]^0 \cong \mu_p \otimes \mathcal{D}_L^{-1}$ then the same computation giving the q-expansion shows that for the canonical non-vanishing differential ω on $\mu_p \otimes \mathcal{D}_L^{-1}$ we get $h_i(\underline{A}, \omega) = 1$ for every i.

Conversely, let $\omega \in \Omega^1_{A/k}$ be a non-vanishing differential such that $h_i(\underline{A}, \omega) = 1$ for every i. We identify $H^0(A, \Omega^1_{A/k})$ with the Dieudonné module of $A[p]^0$ (see Appendix A). The Dieudonné module of $\mu_p \otimes \mathcal{D}_{L/\mathbb{Q}}^{-1}$ is canonically $k \otimes \mathcal{O}_L$, where the $\mathcal{O}_L$ action is the evident one, Fr is the zero map, and Ver is the map determined by $x^p \otimes m \mapsto x \otimes m$.

We define a map

$$\psi\colon H^0(A, \Omega^1_{A/k}) \to k \otimes \mathcal{O}_L \tag{3.39}$$

by

$$\left(\sum x_i \otimes m_i\right)\omega \mapsto \sum x_i \otimes m_i. \tag{3.40}$$

It is obviously an isomorphism of $k \otimes \mathcal{O}_L$ modules. Since Fr is zero on $H^0(A, \Omega^1_A)$ the only question remaining is whether ψ is Ver-equivariant. But this just boils down to the assertion that $\mathrm{Ver}(e_i\omega) = e_{i-1}\omega$, which in turn holds iff $h_i(\underline{A}, \omega) = 1$. $\square$

4. The kernel of the q-expansion map

Similar to the case of $g = 1$, the identification of the kernel of the q-expansion map is based on comparing *modular forms* of μ_N-level with *functions* of μ_{Np}-level. To simplify the exposition, *we assume that p is inert in L.* This is merely a technical requirement if one assumes that p is unramified in L. See [**48**] for an explanation of the yoga. We let $\mathbb{F} = \mathcal{O}_L/(p)$ be the residue field of p^g elements.

Let $N \geq 4$ be a fixed integer prime to p. As before, we denote by $\mathfrak{M}(\mathbb{F}, \mu_N)$ the fine moduli scheme of quadruples

$$\underline{A} = (A, \iota, \lambda, \beta_N)_{/S} \tag{4.1}$$

of polarized abelian varieties A with real multiplication ι and μ_N-level structure $\beta_N\colon \mu_N \otimes \mathcal{D}_L^{-1} \hookrightarrow A$ over $\mathbb{F}$-schemes S. We let $\mathfrak{M}^*(\mathbb{F}, \mu_N)$ be the minimal compactification obtained by adjoining the cusps. We let $\mathfrak{M}(\mathbb{F}, \mu_{Np})$ (resp. $\mathfrak{M}^*(\mathbb{F}, \mu_{Np})$) be the moduli space with μ_{Np}-structure, which we write as $\beta_N \times \beta_p$ (resp. with the cusps added). The morphisms

$$\mathfrak{M}(\mathbb{F}, \mu_{Np}) \to \mathfrak{M}(\mathbb{F}, \mu_N)^{\mathrm{ord}}, \quad \mathfrak{M}^*(\mathbb{F}, \mu_{Np}) \to \mathfrak{M}^*(\mathbb{F}, \mu_N)^{\mathrm{ord}}, \tag{4.2}$$

are Galois coverings (with Galois group isomorphic to $\left(\mathcal{O}_L/(p)\right)^\times$) of the open scheme consisting of the ordinary locus of $\mathfrak{M}(\mathbb{F}, \mu_N)$ and $\mathfrak{M}^*(\mathbb{F}, \mu_N)$, respectively (the cusps are considered ordinary points and the covering is étale outside the cusps). The components of $\mathfrak{M}^*(\mathbb{F}, \mu_{Np})$ are the same as those of $\mathfrak{M}^*(\mathbb{F}, \mu_N)^{\mathrm{ord}}$. This follows from Theorem 6.19 in Chapter 3.

THEOREM 4.1. *Fix a polarization module. Let $(\mathfrak{c}, \mathfrak{c}^+)$ be a representative of the polarization module. Let $\mathbf{T}_\mathfrak{c}(\underline{q})$ be a standard Tate object for this ideal $\mathfrak{c}$ as in* §2. *The kernel of the q-expansion map,*

$$\bigoplus_{\chi \in X(\mathbb{T}_\mathbb{F})} \mathcal{M}(\mathbb{F}, \chi, \mu_N) \to \mathbb{F}[\![q^\nu : \nu \in \mathfrak{c}^{-1,+}]\!] \tag{4.3}$$

is the ideal

$$(h_1 - 1, \dots, h_g - 1). \tag{4.4}$$

PROOF. We consider $\mathbf{T}_{\mathfrak{c}}(\underline{q})$ as a cusp on $\mathfrak{M}(\mathbb{F}, \mu_{Np})$. Let R be the ring of regular functions on the component of $\mathfrak{M}(\mathbb{F}, \mu_{Np})$ corresponding to the polarization module $\mathfrak{c}$. We define a surjective map,

$$r\colon \bigoplus_{\chi \in X(\mathbb{T}_{\mathbb{F}})} \mathcal{M}(\mathbb{F}, \chi, \mu_N) \to R, \tag{4.5}$$

such that the composition,

$$\bigoplus_{\chi \in X(\mathbb{T}_{\mathbb{F}})} \mathcal{M}(\mathbb{F}, \chi, \mu_N) \to R \hookrightarrow \left(\mathcal{O}_{\mathfrak{M}^*(\mathbb{F},\mu_{Np}),\mathbf{T}_{\mathfrak{c}}(\underline{q})}\right)^{\widehat{}}, \tag{4.6}$$

is the q-expansion map. We then prove that

$$\operatorname{Ker}(r) = (h_1 - 1, \dots, h_g - 1). \tag{4.7}$$

This, together with the surjectiveness of r, would prove the Theorem and also give Corollary 4.4.

To define the map r we first need a construction.

LEMMA 4.2. *For every $\chi \in X(\mathbb{T}_{\mathbb{F}})$ there exists a canonical modular form $a(\chi)$ on $\mathfrak{M}^*(\mathbb{F}, \mu_{Np})$ such that:*

- *$a(\chi)$ has weight χ.*
- *$a(\chi)$ has q-expansion 1 at $\mathbf{T}_{\mathfrak{c}}(\underline{q})$.*
- *$a(\chi)a(\chi') = a(\chi\chi')$.*
- *$a(\chi)$ does not vanish.*

PROOF. Given $(\underline{A}, \beta_p)_{/R}$ over a ring R, we have an induced isomorphism

$$\beta_{p*}\colon \mathbf{t}^*_{\mu_p \otimes \mathcal{D}_L^{-1}} \to \mathbf{t}^*_{A/R}. \tag{4.8}$$

The image of dt/t is a non-vanishing differential ω on A/R. We let

$$a(\chi_i) = e_i\omega. \tag{4.9}$$

For $\chi = \chi_1^{a_1} \cdots \chi_g^{a_g}$ we let

$$a(\chi) = a(\chi_1)^{a_1} \cdots a(\chi_g)^{a_g}. \tag{4.10}$$

The properties of the $a(\chi)$'s are easily verified. □

We now define

$$r(f) = \frac{f}{a(\chi)}, \quad f \in \mathcal{M}(\mathbb{F}, \chi, \mu_N), \tag{4.11}$$

and extend it by linearity to

$$r\colon \bigoplus_{\chi \in X(\mathbb{T}_{\mathbb{F}})} \mathcal{M}(\mathbb{F}, \chi, \mu_N) \to R. \tag{4.12}$$

The properties of the $a(\chi)$'s ensure it is a ring homomorphism. Furthermore, the composition in (4.6) is the q-expansion map. First, note that the q-expansion of $r(f)$ is the same as that of f. Second, it follows from Theorem 2.9 that the q-expansion of $r(f)$ with respect to a Tate object is none other than its expression in the local ring of the cusp corresponding to that Tate object.

At this point we may conclude:

- $I := \mathrm{Ker}(r)$ is the kernel of the q-expansion map.
- $I \supseteq (h_1 - 1, \ldots, h_g - 1)$.
- The q-expansion map is injective on every fixed weight, because r restricted to any $\mathcal{M}(\mathbb{F}, \chi, \mu_N)$ is injective!

The next ingredient is defining a

- Galois action. Let $\mathbb{F}^\times = \big(\mathcal{O}_L/(p)\big)^\times$ act on
  - $f \in \mathcal{M}(\mathbb{F}, \chi, \mu_N)$ by $[\alpha]f = \chi(\alpha)f$.
  - $g \in R$ by $([\alpha]g)(\underline{A}, \beta_p) = g\big(\underline{A}, \beta_p \circ (1 \otimes \alpha)\big)$.

One verifies that r is equivariant for this Galois action and that in fact $\mathbb{F}^\times = \mathrm{Gal}(\mathfrak{M}^*(\mathbb{F}, \mu_{Np})/\mathfrak{M}^*(\mathbb{F}, \mu_N)^{\mathrm{ord}})$. Let us note that $\mathbb{F}^\times \hookrightarrow (\mathbb{F} \otimes \mathbb{F})^\times = \mathbb{T}_\mathbb{F}(\mathbb{F})$. Therefore, every character $\chi \in X(\mathbb{T}_\mathbb{F})$ gives a homomorphism $\chi \colon \mathbb{F}^\times \to \mathbb{F}^\times$. We let

$$\begin{aligned} X(1)_p &= \{\chi \in X(\mathbb{T}_\mathbb{F}) : \chi(\mathbb{F}^\times) = 1\} \\ &= \langle \chi_1^p \chi_2^{-1}, \ldots, \chi_i^p \chi_{i+1}^{-1}, \ldots, \chi_g^p \chi_1^{-1} \rangle. \end{aligned} \tag{4.13}$$

Kummer theory gives the equality

$$R = \bigoplus_{\chi \in X/X(1)_p} R^\chi. \tag{4.14}$$

LEMMA 4.3. *We have the identity* $[\alpha]a(\chi) = \chi^{-1}(\alpha)a(\chi)$.

PROOF OF LEMMA. It is enough to prove that

$$[\alpha]a(\chi_i) = \chi_i^{-1}(\alpha)a(\chi_i). \tag{4.15}$$

Now, $a(\chi_i)(\underline{A}, \beta_p) = e_i \beta_{p*}(dt/t) \otimes 1$. Note that $1 \otimes \alpha$ acts by $1 \otimes \alpha$ on the tangent space of $\mu_p \otimes \mathcal{D}_{L/\mathbb{Q}}^{-1}$ and $(1 \otimes \alpha)_*$ acts on the cotangent space by $(1 \otimes \alpha)^{-1}$ whenever this is defined. Thus,

$$\begin{aligned} [\alpha]a(\chi_i)(\underline{A}, \beta_p) &= e_i\big(\beta_p \circ (1 \otimes \alpha)\big)_* \frac{dt}{t} \otimes 1 \\ &= \beta_{p*} e_i (1 \otimes \alpha)_* \frac{dt}{t} \otimes 1 \\ &= \chi_i(\alpha)^{-1} \beta_{p*} e_i \frac{dt}{t} \otimes 1 \\ &= \chi_i(\alpha)^{-1} a(\chi_i)(\underline{A}, \beta_p). \end{aligned} \tag{4.16}$$

□

- r is surjective. Given $g \in R^\chi$ let

$$f = g \cdot a(\chi) \cdot H^n, \quad n \gg 0. \tag{4.17}$$

Then f is a holomorphic modular form on $\mathfrak{M}^*(\mathbb{F}, \mu_{Np})$ which is $\mathbb{F}^\times$-invariant. Hence f is a meromorphic modular form on $\mathfrak{M}^*(\mathbb{F}, \mu_N)$, which is holomorphic if $n \gg 0$. Clearly $r(f) = g$.

- $\mathrm{Ker}(r) = (h_1 - 1, \ldots, h_g - 1)$. Let $f_{\psi_1} + \cdots + f_{\psi_r}$ be in the kernel of r (with f_{ψ_i} of weight $\psi_i \in X(\mathbb{T}_\mathbb{F})$). We may assume without loss of generality that for $i \neq j$, $\psi_i \neq \psi_j$.

Replacing f_{ψ_i} by $f_{\psi_i} + f_{\psi_i}(h_j - 1)$, sufficiently many times and for appropriate i's and j's, we may further assume that

$$i \neq j \implies \psi_i \neq \psi_j \bmod X(1)_p. \tag{4.18}$$

Since r is equivariant, every $r(f_{\psi_i})$ "falls" in a different eigenspace in the decomposition $R = \bigoplus_{\chi \in X/X(1)_p} R^\chi$. Thus every $r(f_{\psi_i}) = 0$. But r is injective on every $\mathcal{M}(\mathbb{F}, \chi, \mu_N)$. Thus, $f_{\psi_i} = 0$ for every i. □

COROLLARY 4.4. *Fix a polarization module $(\mathfrak{c}, \mathfrak{c}^+)$. We have*

$$\bigoplus_{\chi \in X(\mathbb{T}_\mathbb{F})} \mathcal{M}(\mathbb{F}, \chi, \mu_N)/(h_1 - 1, \dots, h_g - 1) \cong R. \tag{4.19}$$

COROLLARY 4.5. *Fix a fractional ideal $\mathfrak{c}$. The kernel of the q-expansion*

$$\bigoplus_{k \in \mathbb{Z}} \mathcal{M}(\mathbb{F}, \mathrm{Norm}^k, \mu_N) \to \mathbb{F}[\![q^\nu : \nu \in \mathfrak{c}^{-1,+}]\!] \tag{4.20}$$

is $(H - 1)$.

REMARK 4.6. Theorem 4.1 and its corollaries hold, with minor modifications, for any cusp.

5. Applications

COROLLARY 5.1. *The q-expansion map over $\mathbb{C}$ is injective.*

PROOF. The kernel of the q-expansion is $\mathrm{Gal}(L^{\mathrm{alg}}/L)$ equivariant and hence the kernel is generated by modular forms with coefficients in L. Let $\sum_{i=1}^t f_{\psi_i} \in \bigoplus \mathcal{M}(\mathbb{C}, \chi, \mu_N)$ be such a sum of modular forms with q-expansion equal to zero. We assume that for $i \neq j$ we have $\psi_i \neq \psi_j$. Now, choose a (inert[7]) prime p such that

- The q-expansion of every f_{ψ_i} is p-integral and is non-zero modulo p.
- The weights $\psi_1, \dots, \psi_t$ are non-congruent modulo $X(1)_p$.

Then the reduction of the forms f_{ψ_i} modulo p is well-defined, $f_{\psi_i} \not\equiv 0 \bmod p$ and the weights are distinct in $X/X(1)_p$. It follows from Theorem 4.1 that each $f_{\psi_i} \bmod p$ is zero—a contradiction. □

EXERCISE 5.2. *Give a direct proof of Corollary* 5.1.

Recall Siegel's construction. For $g > 1$ and $k \geq 2$ (or $g = 1$ and $k \geq 4$) there exists an Eisenstein series $E_k^* = E_k^{L,*}$ (and a normalized Eisenstein series $E_k = E_k^L$) of level 1, weight Norm^k and q-expansion

$$E_k^* = 2^{-g}\zeta_L(1-k) + \sum_{\nu \in \mathcal{O}_L^+} c_{k,\nu} q^\nu, \quad E_k = 2^g \zeta_L(1-k)^{-1} E_k^*, \tag{5.1}$$

where

$$c_{k,\nu} = \sum_{\mathcal{O}_L \supseteq \mathfrak{c} \supseteq (\nu)} \mathrm{Norm}(\mathfrak{c})^{k-1}. \tag{5.2}$$

COROLLARY 5.3. *We have the following bound on denominators:*

$$k \not\equiv 0 \bmod p - 1 \implies \zeta_L(1-k) \textit{ is } p \textit{ integral.} \tag{5.3}$$

[7] If a number field has one inert prime, it has infinitely many inert primes. We insist on inert primes only because we formulated the q-expansion kernel for inert primes. But, in fact, the argument easily generalizes to any totally real number field and non-ramified primes.

PROOF. Assume that p divides the denominator of $\zeta_L(1-k)$. Then we have the following congruence of q-expansions:

$$E_k(q) - 1 \equiv 0 \bmod p. \tag{5.4}$$

However, the ring $\bigoplus \mathcal{M}(\mathbb{F}, \chi, \mu_N)/(h_1 - 1, \ldots, h_g - 1)$ (for a fixed polarization module $(\mathfrak{c}, \mathfrak{c}^+)$) is $X/X(1)_p$ graded. Thus, the homogeneous parts of the relation $E_k(q) - 1 \equiv 0 \bmod p$ also belong to the ideal $(h_1 - 1, \ldots, h_g - 1)$. Since $1 \notin (h_1 - 1, \ldots, h_g - 1)$, it follows that $E_k(q) - 1$ is a homogeneous element. That is

$$\mathrm{Norm}^k \equiv \mathbf{1} \bmod X(1)_p. \tag{5.5}$$

That is, the map $\mathrm{Norm}^k \colon \mathbb{F}^\times \to \mathbb{F}^\times$ is the trivial character. Hence, $(p-1) \mid k$. □

COROLLARY 5.4. *We have the following congruences between values of zeta functions. Suppose that* $k \not\equiv 0 \bmod p-1$ *then*

$$k \equiv k' \bmod p - 1 \implies \zeta_L(1-k) \equiv \zeta_L(1-k') \bmod p. \tag{5.6}$$

PROOF. Let $\alpha = 2^{-g}\big(\zeta_L(1-k) - \zeta_L(1-k')\big)$. Consider the congruence

$$E_k^* - E_{k'}^* - \alpha = 0 \bmod p. \tag{5.7}$$

Using a grading argument, we infer from $\mathrm{Norm}^k \equiv \mathrm{Norm}^{k'} \not\equiv \mathbf{1} \bmod X(1)_p$ that $\alpha \equiv 0 \bmod p$. That is, $\zeta_L(1-k) \equiv \zeta_L(1-k') \bmod p$. □

This method of obtaining bounds on the denominator of the zeta function of L and congruences between its values at negative integers can be generalized considerably to cover the cases $k \equiv 0 \bmod p-1$ and to give congruences modulo p^m. The idea, following Serre and Katz, is to consider the moduli scheme modulo p^m, and use the easy congruences on the higher coefficients of suitable Eisenstein series to obtain information on the constant term. The details may be found in [**49**]. The same results may be deduced from Deligne–Ribet [**27**]. We include here the results for completeness and for the reader's convenience. It seems they are sharp.

PROPOSITION 5.5. *Let* p *be inert in* L. *Let* $k \geq 2$ *be an integer.*

(1) *Let* $p \neq 2$, *then if* $k \equiv 0 \bmod p-1$

$$\mathrm{val}_p\big(\zeta_L(1-k)\big) \geq -1 - \mathrm{val}_p(k), \tag{5.8}$$

and $\zeta_L(1-k)$ *is* p*-integral if* $k \not\equiv 0 \bmod p-1$.

(2) *If* $p = 2$, *then*

$$\mathrm{val}_2\big(\zeta_L(1-k)\big) \geq g - 2 - \mathrm{val}_2(k). \tag{5.9}$$

PROPOSITION 5.6. *Let* p *be inert in* L. *Let* $k, k' \geq 2$ *be integers such that* $k \equiv k' \bmod (p-1)p^m$.

(1) *If* $k \not\equiv 0 \bmod p-1$ *then*

$$(1 - p^{g(k-1)})\zeta_L(1-k) \equiv (1 - p^{g(k'-1)})\zeta_L(1-k') \bmod p^{m+1}. \tag{5.10}$$

(2) *If* $k \equiv 0 \bmod p-1$ *but* $p \neq 2$, *then*

$$(1 - p^{g(k-1)})\zeta_L(1-k) \equiv (1 - p^{g(k'-1)})\zeta_L(1-k') \bmod p^{m-1-\mathrm{val}_p(k\cdot k')}. \tag{5.11}$$

(3) *If* $p = 2$, *then*

$$(1 - 2^{g(k-1)})\zeta_L(1-k) \equiv (1 - 2^{g(k'-1)})\zeta_L(1-k') \bmod 2^{m+g-2-\mathrm{val}_2(k\cdot k')}. \tag{5.12}$$

PROPOSITION 5.7. *There exists a notion of* filtration *on Hilbert modular forms. Given a q-expansion $b(q)$, which is a q-expansion of some $\mathfrak{c}$-polarized Hilbert modular form, there exists a unique $\mathfrak{c}$-polarized modular form f_0 such that the set of all modular forms with q-expansion $b(q)$ is the set*

$$\left\{ f_0 \cdot \prod_{i=1}^{g} h_i^{a_i} : a_i \geq 0 \right\}. \tag{5.13}$$

We call the weight of f_0 the filtration *of the q-expansion $b(q)$.*

PROOF. If f and g have the same q-expansion then $r(f) = r(g)$, and vice versa. We are given that $b(q)$ is a q-expansion of some Hilbert modular form of weight, say, χ. Let f' be a function on $\mathfrak{M}(\mathbb{F}, \mu_{Np})$ such that $f' \in R^\chi$ and in the local ring of the appropriate cusp $f' = b(q)$. Then all the meromorphic modular forms having q-expansion $b(q)$ are of the form $f' \cdot a(\chi) \cdot \prod h_i^{a_i}$ where the $a_i \in \mathbb{Z}$. But the divisor of h_i is a reduced effective divisor $W_{\{i\}}$. Therefore, there is a choice $a_1^*, \ldots, a_g^*$ such that $f_0 = f' \cdot a(\chi) \cdot \prod h_i^{a_i^*}$ is holomorphic and non-vanishing on some component of every $W_{\{i\}}$. It follows that every other holomorphic form with the same q-expansion is a multiple $f_0 \cdot \prod_{i=1}^g h_i^{a_i}$ with $a_i \geq 0$. □

As in Chapter 4, one can define for modular forms of μ_N-level (and any weight) Hecke and diamond operators, U and V operators and theta operators. All those operators can also defined on functions with μ_{Np}-level structures and the map r constructed in Theorem 4.1 is equivariant for these operators. A detailed exposition will appear elsewhere. Here we just note two corollaries that are immediate from the results above: (i) If f has filtration χ then Vf has filtration $\chi^{(p)} = \chi \circ \mathrm{Fr}$; (ii) One can define a theta operator on modular forms of parallel weight $\underline{k}$ and level μ_N by the formula $\theta\big(r(f)\big) = dr(f)/a(\mathrm{Norm}^2)$. This gives a well-defined modular form of parallel weight $\underline{k+p+1}$.

6. p-adic Hilbert modular forms

This section is modeled after Chapter 4, §6. Many of the definitions and proofs extend verbatim, with additional supporting arguments needed every once in a while. We shall therefore be brief and let the reader refer back to loc. cit. whenever needed. For simplicity we shall discuss only Hilbert modular forms of parallel level. The totally real field is of course fixed and is denoted by L; the prime p is unramified and for simplicity of exposition assumed to be inert in L. We furthermore fix a fractional ideal $\mathfrak{c}$ of L with its natural notion of positivity. All definitions and objects should be understood relative to this fixed choice.

6.1. Test objects and overconvergent forms. We fix a Hilbert modular form E_ℓ with respect to L with the following properties:

- E_ℓ has level 1.
- E_ℓ has q-expansion $E_\ell(q)$ with coefficients in $\mathcal{O}_{L_p}$.
- $E_\ell(q)$ is congruent to 1 modulo p.
- E_ℓ has weight Norm^ℓ and $(p-1) \mid \ell$.

REMARK 6.1. (1) Such a modular form always exists. Indeed, for $n \gg 0$ the modular form H^n over $\mathbb{F}$ lifts to a modular form of level 1 over $\mathcal{O}_{L_p}$.

(2) Given Leopoldt's conjecture one can choose E_ℓ to be the Eisenstein series E_ℓ given in (5.1).

We fix:

- B—a p-adic ring.
- r—an element of B.

DEFINITION 6.2. Let C be a p-adic ring which is a B-algebra. A *test object* of:

- level μ_N,
- growth condition r,
- over B,

is a quadruple:

$$(\underline{A}, \omega, \beta_N, Y)_{/C}, \tag{6.1}$$

such that:

- $\underline{A}_{/C} = (A, \iota, \lambda)_{/C}$ is a polarized abelian variety with real multiplication defined over C, such that condition (R) holds;
- $\omega \in \mathfrak{t}^*_{\underline{A}/C}$ is a relative non-vanishing differential;
- $\beta_N : \mu_{N/C} \otimes \mathcal{D}_L^{-1} \hookrightarrow \underline{A}$ is an embedding of $\mathcal{O}_L$-group schemes over Spec(C);
- $Y \in C$ satisfies $Y \cdot E_\ell(\underline{A}, \omega) = r$.

If $(\underline{A}, \omega, \beta_N, Y)_{/C}$ is a test object, so is $(\underline{A}, \lambda\omega, \beta_N, \mathrm{Norm}(\lambda)^\ell \cdot Y)_{/C}$ for $\lambda \in (\mathcal{O}_L \otimes C)^\times$. Indeed,

$$(\mathrm{Norm}(\lambda)^\ell \cdot Y) \cdot \big(E_\ell(\underline{A}, \lambda\omega)\big) = (\mathrm{Norm}(\lambda)^\ell \cdot Y) \cdot \mathrm{Norm}(\lambda)^{-\ell} \cdot E_\ell(\underline{A}, \omega) = r. \tag{6.2}$$

Equivalently, a test object could be thought of as

$$(\underline{A}, \beta_N, Y)_{/C}, \quad Y \in (\det \mathfrak{t}^*_{\underline{A}/C})^{-\ell}, \tag{6.3}$$

such that $Y \cdot E_\ell = r$, where E_ℓ is interpreted as a rule associating to $\underline{A}/C$ a section of $(\det \mathfrak{t}^*_{\underline{A}/C})^\ell$.

DEFINITION 6.3. A *p-adic Hilbert modular form* (à la Katz) of:

- weight Norm^k, $k \in \mathbb{Z}$,
- level μ_N,
- growth r,
- defined over B,

is a rule associating to each test object $(\underline{A}, \omega, \beta_N, Y)_{/C}$ an element $f(\underline{A}, \omega, \beta_N, Y) \in C$ such that:

- $f(\underline{A}, \omega, \beta_N, Y)$ depends only on the isomorphism class of the test object $(\underline{A}, \omega, \beta_N, Y)_{/C}$;
- the rule f commutes with base change;
- for every $\lambda \in (\mathcal{O}_L \otimes C)^\times$ and every test object $(\underline{A}, \omega, \beta_N, Y)_{/C}$

$$f(\underline{A}, \lambda\omega, \beta_N, \mathrm{Norm}(\lambda)^\ell \cdot Y) = \mathrm{Norm}(\lambda)^{-k} \cdot f(\underline{A}, \omega, \beta_N, Y). \tag{6.4}$$

DEFINITION 6.4. The space of Hilbert modular forms over B, of μ_N-level, weight Norm^k and growth condition r is denoted by $\mathbb{F}(B, k, \mu_N; r)$. If $r \notin B^\times$, they are called *overconvergent* modular forms.

Again, the study of modular forms of growth $r \in B^\times$ amounts to the case $r = 1$. In that case we discard all abelian varieties having non-ordinary reduction.

Every classical modular form f of weight Norm^k, level μ_N, over B defines a p-adic modular form (still denoted f) in $\mathbb{F}(B, k, \mu_N; r)$:

$$f(\underline{A}, \omega, \beta_N, Y)_{/C} := f(\underline{A}, \omega, \beta_N)_{/C}. \tag{6.5}$$

As an example of a truly p-adic modular form, consider

$$f(\underline{A}, \omega, \beta_N, Y) = Y. \tag{6.6}$$

By definition,

$$f(\underline{A}, \lambda\omega, \beta_N, \mathrm{Norm}(\lambda)^{\ell} \cdot Y) = \mathrm{Norm}(\lambda)^{\ell} \cdot Y = \mathrm{Norm}(\lambda)^{\ell} \cdot f(\underline{A}, \omega, \beta_N, Y). \tag{6.7}$$

Therefore, it is a modular form in $\mathbb{F}(B, \mathrm{Norm}^{-\ell}, \mu_N; r)$.

6.2. q-expansion for p-adic modular forms. If $E_\ell(\underline{A}, \omega)$ is invertible, then in any test object $(\underline{A}, \omega, \beta_N, Y)$ we must have $Y = r \cdot E_\ell(\underline{A}, \omega)^{-1}$. This applies in particular to the standard Tate object $\mathbf{T}_{\mathfrak{c}}(\underline{q}) = \mathbb{G}_m \otimes \mathcal{D}_L^{-1}/\underline{q}(\mathfrak{c}^{-1})$ over $S_B = \mathbb{Z}[\![\mathfrak{c}^{-1}; \Delta]\!] \otimes B$. The Tate object carries a canonical μ_N-level:

$$\beta_{\mathrm{can}} : \mu_N \otimes \mathcal{D}_L^{-1} \hookrightarrow \mathbb{G}_m \otimes \mathcal{D}_L^{-1} \to \mathbf{T}_{\mathfrak{c}}(\underline{q}), \tag{6.8}$$

and a canonical differential ω_{can} induced from the differential $dt/t \otimes 1$ on $\mathbb{G}_m \otimes \mathcal{D}_L^{-1}$.

DEFINITION 6.5. Let $f \in \mathbb{F}(B, k, \mu_N; r)$ be a p-adic Hilbert modular form. The q-expansion of f at the cusp $\mathbf{T}_{\mathfrak{c}}(\underline{q})$ is

$$f\left(\mathbf{T}_{\mathfrak{c}}(\underline{q}), \omega_{\mathrm{can}}, \beta_{\mathrm{can}}, \frac{r}{E_\ell(\mathbf{T}(q), \omega_{\mathrm{can}})}\right) \in \mathbb{Z}[\![q^\nu : \nu \in \mathfrak{c}^{-1+}]\!]^{U_N^2} \otimes_{\mathbb{Z}} B. \tag{6.9}$$

Similarly, for any cusp $(\mathbf{T}_{\mathfrak{a},\mathfrak{b},j}(\underline{q}), \beta_N)$ we define the q-expansion of f by

$$f\left(\mathbf{T}_{\mathfrak{a},\mathfrak{b},j}(\underline{q}), \omega_{can,j}, \beta_{N,j}, \frac{r}{E_\ell(\mathbf{T}_{\mathfrak{a},\mathfrak{b},j}(\underline{q}), \omega_{\mathrm{can}})}\right) \in \mathbb{Z}[\![q^\nu : \nu \in (\mathfrak{a}\mathfrak{b})^{+}]\!]^{U_N^2} \otimes_{\mathbb{Z}} B. \tag{6.10}$$

We call f a *holomorphic* modular form if for every cusp $(\mathbf{T}_{\mathfrak{a},\mathfrak{b},j}(\underline{q}), \beta_N)$ its q-expansion lies in $B[\![q^\nu : \nu \in (\mathfrak{a}\mathfrak{b})^+]\!]$. We call f a *cusp* form if it is holomorphic, and for every cusp $(\mathbf{T}_{\mathfrak{a},\mathfrak{b},j}(\underline{q}), \beta_N)$ its q-expansion has a vanishing constant coefficient. We denote the holomorphic (resp. cusp) forms by

$$\mathbb{M}(B, k, \mu_N; r) \quad (\text{resp. } \mathbb{S}(B, k, \mu_N; r)). \tag{6.11}$$

PROPOSITION 6.6. (1) *Take any* $\mathbb{X} \in \{\mathbb{F}, \mathbb{M}, \mathbb{S}\}$. *Then*

$$\mathbb{X}(B, k, \mu_N; r) = \varprojlim \mathbb{X}(B/p^n B, k, \mu_N; r). \tag{6.12}$$

(2) (Köcher's principle). *For* $g > 1$,

$$\mathbb{F}(B, k, \mu_N; r) = \mathbb{M}(B, k, \mu_N; r). \tag{6.13}$$

PROOF. All our objects are p-adic. The same proof as in Chapter 4, Proposition 6.7 works.

Köcher's principle follows from the fact that the Tate objects are defined over normal local rings of dimension greater than 1 (See §2). Since divisors have codimension 1 we conclude that a modular form, whose q-expansion is a rational function on the base with no poles outside the closed point, is regular. □

We offer the same remarks as in p. 150.

(1) The q-expansion of a p-adic modular form f' induced from a classical modular form f as in (6.5) is the same q-expansion as of f.
(2) The q-expansion map is injective. See Corollary 6.10.

6.3. The case when p is nilpotent or r is a unit. In this section we let p be a prime and $N > 4$ an integer prime to p. As before, let B be a p-adic ring. We now assume furthermore that p is nilpotent in B. We use our usual notation,

$$\mathcal{M}(B, k, \mu_N), \tag{6.14}$$

for classical Hilbert holomorphic modular forms over B, of μ_N-level structure and weight Norm^k. We shall assume that $g \geq 2$. Thus,

$$\mathcal{F}(B, k, \mu_N) = \mathcal{M}(B, k, \mu_N), \tag{6.15}$$

where $\mathcal{F}(B, k, \mu_N)$ are the classical modular forms over B, of μ_N-level structure and weight k with possible poles at infinity. As above, $\mathbb{M}(B, k, \mu_N; r)$ is the space of holomorphic p-adic Hilbert modular forms with growth r.

Given $j \geq 0$, define a map:

$$\mathcal{M}(B, k + j\ell, \mu_N) \to \mathbb{M}(B, k, \mu_N; r), \quad f \mapsto \tilde{f}, \tag{6.16}$$

where

$$\tilde{f}(\underline{A}, \omega, \beta_N, Y)_{/C} = Y^j f(\underline{A}, \omega, \beta_N). \tag{6.17}$$

We claim that $\tilde{f}$ is a p-adic Hilbert modular form of weight Norm^k. Indeed,

$$\begin{aligned} \tilde{f}(\underline{A}, \lambda\omega, \beta_N, \mathrm{Norm}(\lambda)^\ell \cdot Y) &= (\mathrm{Norm}(\lambda)^\ell \cdot Y)^j f(\underline{A}, \lambda\omega, \beta_N) \\ &= \mathrm{Norm}(\lambda)^{j\ell} \cdot Y^j \cdot \mathrm{Norm}(\lambda)^{-(k+j\ell)} \cdot f(E, \omega, \beta_N) \\ &= \mathrm{Norm}(\lambda)^{-k} \cdot \tilde{f}(\underline{A}, \omega, \beta_N, Y). \end{aligned} \tag{6.18}$$

Under this map, $E_\ell f$ is sent to $\widetilde{E_\ell f}$, and

$$\begin{aligned} \widetilde{E_\ell f}(\underline{A}, \omega, \beta_N, Y) &= Y^{j+1} f(\underline{A}, \omega, \beta_N) \cdot E_\ell(\underline{A}, \omega, \beta_N) \\ &= r \cdot Y^j f(\underline{A}, \omega, \beta_N) \\ &= r \cdot \tilde{f}. \end{aligned} \tag{6.19}$$

Therefore, we have obtained a well-defined homomorphism of B-modules as follows:

$$\Big(\bigoplus_{j \geq 0} \mathcal{M}(B, k + j\ell, \mu_N)\Big)/(E_\ell - r) \to \mathbb{M}(B, k, \mu_N; r). \tag{6.20}$$

The analogue of the isomorphism

$$\Big(\bigoplus_{j \geq 0} \mathcal{F}(B, k + j\ell, \mu_N)\Big)/(E_\ell - r) \xrightarrow{\cong} \mathbb{F}(B, k, \mu_N; r) \tag{6.21}$$

existing for $g = 1$ (Chapter 4, Proposition 6.12) is probably not true. The proof there, recall, was based on the fact that the modular curve with no cusps is affine. However, some information along these lines is available (e.g., Equation (6.29)) in Chapter 4 has an analogue). Moreover, in analogue to the elliptic case (Chapter 4, Proposition 6.14) we do have

PROPOSITION 6.7. *Let B any p-adic ring (p not necessarily nilpotent). Let r be a p-adic unit in B. Then*

$$\mathbb{M}(B, k, \mu_N; r) \cong \Big(\bigoplus_{j=0}^{\infty} \mathcal{M}(B, k + j\ell, \mu_N)\Big)/(E_\ell - r), \tag{6.22}$$

and

$$\mathbb{S}(B,k,\mu_N;r) \cong \Big(\bigoplus_{j=0}^{\infty} \mathcal{S}(B,k+j\ell,\mu_N)\Big)/(E_\ell - r). \tag{6.23}$$

As in passing in Chapter 4 from Proposition 6.14 to Theorem 6.15, one deduces the general case from the case where p is nilpotent.

Assume that p is nilpotent. The proof for holomorphic modular forms is practically the same as in the case $g = 1$. One needs to use that the ordinary locus $\mathfrak{M}(\mathbb{F},\mu_N)^{\mathrm{ord}}$ (the cusps are ordinary) is affine. This follows from the fact that its complement is (H) and H is a section of the ample line bundle $(\det \underline{\omega})^{p-1}$.

For the case of cusps forms one needs that cusp forms are sections of a suitable quasi-coherent sheaf (and thus its higher cohomology on an affine scheme vanishes). This is a bit delicate (and technically demanding). We refer the interested reader to [**49**].

6.4. Katz's expansion. In this section k is large enough so that base change property hold. Sufficient is the vanishing of $H^1(\mathfrak{M}^*(\mathbb{F},\mu_N),\underline{\omega}^k)$ for any field $\mathbb{F}$ which is an $\mathcal{O}_L$-algebra. It would be interesting to know k explicitly.

Consider the map:

$$\mathcal{M}(\mathcal{O}_{Lp},k+j\ell,\mu_N) \xrightarrow{\times E_\ell} \mathcal{M}(\mathcal{O}_{Lp},k+(j+1)\ell,\mu_N). \tag{6.24}$$

The map in (6.24) is injective and splits. We choose complements:

$$\begin{aligned}\mathcal{M}(\mathcal{O}_{Lp},&k+(j+1)\ell,\mu_N)\\ &= E_\ell\cdot\mathcal{M}(\mathcal{O}_{Lp},k+j\ell,\mu_N)\oplus A(\mathcal{O}_{Lp},k+(j+1)\ell,\mu_N).\end{aligned} \tag{6.25}$$

(We let $A(\mathcal{O}_{Lp},k,\mu_N) = \mathcal{M}(\mathcal{O}_{Lp},k,\mu_N)$.) We may tensor with B and we get the same equality with B-coefficients. Then

$$\bigoplus_{a=0}^{j} A(B,k+a\ell,\mu_N) \cong \mathcal{M}(B,k+j\ell,\mu_N); \tag{6.26}$$

the map given by

$$(f_0,\ldots,f_j) \mapsto \sum_{a=0}^{j} f_a\cdot E_\ell^{j-a}. \tag{6.27}$$

Consider the p-adically complete B-module:

$$\begin{aligned}&A^{\mathrm{rigid}}(B,k,\mu_N)\\ &\quad= \Big\{\sum_{a=0}^{\infty} b_a : b_a\in A(B,k+a\ell,\mu_N), b_a\to 0 \text{ p-adically uniformly}\Big\}.\end{aligned} \tag{6.28}$$

By the same methods of Chapter 6.6 we prove:

PROPOSITION 6.8 (Katz's expansion). *Let r be a p-adic unit. There exists an isomorphism:*

$$A^{\mathrm{rigid}}(B,k,\mu_N) \xrightarrow[\psi]{\sim} \mathbb{M}(B,k,\mu_N;r) \tag{6.29}$$

given by

$$\sum_{a=0}^{\infty} b_a \mapsto \left\langle\!\!\left\langle \sum_{a=0}^{\infty} r^a b_a / E_\ell^a \right\rangle\!\!\right\rangle, \tag{6.30}$$

where the right-hand side stands for the p-adic Hilbert modular form whose value on a test object $(\underline{A}, \omega, \beta_N, Y)_{/A}$ *(where* $Y \cdot E_\ell(\underline{A}, \omega) = r$*) is:*

$$\left\langle\!\!\left\langle \sum_{a=0}^{\infty} b_a / E_\ell^a \right\rangle\!\!\right\rangle (\underline{A}, \omega, \beta_N, Y) = \sum_{a=0}^{\infty} Y^a b_a(\underline{A}, \omega, \beta_N). \tag{6.31}$$

6.5. Properties of q-expansions of p-adic modular forms.

PROPOSITION 6.9. *Let* $b \in B$ *be an element dividing a positive power of* p*. Let* $f \in \mathbb{M}(B, k, \mu_N; 1)$*. The followings assertions are equivalent:*

(1) $f \in b \cdot \mathbb{M}(B, k, \mu_N; 1)$.
(2) *The q-expansion of f lies in* $b \cdot B[\![q]\!]$.

The proof is the same as the proof of Proposition 6.19 in Chapter 4.

COROLLARY 6.10. *The q-expansion map on p-adic modular forms is injective.*

REMARK 6.11. If r divides s there is an injective homomorphism that commutes with taking q-expansions

$$\mathbb{M}(B, k, \mu_N; s) \to \mathbb{M}(B, k, \mu_N; r). \tag{6.32}$$

It follows that it is enough to prove Corollary 6.10 for $r = 1$.

THEOREM 6.12. *Let* $f(q) \in B[\![q]\!]$ *be a power series. The following assertions are equivalent:*

(1) $f(q)$ *is the q-expansion of an element* $f \in \mathbb{M}(B, k, \mu_N; 1)$.
(2) *For all n, there exists a positive integer* $M(n)$, $M(n) \equiv 0 \bmod p^{n-1}$*, and a classical modular form* $g_n \in \mathcal{M}(B, k + M(n)\ell, \mu_N)$ *such that the q-expansion* $g_n(q) \equiv f(q) \bmod p^n$.

PROOF. One just repeats the proof of Theorem 6.21 in Chapter 4. First, let us show that (2) $\Longrightarrow$ (1). Writing $E_\ell(q) = 1 + px$, we see that $E_\ell^{p^{n-1}} \equiv 1 \mod p^n$. Now, multiplication of g_n by $E_\ell^{p^{n-1}}$ changes the weight by ℓp^{n-1}, so we can assume $M(n)$ is increasing. Let $\Delta(n) = M(n+1) - M(n)$, so

$$g_{n+1} - g_n \cdot E_\ell^{\Delta(n)} \in p^n \cdot \mathcal{M}(B, k + \ell M(n+1), \mu_N), \tag{6.33}$$

(since $\Delta(n) \equiv 0 \bmod p^{n-1}$).

Hence, $g_0 + \sum_{a=0}^{\infty}(g_{a+1} - g_a \cdot E_\ell^{\Delta(a)}) \in \mathbb{M}(B, k, \mu_N; 1)$. Modulo p^n this sum is

$$g_0 + (g_1 - g_0 E_\ell^{\Delta(0)}) + \cdots + (g_n - g_{n-1} \cdot E_\ell^{\Delta(n-1)}). \tag{6.34}$$

But $E_\ell = 1$ in $\mathbb{M}(B, k, \mu_N; 1)$, so the telescopic sum is equal to g_n. Hence, the q-expansion is $\lim g_n(q) = f(q)$.

The implication (1) $\Longrightarrow$ (2) can be proved as follows: Let $f \in \mathbb{M}(B, k, \mu_N; 1)$. Then

$$f = \psi\left(\sum_{a=0}^{\infty} b_a\right), \quad b_a \in A(B, k + a\ell, \mu_N). \tag{6.35}$$

Consider $c_n = \psi(\sum_{a=0}^{n} b_a) = \phi(\eta(\sum_{a=0}^{n} b_a)) \in \mathcal{M}(B, k + n\ell, \mu_N)$. Take $M(n)$ to be suitably increasing powers of p and $g_n = c_{M(n)}$. □

We didn't define Serre modular forms. But the definition is rather obvious.

DEFINITION 6.13. A p-adic Hilbert modular form (à la Serre) over L_p, of level μ_N, is a uniform p-adic limit of Hilbert modular forms over L, of level μ_N.

Such a modular form would have a limit weight which is a weight in $\mathbb{Z}_p$. This follows from the results in [**49**].

COROLLARY 6.14. *Serre's p-adic modular forms of weight* Norm^k, $k \in \mathbb{Z}$ *are the same as Katz's p-adic modular forms* $\mathbb{M}(B, k, \mu_N; 1)$ *of growth condition* 1.

CHAPTER 6

Deformation Theory of Abelian Varieties

The goal of this chapter is to explain the local deformation theory of abelian varieties in characteristic p, concentrating on the case of abelian varieties with real multiplication. Some of the applications we have in mind are proving that the divisor of the Hasse invariant is a reduced normal crossing divisor.

Most attributions will be omitted. The results are due foremost to Kodaira, Spencer, Grothendieck, Serre, Tate, Dieudonné, Cartier, Lazard, Mumford, Norman, Deligne, Rapoport, Oort, Zink and others.

Here is a rough diagram of the strategy for studying deformations of abelian varieties we are going to present:

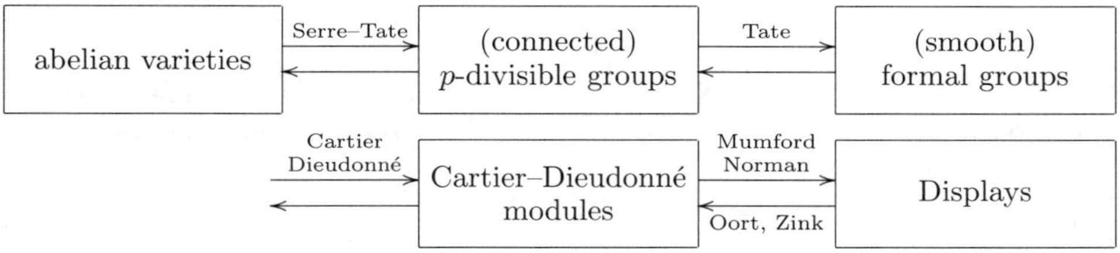

1. Fine moduli schemes

Fix a rational prime p, an integer $g \geq 1$, and an integer $n \geq 3$ such that $(n,p) = 1$. We shall assume we are in one of the following situations:

I. We consider principally polarized abelian schemes $(A, \lambda, \alpha)/_S$ of relative dimension g, and full level-n-structure. Thus, $A \to S$ is an abelian scheme of relative dimension g, $\lambda\colon A \to A^t$ is an isomorphism, defined étale locally on S by a relatively ample line bundle, and $\alpha\colon (\mathbb{Z}/n\mathbb{Z})^{2g} \to A[n]$ is an isomorphism of constant S-group schemes.

There exists a fine moduli space $\mathcal{A}_{g,n} \to \operatorname{Spec}(\mathbb{Z})$ for such data [**113**, Chapter 7, §§2–3 and Appendix]. That is, there exists a scheme $\mathcal{A}_{g,n} \to \operatorname{Spec}(\mathbb{Z})$ representing the functor

$$S \mapsto \{S\text{-isomorphism classes of } (A, \lambda, \alpha)/_S\}. \tag{1.1}$$

DEFINITION 1.1. Let $S \to T$ be a morphism of schemes and $(A, \lambda, \alpha)/_S$ be given. A *deformation* of $(A, \lambda, \alpha)/_S$ to T is a triple $(A', \lambda', \alpha')/_T$ together with an isomorphism over S

$$(A', \lambda', \alpha') \times_T S \cong (A, \lambda, \alpha). \tag{1.2}$$

Two deformations, (A', λ', α') and $(A'', \lambda'', \alpha'')$, are isomorphic if there exists an isomorphism of abelian schemes over T, say $\varphi\colon A' \to A''$, such that $\varphi|_S$ is the identity, $\varphi^*\lambda'' = \lambda'$, and $\varphi \circ \alpha' = \alpha''$.

Let k be a field. Let $\Lambda = k$ if k has characteristic zero and $\Lambda = W(k)$ (infinite Witt vectors) if k has characteristic p. Let $\mathcal{C}_k$ be the category of local artinian rings Λ-algebras $(R, \mathfrak{m}_R)$ together with a *given* isomorphism $R/\mathfrak{m}_R \cong k$. Morphisms are local homomorphisms of rings inducing the identity on k.

Fix an object $(A, \lambda, \alpha)/_{\mathrm{Spec}(k)}$ and consider the functor $\mathcal{C}_k \to \mathbf{Sets}$ given by

$$R \mapsto \{\text{isomorphism classes of deformations over } \mathrm{Spec}(R) \text{ of } (A, \lambda, \alpha)/_k\}. \tag{1.3}$$

This functor is known to be pro-representable by a ring[1], say R^U, and if $x \in \mathcal{A}_{g,n}(k)$ is the moduli point corresponding to $(A, \lambda, \alpha)/_S$, then

$$R^U \cong \widehat{\mathcal{O}}_{\mathcal{A}_{g,n},x}. \tag{1.4}$$

We shall say we are in the situation $\mathbf{I}_p$ if all schemes are $\mathbb{F}_p$-schemes.

II. We are given a totally real number field L of degree g over $\mathbb{Q}$. We denote by $\mathcal{O}_L$ its ring of integers, by $\mathcal{D}_L$ its different ideal relative to $\mathbb{Q}$ and by d_L its discriminant. We fix representatives for the strict ideal classes $(\mathfrak{a}, \mathfrak{a}^+), \dots,$ and we assume that $p \nmid d_L$. Write p as a product of prime ideals

$$p = \mathfrak{p}_1 \cdot \dots \cdot \mathfrak{p}_r \tag{1.5}$$

in $\mathcal{O}_L$, and let

$$f_i = \deg(\mathfrak{p}_i/p), \quad f = \mathrm{lcm}\{f_i\}. \tag{1.6}$$

Fix a field $\mathbb{F}$ of p^f elements. Note that if we let σ denote the Frobenius automorphism on $\mathbb{F}$, as well as on the Witt vectors $W(\mathbb{F})$, then the embeddings of L in $W(\mathbb{F}) \otimes_{\mathbb{Z}} \mathbb{Q}$ can be indexed by

$$\sigma_{(j,i)}, \quad 1 \le j \le r, 1 \le i \le f_j, \tag{1.7}$$

in such a way that

$$\sigma \circ \sigma_{(j,i)} = \sigma_{(j,i+1)}. \tag{1.8}$$

We consider quadruples $(A, \iota, \lambda, \beta_N)/_S$ where S is a $\mathbb{Z}[{d_L}^{-1}]$-scheme, consisting of an abelian variety $A \to S$ of relative dimension g, an embedding of rings $\iota \colon \mathcal{O}_L \to \mathrm{End}(A/S)$ such that condition (R) holds (Chapter 3, §5), an isomorphism of $\mathcal{O}_L$ modules with a notion of positivity $\lambda \colon (\mathcal{M}_A, \mathcal{M}_A^+) \cong (\mathfrak{a}, \mathfrak{a}^+)$, and an $\mathcal{O}_L$-equivariant embedding $\beta_N \colon \mu_N \otimes \mathcal{D}_L^{-1} \to A$.

There exists a fine moduli scheme $\mathfrak{M}(\mu_N) \to \mathbb{Z}[d_L^{-1}]$ for such data (see [**135**] for similar statements). The same remarks concerning deformations, deformations over $\mathcal{C}_k$, etc. apply here with the obvious modifications. We shall say we are in case $\mathbf{II}_p$ if all schemes are $\mathbb{F}$-schemes.

Most local deformations theories over p-adic rings for case $\mathbf{I}_p$ or $\mathbf{II}_p$ (or similar scenarios) pass through the *p-divisible group*,

$$A(p) = \varinjlim_n A[p^n], \tag{1.9}$$

of the abelian variety A. See [**11**] for a review of the available tools; [**144, 161**] for essentials on p-divisible groups, and [**65**] (that should be updated using [**75**]) for interesting applications; we content ourselves with the mere definitions.

[1]This follows from the existence of the global moduli space $\mathcal{A}_{g,n}$—cf. Chapter 3, Section 4.1. It can also be proven directly using Schlessinger's theory. See [**127**, §2.3].

DEFINITION 1.2. Let S be a scheme. A p-divisible group $\mathcal{G}$ over S is an inductive system of finite flat commutative group schemes $\{G_i : i \in \mathbb{N}\}$ and closed immersions

$$0 = G_0 \hookrightarrow G_1 \hookrightarrow G_2 \hookrightarrow G_3 \hookrightarrow \cdots , \tag{1.10}$$

such that for every $i \geq 1$ we have an exact sequence

$$0 \to G_1 \to G_{i+1} \xrightarrow{[p]} G_i \to 0. \tag{1.11}$$

REMARK 1.3. Some authors call p-divisible groups *Barsotti–Tate* groups.

EXERCISE 1.4. *Prove that for $i, j \geq 0$ we have $G_i = \mathrm{Ker}([p^i]\colon G_{i+j} \to G_{i+j})$ and in fact $[p^i]G_{i+j} = G_j$. Prove that if S is connected there exists an integer h, called the* height *of $\mathcal{G}$, such that G_i has rank p^{hi}.*

Given a p-divisible group $\mathcal{G} = \{G_i\}$, multiplication by p gives surjective morphisms $[p]\colon G_{i+1} \to G_i$ for every $i \geq 1$ that induce closed embeddings of the dual group schemes $G_i^t \xrightarrow{[p]^*} G_{i+1}^t$.

DEFINITION 1.5. Let $\mathcal{G} = \{G_i\}$ be a p-divisible group. Define the *dual p-divisible group* $\mathcal{G}^t$ to be the inductive system $\{G_i^t : i \in \mathbb{N}\}$ with closed immersion induced from the map $[p]$ on $\mathcal{G}$, as above.

EXERCISE 1.6. *Prove that* $\mathcal{G}^t$ *is a p-divisible group of the same height as $\mathcal{G}$.*

If R is a complete Noetherian local ring of residue characteristic p, one can define the dimension of a p-divisible group $\mathcal{G}$ over R in the sense of formal schemes. (Via Tate's Theorem—Theorem 3.6—this is the dimension of the associated formal group.) A fundamental formula relating the dimension and the height is

$$\dim(\mathcal{G}) + \dim(\mathcal{G}^t) = \mathrm{height}(\mathcal{G}). \tag{1.12}$$

As an illustration consider the p-divisible group of height one $\widehat{\mathbb{G}}_m$. It is the inductive system

$$1 \hookrightarrow \mu_p \hookrightarrow \mu_{p^2} \hookrightarrow \mu_{p^3} \hookrightarrow \cdots . \tag{1.13}$$

The system of dual group schemes is

$$0 \hookrightarrow \underline{\mathbb{Z}/p\mathbb{Z}} \hookrightarrow \underline{\mathbb{Z}/p^2 Z} \hookrightarrow \underline{\mathbb{Z}/p^3 Z} \hookrightarrow \cdots , \tag{1.14}$$

where the map $\underline{\mathbb{Z}/p^i\mathbb{Z}} \to \underline{\mathbb{Z}/p^{i+1}\mathbb{Z}}$ is induced from the map of residue classes taking $a \bmod p^i$ to $pa \bmod p^{i+1}$. We therefore rewrite (1.14) as

$$0 \hookrightarrow \underline{p^{-1}\mathbb{Z}/\mathbb{Z}} \hookrightarrow \underline{p^{-2}\mathbb{Z}/\mathbb{Z}} \hookrightarrow \underline{p^{-3}\mathbb{Z}/\mathbb{Z}} \hookrightarrow \cdots \tag{1.15}$$

(with the natural inclusions). Thus, the dual of $\widehat{\mathbb{G}}_m$ is $\underline{\mathbb{Q}_p/\mathbb{Z}_p}$.

Let k be a fixed field of characteristic p. Let A/k be an abelian variety.

THEOREM 1.7 (Serre–Tate). *For every ring R in $\mathcal{C}_k$ the functor*

$$\mathbb{A} \mapsto \mathbb{A}(p), \tag{1.16}$$

induces an equivalence of categories between the category of deformations $\mathbb{A}$ of A over R with morphisms being morphisms of abelian schemes over R, to the category of deformations of $A(p)$ into p-divisible groups $\mathcal{G}$ over R with the morphisms being morphisms of p-divisible groups whose restriction to $A(p)$ comes from an endomorphism of A/k.

One immediately deduces that in case $\mathbf{I}_p$ we have an equivalence of categories

$$\left\{\begin{array}{l}\text{Isomorphism classes of defor-}\\ \text{mations of } (A,\lambda,\alpha)/k \text{ to } R\end{array}\right\} \to \left\{\begin{array}{l}\text{Isomorphism classes of deforma-}\\ \text{tions } \mathcal{G} \text{ of } \big(A(p),\lambda(p)\big)/k \text{ to } R\end{array}\right\}, \tag{1.17}$$

where $\lambda(p)\colon A(p) \to A^t(p) \cong (A(p))^t$ is deformed into a symmetric isomorphism $\lambda\colon \mathcal{G} \to \mathcal{G}^t$.

In case $\mathbf{II}_p$ we have an equivalence of categories

$$\left\{\begin{array}{l}\text{Isomorphism classes of defor-}\\ \text{mations of } (A,\iota,\lambda,\alpha)/k \text{ to } R\end{array}\right\} \to \left\{\begin{array}{l}\text{Isomorphism classes of deforma-}\\ \text{tions } \mathcal{G} \text{ of } \big(A(p),\iota(p)\big)/k \text{ to } R\end{array}\right\}, \tag{1.18}$$

where $\iota(p)\colon \mathcal{O}_L \to \operatorname{End}\big(A(p)\big)$ is deformed to homomorphisms $\iota\colon \mathcal{O}_L \to \operatorname{End}(\mathcal{G})$.

REMARK 1.8. Note that the extra level structure disappears. This follows from rigidity of étale group schemes; i.e., there always exists a way to extend this level structure to any such deformation, and uniquely so. Note also that we do not require condition (R) to hold for a deformation. It is automatically satisfied. This follows from

EXERCISE 1.9. *Let $(R,\mathfrak{m})$ be an object of $\mathcal{C}_k$. If M is a finitely generated $R\otimes\mathcal{O}_L$-module, free as an R-module, such that $M\otimes_R R/\mathfrak{m}$ is a free $R/\mathfrak{m}\otimes\mathcal{O}_L$-module then M is a free $R\otimes\mathcal{O}_L$-module.*

It also turns out that deforming the $\mathcal{O}_L$ structure automatically allows deformation of the polarization structure (uniquely). This follows from the deformation theory we shall develop.

EXAMPLE 1.10. Let $E/\bar{k}$ be an ordinary elliptic curve over $\bar{k}$. Then

$$E(p) \cong \underline{\mathbb{Q}_p/\mathbb{Z}_p}_{/\bar{k}} \oplus (\widehat{\mathbb{G}}_m)_{/\bar{k}}. \tag{1.19}$$

There is a canonical deformation, or *canonical lift*, of $E(p)$ to R, namely,

$$\underline{\mathbb{Q}_p/\mathbb{Z}_p}_{/R} \oplus (\widehat{\mathbb{G}}_m)_{/R}. \tag{1.20}$$

Clearly it has the property that every endomorphism of $E(p)$ extends to it. We conclude that for every $R \in \mathcal{C}_{\bar{k}}$ there exists a canonical lift of E over R, say E^{can}/R with the property

$$\operatorname{End}_R(E^{\mathrm{can}}) \cong \operatorname{End}_{\bar{k}}(E). \tag{1.21}$$

We therefore find a coarser formulation of the Serre–Tate coordinates discussed in Chapter 3, §4.2. The same argument applies to ordinary abelian varieties.

EXERCISE 1.11. *Explain the (many) difficulties in making sense of a "canonical lifting" of an elliptic curve over a nonalgebraically closed field of characteristic p. However, for ordinary elliptic curves (or abelian varieties) over a perfect field, such a theory exists and is not hard to imagine. One uses "rigidity of tori." See, for example,* [**96**, Chapter V, §3].

EXERCISE 1.12. *Let* val *be the valuation on $\overline{\mathbb{Q}_p}$ normalized such that* $\operatorname{val}(p) = 1$. *We say that a j-invariant is CM if the corresponding elliptic curve has CM over $\overline{\mathbb{Q}_p}$.*

Let $j_0 \in \mathbb{Q}_p$ be a j-invariant. Prove that if $\operatorname{val}(j_0) < 0$ *then there is no CM j-invariant j_1 such that* $\operatorname{val}(j_1 - j_0) > 0$, *and if* $\operatorname{val}(j_0) \geq 0$ *and of ordinary reduction then there is a CM j-invariant $j_1 \in \mathbb{Z}_p$ such that* $\operatorname{val}(j - j_1) > 0$.

What happens if we drop the assumption of ordinary reduction? And, are the CM j-invariants dense in the open set of j-invariants of nonnegative valuation?

2. Proof of the Serre–Tate Theorem

We follow Drinfeld's proof as presented in [**83**]. Consider a more general scenario: We are given two rings, R and R_0, and a ring homomorphism $\varphi\colon R \to R_0$ with kernel I, such that there exists a positive integer b with $I^b = 0$ and there exists a positive integer a such that $p^a = 0$ in R_0. We consider two categories:

- $\mathbf{Ab}_R$ = Category of abelian schemes over R with morphisms being homomorphisms of abelian schemes.
- $\mathbf{Def}_R$ = Category of triples $(A/R_0, G/R, \iota)$, where A/R_0 is an abelian scheme, G/R is a p-divisible group and $\iota\colon A(p) \to G|_{R_0}$ is an isomorphism. An arrow $(A/R_0, G/R, \iota) \to (A'/R_0, G'/R, \iota')$ is a homomorphism $f\colon G \to G'$ whose restriction $f_0\colon A(p) \to A'(p)$, induced via ι and ι', is induced also from a homomorphism of abelian schemes $A \to A'$.

We have a natural functor

$$\Phi\colon \mathbf{Ab}_R \to \mathbf{Def}_R; \quad \Phi(A) = (A \otimes_R R_0, A(p)/R, \iota_{\mathrm{can}}). \tag{2.1}$$

THEOREM 2.1 (Serre–Tate). *The functor Φ is an equivalence of categories.*

PROOF. We first make a

DEFINITION 2.2. Let G be either a finite flat group scheme over R or an abelian scheme over R. Let J be an ideal of R. Define a group functor G_J on the category R-**Alg** by

$$G_J(S) = \mathrm{Ker}\big(G(S) \to G(S/J)\big), \tag{2.2}$$

where here and below S/J stands for $S \otimes_R (R/J)$.

LEMMA 2.3. *Let $q = p^{ab}$. Then*

$$[q]G_I = 0. \tag{2.3}$$

PROOF. We assume that G is an abelian scheme. After localizing on R, we may assume that $G = \mathrm{Spf}\, R[\![x_1, \dots, x_d]\!]$ with augmentation ideal topologically generated by $x_1, \dots, x_d$ (for a finite flat group scheme, write $G = \mathrm{Spec}\, R[x_1, \dots, x_d]/\mathfrak{a}$ for a suitable ideal $\mathfrak{a}$ and augmentation ideal generated by $x_1, \dots, x_d$).

Let $S \in R$-**Alg** and $\alpha \in G_J(S)$. Then we may identify α with a vector $(\alpha_1, \dots, \alpha_d)$ with $\alpha_i \in JS$. Writing

$$([p^a]\alpha)_i = p^a\alpha_i + \text{higher order terms in } \alpha_1, \dots, \alpha_d, \tag{2.4}$$

we see that $[p^a]\alpha \in G_{IJ+J^2}$. Applying this to the ideals $J = I, I^2, I^3, \dots$, we see that $[q]G_I = 0$. ◻

Note that the lemma also holds for a p-divisible group over R.

PROPOSITION 2.4. *Let G and H be either p-divisible groups or abelian schemes over R. Let G_0 and H_0 be the objects obtained from G and H, respectively, by base change to R_0. Let $q = p^{ab}$.*

1. *The groups $\mathrm{Hom}_R(G, H)$ and $\mathrm{Hom}_{R_0}(G_0, H_0)$ are torsion free.*
2. *The homomorphism $\mathrm{Hom}_R(G, H) \to \mathrm{Hom}_{R_0}(G_0, H_0)$ is injective.*
3. *Given f_0 in $\mathrm{Hom}_{R_0}(G_0, H_0)$ there exists a unique homomorphism $\widetilde{qf_0}$ in $\mathrm{Hom}_R(G, H)$ lifting qf_0.*
4. *The homomorphism f_0 lifts to a homomorphism $\widetilde{f_0}$ in $\mathrm{Hom}_R(G, H)$ if and only if $G[q] \subset \mathrm{Ker}(\widetilde{qf_0})$.*

PROOF. Part (1) is clear. For (2), we note that

$$\mathrm{Ker}\big(\mathrm{Hom}_R(G,H) \to \mathrm{Hom}_{R_0}(G_0,H_0)\big) = \mathrm{Hom}_R(G,H_I). \tag{2.5}$$

But H_I is torsion and $\mathrm{Hom}_R(G,H)$ is torsion free!

(3) Let $\alpha \in G(S)$ and let $\bar{\alpha}$ denote its image in $G(S/I)$. Regarding $\bar{\alpha}$ as an element of $G_0(S/I)$, we write $f_0(\bar{\alpha})$ for the image in $H_0(S/I)$. Since the map $H(S) \to H(S/I)$ is surjective, we may lift $f_0(\bar{\alpha})$ to some $\alpha' \in H(S)$. Let

$$\widetilde{qf_0}(\alpha) = q\alpha'. \tag{2.6}$$

This is well-defined, because any other choice of lift, say α'' differs from α' by an element of $H_I(S)$, hence killed by q. Immediate verification shows that $\widetilde{qf_0}$ is a homomorphism and its uniqueness follows from part 2.

(4) Clearly if f_0 lifts to some $\widetilde{f_0}$ then $q\widetilde{f_0}$ lifts $\widetilde{qf_0}$ and so, by uniqueness, $\widetilde{qf_0} = q\widetilde{f_0}$. Therefore, $\widetilde{qf_0}$ kills $G[q]$. Conversely, if $\widetilde{qf_0}$ kills $G[q]$ we may factor it as $\widetilde{qf_0} = qg$ for some $g \in \mathrm{Hom}_R(G,H)$. Let $g_0 \in \mathrm{Hom}_{R_0}(G_0,H_0)$ be the homomorphism induced by g. Then $qg_0 = qf_0$ as both are obtained by restricting $\widetilde{qf_0}$. Since $\mathrm{Hom}_{R_0}(G_0,H_0)$ is torsion free, $g_0 = f_0$. □

Let us show now that Φ is fully-faithful, i.e., that

$$\mathrm{Hom}_R(A,B) \to \mathrm{Hom}\Big(\big(A_0, A(p)\big), \big(B_0, B(p)\big)\Big) \tag{2.7}$$

is an isomorphism. We know it is injective by Proposition 2.4. Let, therefore, f be in $\mathrm{Hom}(A_0,B_0)$ and ϕ be in $\mathrm{Hom}\big(A(p),B(p)\big)$ such that

$$f(p) = \phi_0, \tag{2.8}$$

where here and below "(p)" denotes passing to the p-divisible and the subscript "$_0$" denotes base change to R_0. Using Proposition 2.4, we see that the homomorphism qf lifts uniquely to a homomorphism $\widetilde{qf} \in \mathrm{Hom}_R(A,B)$ and f lifts to $\mathrm{Hom}_R(A,B)$ if and only if $\widetilde{qf}$ kills $A[q]$. But $\widetilde{qf}$ kills $A[q]$ if and only if $\widetilde{qf}(p)$ kills $A[q]$. Since $\widetilde{qf}(p)$ is a lift of $q\phi_0$ to $A(p)$ and since ϕ_0 *does* lift to $A(p)$, we have, by part (4) of Proposition 2.4, that $\widetilde{qf}(p)$ kills $A[q]$. Therefore, f lifts.

It remains to prove essential surjectivity of Φ. Let (A_0, G, ι) be an object of $\mathbf{Def}_R$. Because $R \to R_0$ is a nilpotent thickening there exists *some* abelian scheme B lifting A_0 to R. This can be deduced for example from [**113**, §6.3]. Let α_0 be the isomorphism $B(p)_0 \to G_0$. Let $\widetilde{q\alpha_0} \in \mathrm{Hom}_R(B(p),G)$ be the canonical lift of $q\alpha_0$. Let $K = \mathrm{Ker}(\widetilde{q\alpha_0})$. Then K is a finite flat group scheme such that $K_0 = A_0[q]$, and $D := B/K$ is an abelian scheme. The map $\widetilde{q\alpha_0}$ induces an isomorphism between $D(p)$ and G, which under the identification $D_0 = A_0/A_0[q]$ is equal to α_0. □

REMARK 2.5. The question of finding the "largest" infinitesimal neighborhood to which an endomorphism lifts is very interesting and probably much remains to be discovered here. We content ourselves by mentioning the following intriguing formula of Gross–Zagier [**59**, Prop. 2.3]:

Let W be a complete d.v.r. with uniformizer π and of mixed characteristic. Normalize the valuation on W so that $\mathrm{val}(\pi) = 1$. Assume that the residue field is algebraically closed. Let E, E' be two elliptic curves over W with good reduction and let j, j' be the corresponding j-invariants. Let $2i(n)$ be the number of isomorphisms $E_{/W/(\pi^n)} \to E'_{/W/(\pi^n)}$, i.e., the number of isomorphisms from the curve E

to the curve E' considered as curves modulo π^n. Then

$$\mathrm{val}(j - j') = \sum_{n=1}^{\infty} i(n). \tag{2.9}$$

3. Deformation of p-divisible groups

Let R be a ring. By an *n-dimensional smooth commutative formal group over R*, or simply, *n-dimensional formal group*, we mean the following: It is a power series ring $R[\![x_1, \ldots, x_n]\!]$ together with n power series

$$F(\underline{x}, \underline{y}) = \big(F_1(\underline{x}, \underline{y}), \ldots, F_n(\underline{x}, \underline{y})\big) \tag{3.1}$$

in the $2n$ variables $\underline{x} = (x_1, \ldots, x_n)$ and $\underline{y} = (y_1, \ldots, y_n)$. The following are required to hold for every i:

- $F_i(\underline{x}, 0) = x_i$, $F_i(0, \underline{y}) = y_i$.
- $F_i(\underline{x}, \underline{y}) = F_i(\underline{y}, \underline{x})$.
- $F_i(F(\underline{x}, \underline{y}), \underline{z}) = F_i\big(\underline{x}, F(\underline{y}, \underline{z})\big)$.

Define inductively the multiplication by m maps $[m]$:

$$[1](\underline{x}) = \underline{x}; \quad [m](\underline{x}) = F([m-1](\underline{x}), \underline{x}). \tag{3.2}$$

Note that $[m](\underline{x}) = \big([m]_1(\underline{x}), \ldots, [m]_n(\underline{x})\big)$ defines an endomorphism of $R[\![\underline{x}]\!]$ by

$$x_i \mapsto [m]_i(\underline{x}). \tag{3.3}$$

DEFINITION 3.1. Let R be a p-adic ring, i.e., $R \cong \varprojlim R/p^m R$. We say that the formal group over R is *p-divisible* if $R[\![\underline{x}]\!]$, considered as an $R[\![\underline{x}]\!]$ module via $[p](\underline{x})$, is a finitely generated module.

EXAMPLE 3.2. Let $n = 1$. Define the *additive formal group* $(\widehat{\mathbb{G}}_a)_{/R}$ over any ring R by

$$F_a(x, y) = x + y. \tag{3.4}$$

Thus,

$$[p](x) = px. \tag{3.5}$$

One verifies that $\widehat{\mathbb{G}}_a$ is never p-divisible. Note that $\widehat{\mathbb{G}}_a/R$ is the completion of $\mathbb{G}_a/R = \mathrm{Spec}(R[x])$ along the zero section defined by the ideal sheaf (x).

EXAMPLE 3.3. Let $n = 1$. Define the *multiplicative formal group* $(\widehat{\mathbb{G}}_m)_{/R}$ over any ring R by

$$F_m(x, y) = x + y - xy. \tag{3.6}$$

Equivalently, $1 - F_m(x, y) = (1 - x)(1 - y)$. Thus,

$$1 - [p](x) = (1 - x)^p. \tag{3.7}$$

One verifies that $\widehat{\mathbb{G}}_m$ is p-divisible for every p. In fact, $\{1, x, \ldots, x^{p-1}\}$ span $R[\![x]\!]$ as a module (via $[p](x)$) over $R[\![x]\!]$. Note that $\widehat{\mathbb{G}}_m/R$ is the completion of $\mathbb{G}_m/R = \mathrm{Spec}(R[t, t^{-1}]) = \mathrm{Spec}(R[1 - x, (1 - x)^{-1}])$, where $x = 1 - t$, along the zero section defined by the ideal sheaf (x).

Note that F_m is the group law of $1 - \exp(-x)$. That is

$$1 - \exp\big(-(x + y)\big) = F_m\big(1 - \exp(-x), 1 - \exp(-y)\big). \tag{3.8}$$

EXAMPLE 3.4. Let ϵ and δ be free variables. Define a power series $F_{\epsilon,\delta}(x,y)$ in $\mathbb{Z}[\frac{1}{2}][\![\epsilon,\delta]\!]$ by

$$F_{\epsilon,\delta}(x,y) = \frac{x\sqrt{\ell(y)} + y\sqrt{\ell(x)}}{1-\epsilon x^2y^2}, \quad \ell(x) = 1 - 2\delta x^2 + \epsilon x^4. \tag{3.9}$$

The fact that this is a formal group law is quite tedious to check directly. A more insightful approach is to prove that this is the formal group associated to the elliptic curve $E\colon y^2 = \ell(x)$ with the zero point $(0,1)$. This formal group plays a special role in elliptic cohomology. The paper [**136**] is a good starting point.

One defines the Legendre polynomials $P_n(t) \in \mathbb{Z}[\frac{1}{2}][t]$ by the identity

$$\frac{1}{\sqrt{1-2tx+x^2}} = \sum_{n=0}^{\infty} P_n(t)x^n, \tag{3.10}$$

and the homogeneous Legendre polynomials by the formula $P_n(\delta,\epsilon) = P_n(\delta/\sqrt{\epsilon})\epsilon^{n/2}$. The logarithm $\log_{\epsilon,\delta}$ of the formal group $F_{\epsilon,\delta}(x,y)$ is given by the power series

$$\log_{\epsilon,\delta}(x) = \sum_{n=0}^{\infty} \frac{P_n(\delta,\epsilon)}{2n+1}x^{2n+1}. \tag{3.11}$$

As a special case, take ϵ and δ equal to 1 (this is related to the signature of a manifold). Then the formal group is simply

$$F(x,y) = \frac{x+y}{1+xy}. \tag{3.12}$$

It is immediate to verify the axioms in this case. Note that $F(x,y)$ is the group law of the hyperbolic tangent

$$\tanh(x) = \frac{e^x - e^{-x}}{e^x + e^{-x}}. \tag{3.13}$$

That is,

$$\tanh(x+y) = F\big(\tanh(x), \tanh(y)\big). \tag{3.14}$$

EXERCISE 3.5. *Work out the case of* $\delta = -\frac{1}{8}$ *and* $\epsilon = 0$ *(this is related to the* $\hat{A}$*-genus of a manifold).*

THEOREM 3.6 (Tate [**161**, Prop. 1]). *Let* $(R, \mathfrak{m})$ *be a complete Noetherian local ring with residue field* $k = R/\mathfrak{m}$ *of characteristic* p. *The category of connected* p*-divisible groups over* R *is equivalent to the category of* p*-divisible smooth commutative formal groups over* R.

Note that every R in $\mathcal{C}_k$ is of this form. Thus, to study local deformations of data $\mathbf{I}_p$ or $\mathbf{II}_p$, one is led to consider p-divisible formal groups over local artinian rings. We remark that the situation that will ultimately interest us is when both the p-divisible group and its dual are connected. To deal with the general situation one uses the theory of mixed extensions (see [**11**]). We further remark that certain strengthenings of Tate's Theorem are known. For example, if one generalizes the notion of a formal group to objects "looking locally like a formal group," then Tate's Theorem holds for p-adic rings R (i.e., $R \cong \varprojlim R/p^nR$)—see [**96**, Chapter II].

4. Commutative smooth formal groups

We follow Lazard [**91**] in our exposition of commutative smooth formal groups, or simply, formal groups, via their Cartier-Dieudonné modules. We shall refer for most details to loc. cit. Again, the scope of the text compels us to omit most proofs. However, the reader is encouraged to work them out himself. Most of them are not very difficult.

Before embarking, we remark that while Cartier–Dieudonné theory works for *any* (commutative, smooth) formal group over *any* commutative associative ring with 1, the *applications* to studying deformations of p-divisible groups are limited to specific rings, e.g., artinian local rings.

Let K be any commutative associative ring with 1 and denote by $\mathbf{nil}(K)$ the category of commutative nil-algebras A over K. This means that every element in A is nilpotent. Note that A has no identity element. We also let $\mathbf{nil}(K,n)$ be the full subcategory of $\mathbf{nil}(K)$ consisting of nil-algebras A such that for every $x_1,\dots,x_{n+1}$ in A we have $x_1\cdot\dots\cdot x_{n+1}=0$.

We are interested in three closely related functors from the category $\mathbf{nil}(K)$ to the category of pointed sets $^*\mathbf{Sets}$. They are the following:

Models: These are the functors

$$D^{(I)}\colon \mathbf{nil}(K)\to {}^*\mathbf{Sets};\quad D^{(I)}(A)=A^{(I)}, \tag{4.1}$$

where I is any index set and

$$A^{(I)}=\{(a_\alpha)_{\alpha\in I} : a_\alpha\in A \text{ and } a_\alpha=0 \text{ except for finitely many } \alpha\}. \tag{4.2}$$

Formal modules: For any *free* K-module L, define a functor

$$L^+\colon \mathbf{nil}(K)\to {}^*\mathbf{Sets};\quad L^+(A)=A\otimes_K L. \tag{4.3}$$

Formal varieties: A formal variety over K is a functor,

$$\mathcal{V}\colon \mathbf{nil}(K)\to {}^*\mathbf{Sets}, \tag{4.4}$$

isomorphic to $D^{(I)}$ for some I.

Thus, models and formal modules are examples of formal varieties. For the latter, a choice of a K-basis to L provides such an isomorphism. Formal varieties appear naturally in algebraic geometry in the following way:

Let V be an n-dimensional algebraic variety over K and let $v\in V$ be a non-singular K-rational point. The completion of the local ring of v is isomorphic, by a choice of local coordinates, to $K[\![x_1,\dots,x_n]\!]$. The functor defined on $\mathbf{nil}(K)$ by

$$A\mapsto \mathrm{Hom}_K^{\mathrm{Cont}}(K[\![x_1,\dots,x_n]\!],K\oplus A), \tag{4.5}$$

where

$$\mathrm{Hom}_K^{\mathrm{Cont}}(K[\![x_1,\dots,x_n]\!],K\oplus A):=\varinjlim \mathrm{Hom}_k(K[\![x_1,\dots,x_n]\!]/I^j,K\oplus A) \tag{4.6}$$

and I is the maximal ideal of $K[\![x_1,\dots,x_n]\!]$, is identified with D^n. More canonically, the completion $\widehat{V}$ of V along v defines a formal variety taking every nil-K-algebra A to $\mathrm{Mor}(\mathrm{Spec}(K\oplus A),\widehat{V})$ (morphisms in the category of formal varieties).

A particular case of great importance is when V is an abelian variety (resp. scheme) and v is its zero point (resp. section). This includes Example 3.4 as a special case. The case when V is the additive group (resp. multiplicative group) leads to Example 3.2 (resp. Example 3.3).

Formal varieties form a category whose morphisms are functorial morphisms (natural transformations) of functors. If $\mathcal{V}$ and $\mathcal{W}$ are formal varieties, we denote by $\mathrm{Mor}(\mathcal{V}, \mathcal{W})$ the morphisms from $\mathcal{V}$ to $\mathcal{W}$.

We shall use multi-index notation. If $\alpha \in \mathbb{N}^{(I)}$ is the vector $(\alpha_i)_{i\in I}$ then we let $x^\alpha = \prod_{i\in I} x^{\alpha_i}$ and $|\alpha| = \sum_{i\in I} \alpha_i$. In general, for $\alpha \in A^{(I)}$ the support of α is the finite set $\mathrm{supp}(\alpha) = \{i \in I : \alpha_i \neq 0\}$.

LEMMA 4.1 (Morphism Lemma, [**91**, p. 10]). *The functorial morphisms*

$$f\colon D^{(I)} \to D, \tag{4.7}$$

are in bijection with formal power series $\sum_{\alpha\in\mathbb{N}^{(I)}} c_\alpha x^\alpha$ *with* $c_0 = 0$. *A morphism* $f\colon D^{(I)} \to D^{(J)}$ *is defined by its components* $f_j\colon D^{(I)} \to D$, $j \in J$.

PROOF. The only point here is that any morphism $f\colon D^{(I)} \to D$ can be expressed as a power series. Given $\lambda = (J, n)$ with J a finite subset of I and n an integer, let A_λ be freely generated in $\mathbf{nil}(K, n)$ by elements $\{x_{\lambda,i} : i \in J\}$.

Consider the point x_λ in $D^{(I)}(A_\lambda)$ given by $x_\lambda = (x_{\lambda,i})_{i\in I}$, where $x_{\lambda,i} = 0$ for $i \notin J$. Then $f_{A_\lambda}(x_\lambda)$ is an element of A_λ and can thus be written as

$$f_{A_\lambda}(x_\lambda) = \sum_{\substack{\alpha\in\mathbb{N}^{(I)}\\ 0<|\alpha|\leq n\\ \mathrm{supp}(\alpha)\subset J}} c_{\lambda,\alpha} x^\alpha \tag{4.8}$$

for uniquely determined $c_{\lambda,\alpha} \in K$. But for any $A \in \mathbf{nil}(K)$ and any $x \in A$ we can find $\lambda = (J, n)$ and $\phi\colon A_\lambda \to A$ such that $x = \phi(x_\lambda)$. By functoriality

$$f_A(x) = \sum_{\substack{\alpha\in\mathbb{N}^{(I)}\\ 0<|\alpha|\leq n\\ \mathrm{supp}(\alpha)\subset J}} c_{\lambda,\alpha} x^\alpha. \tag{4.9}$$

Finally, applying this consideration for the case $x = x_{\lambda'} \in A_{\lambda'}$ for $\lambda' = (J', n')$ and $\lambda = (J, n)$ such that $J' \subset J$ and $n' \leq n$, one sees that $c_{\lambda,\alpha}$ depend only on α. □

Let S be a scheme over K. Yoneda's Lemma tells us that S is determined by its functor of points $h_S\colon K\text{-}\mathbf{alg} \to \mathbf{Sets}$ given by $h_S(R) = \mathrm{Mor}_{\mathrm{Spec}(K)}(\mathrm{Spec}(R), S)$. That is, the functor $S \mapsto h_S$ is fully-faithful [**111**, II.6], [**168**]. Lemma 4.1 is Yoneda's Lemma for formal varieties, only that on the functor side we only take its values on $\mathbf{nil}(K)$. The functor $D^{(I)}$ is naturally identified with the functor of points of the formal scheme $R[\![x_i : i \in I]\!]$ and note that giving a power series $\sum_{\alpha\in\mathbb{N}^{(I)}} c_\alpha x^\alpha$ with constant term 0 is equivalent to giving a morphism of formal schemes over R from $R[\![x]\!]$ to $R[\![x_i : i \in I]\!]$.

4.1. Curves. Let $\mathcal{V}$ be a formal variety. A *curve* γ in $\mathcal{V}$ is a morphism

$$\gamma\colon D \to \mathcal{V}. \tag{4.10}$$

Put

$$\mathfrak{C}(\mathcal{V}) = \mathrm{Mor}(D, \mathcal{V}) \tag{4.11}$$

—the set of curves in $\mathcal{V}$. One defines $\mathcal{T}\gamma$ to be the induced map

$$\mathcal{T}\gamma\colon D|_{\mathbf{nil}(K,1)} \to \mathcal{V}|_{\mathbf{nil}(K,1)}, \tag{4.12}$$

and calls it the *tangent* to γ. We further let

$$\mathcal{T}\mathcal{V} = \{\mathcal{T}\gamma : \gamma \in \mathfrak{C}(\mathcal{V})\}, \tag{4.13}$$

and call it the *tangent space* to $\mathcal{V}$. We make $\mathcal{T}$ into a functor by defining $\mathcal{T}f$ for $f\colon \mathcal{V} \to \mathcal{W}$ by the formula

$$\mathcal{T}f(\mathcal{T}\gamma) = \mathcal{T}(f \circ \gamma). \tag{4.14}$$

One easily verifies, using the Morphism Lemma, that any curve in $\mathfrak{C}(D^{(I)})$ is of the form $\gamma = \sum_{n\geq 1} a_n t^n$ for some $a_n \in K^{(I)}$ and vice versa. (Thus, the geometric intuition is that of a "germ of a curve through the origin.") Then $\mathcal{T}\gamma$ is just $a_1 \in K^{(I)}$. Beside showing that the definition of tangent agrees with the intuitive one, it also gives a canonical isomorphism of $\mathcal{T}D^{(I)}$ with $K^{(I)}$, and thus $\mathcal{T}D^{(I)}$ is canonically a free K-module. One can show further that given a formal variety $\mathcal{V}$, the K-linear structure on $\mathcal{T}\mathcal{V}$ deduced from a choice of isomorphism $\mathcal{V} \cong D^{(I)}$ is in fact independent of the choice of isomorphism. Furthermore, if $f\colon \mathcal{V} \to \mathcal{W}$ is a morphism then $\mathcal{T}f$ is K-linear. One puts:

$$\dim(\mathcal{V}) = \dim_K(\mathcal{T}\mathcal{V}). \tag{4.15}$$

The concept of curves is crucial to the whole theory we are about to present. The following results are evidence for that.

THEOREM 4.2 (Isomorphism Theorem, [**91**, I.8]). *A morphism of formal varieties $f\colon \mathcal{V} \to \mathcal{W}$ is an isomorphism if and only if $\mathcal{T}f\colon \mathcal{T}\mathcal{V} \to \mathcal{T}\mathcal{W}$ is an isomorphism.*

The proof is not difficult. One direction is immediate. For the other, one passes to models and applies the Morphism Lemma, which reduces the assertion to *formal* inversion of power series.

LEMMA 4.3 (Curves Lemma, [**91**, I.10]). *The functor $\mathcal{V} \mapsto \mathfrak{C}(\mathcal{V})$ is faithful. That is, if f and f' are morphisms from $\mathcal{V}$ to $\mathcal{W}$ and $f \circ \gamma = f' \circ \gamma$ for every $\gamma \in \mathfrak{C}(\mathcal{V})$, then $f = f'$.*

The argument runs as follows: Reduce to the case $\mathcal{V} = D^{(I)}$ and $\mathcal{W} = D$ and use the Morphism Lemma and enough "test curves" to show that $f - f'$ (the difference defined using the power series expression) is actually zero.

4.2. Formal groups. A formal variety $\mathcal{G}$ is called a *formal group* if $\mathcal{G}$ is a commutative group object in the category of formal varieties. This amounts to giving morphisms

$$m\colon \mathcal{G} \times \mathcal{G} \to \mathcal{G}, \quad \text{inv}\colon \mathcal{G} \to \mathcal{G}, \tag{4.16}$$

such that the expected diagrams commute. Still more concretely, it is equivalent to giving for every $A \in \mathbf{nil}(K)$ a group structure on $\mathcal{G}(A)$ that is functorial in A.

We see that if $\mathcal{G} = D^{(I)}$ this agrees with the previous notion of (smooth commutative) formal groups defined in §3 using power series. Note that the coordinates of the formal power series $F(\underline{x}, \underline{y}) = \bigl(F_1(\underline{x}, \underline{y}), \dots, F_n(\underline{x}, \underline{y})\bigr)$ are the coordinates for the morphism $m\colon D^n \times D^n \to D^n$ constructed from F. The simplest examples of formal groups are thus $\widehat{\mathbb{G}}_a = D$ with $F(x, y) = x + y$ and $\widehat{\mathbb{G}}_m = D$ with $F(x, y) = x + y - xy$.

A more interesting example is provided by the completion $\hat{A}$ of an abelian variety A at its identity. We explained above that $\hat{A}$ is a formal variety, and we note that the multiplication morphism $m\colon A \times A \to A$ induces a multiplication map

$$\hat{A} \times \hat{A} \to \hat{A}, \tag{4.17}$$

making it into a formal group. See also Example 3.4.

If $\mathcal{G}$ is a formal group then $\mathcal{V} \mapsto \mathrm{Mor}(\mathcal{V}, \mathcal{G})$ is a contravariant functor from formal varieties to abelian groups. In particular $\mathfrak{C}(\mathcal{G})$ is an abelian group! We shall write the addition in this group as

$$\gamma_1 \underset{\mathcal{G}}{+} \gamma_2. \tag{4.18}$$

4.3. Operators on $\mathfrak{C}(\mathcal{G})$. Given $\varphi \in \mathrm{Mor}(D, D)$ we define a composition operator,

$$\mathrm{comp}(\varphi) \colon \mathfrak{C}(\mathcal{G}) \to \mathfrak{C}(\mathcal{G}), \tag{4.19}$$

by

$$\mathrm{comp}(\varphi) \cdot \gamma = \gamma \circ \varphi. \tag{4.20}$$

Thus, if $\mathcal{G} = D^{(I)}$, $\gamma = \sum_{n \geq 1} a_n t^n$ with $a_n \in K^{(I)}$ and $\varphi(t) = \sum_{n \geq 1} b_n t^n$ with $b_n \in K$, then $\mathrm{comp}(\varphi)\gamma = \sum_{n \geq 1} a_n (\sum_{m \geq 1} b_m t^m)^n$. Note that

$$\mathrm{comp}(\varphi_1 \circ \varphi_2) = \mathrm{comp}(\varphi_2)\,\mathrm{comp}(\varphi_1). \tag{4.21}$$

We define the following composition operators:

- For $n \geq 1$, let V_n denote $\mathrm{comp}(\varphi)$ for $\varphi(t) = t^n$.
- For $c \in K$, let $[c]$ denote $\mathrm{comp}(\varphi)$ for $\varphi(t) = ct$.
- For $n \geq 1$, define an operator F_n as follows: Fix an isomorphism $\mathcal{G} \cong D^{(I)}$. Given $\gamma \in \mathfrak{C}(\mathcal{G})$, let $\sigma_\gamma \colon D^n \to \mathcal{G}$ be defined by

$$\sigma_\gamma(t_1, \ldots, t_n) = \gamma(t_1) \underset{\mathcal{G}}{+} \cdots \underset{\mathcal{G}}{+} \gamma(t_n). \tag{4.22}$$

One can prove ([**91**, I.11]) that σ_γ factors as $s_\gamma \circ \mathrm{sym}_n$, where sym_n is the morphism with components equal to the elementary symmetric polynomials in n variables

$$\mathrm{sym}_n(x_1, \ldots, x_n) = \Big(\sum x_i, \sum_{i<j} x_i x_j, \ldots, x_1 \cdots x_n \Big). \tag{4.23}$$

One defines

$$F_n \cdot \gamma(t) = s_\gamma(0, \ldots, 0, (-1)^{n-1} t). \tag{4.24}$$

Consider $\mathfrak{C}$ as a functor on the category of formal groups, and consider the ring of natural transformations of this functor. Denote it by

$$\mathrm{Cart}(K). \tag{4.25}$$

Note that each element of $\mathrm{Cart}(K)$ is, by definition, additive. That is, if $\phi \in \mathrm{Cart}(K)$ then for every formal group $\mathcal{G}$ the map $\phi_{\mathcal{G}} \colon \mathfrak{C}(\mathcal{G}) \to \mathfrak{C}(\mathcal{G})$ is a group homomorphism. Every composition operator we defined above is an element of $\mathrm{Cart}(K)$. For that one needs to verify that the following statement is true.

EXERCISE 4.4. *The composition operators $[c]$, V_n and F_n are additive.*

THEOREM 4.5. *In $\mathrm{Cart}(K)$ the following relations hold:*

$$\begin{gathered} [1_K] = V_1 = F_1 = 1_{\mathrm{Cart}(K)}; \quad [c][d] = [cd]; \quad F_n F_m = F_{nm}; \quad V_n V_m = V_{nm}; \\ [c] V_n = V_n [c^n]; \quad F_n [c] = [c^n] F_n; \quad F_n V_n = n \cdot 1_{\mathrm{Cart}(K)}; \\ V_n F_m = F_m V_n, \quad \textit{if } (n, m) = 1. \end{gathered} \tag{4.26}$$

Most of the properties are easy to verify from the definition. For the proof see [**91**, I.10, III.3, IV.1] or [**69**, §§15–16]. Note that in general $[c] + [d] \neq [c + d]$, and in particular $n \cdot 1_{\mathrm{Cart}(K)} \neq [n]$. Also, in general, $V_n F_n \neq n \cdot 1_{\mathrm{Cart}(K)}$.

EXAMPLE 4.6 (The formal group $\widehat{\mathbb{G}}_a$; cf. Example 3.2). Let $\gamma(t) = \sum a_n t^n$ and $\gamma' = \sum a'_n t^n$ be two elements of $\mathfrak{C}(\widehat{\mathbb{G}}_a)$. Then

$$\gamma \underset{\widehat{\mathbb{G}}_a}{+} \gamma'(t) = \sum (a_n + a'_n) t^n, \tag{4.27}$$

and

$$[c]\gamma(t) = \sum a_n c^n t^n, \quad V_m\gamma(t) = \sum a_n t^{mn}, \quad F_m\gamma(t) = \sum m a_{mn} t^n. \tag{4.28}$$

In particular, letting $\gamma_{\underline{a}}(t) = t$, we get $F_m \gamma_{\underline{a}} = 0$ for all $m > 1$. See [**91**, III.3].

EXAMPLE 4.7 (The formal group $\widehat{\mathbb{G}}_m$; cf. Example 3.3). Let γ and γ' be as above. Then

$$\gamma \underset{\widehat{\mathbb{G}}_m}{+} \gamma'(t) = \sum a''_n t^n, \tag{4.29}$$

where

$$a''_n = a_n + a'_n - \sum_{0<i<n} a_i a'_{n-i}. \tag{4.30}$$

We also have

$$[c]\gamma(t) = \sum a_n c^n t^n, \quad V_m\gamma(t) = \sum a_n t^{mn}. \tag{4.31}$$

Let $\gamma_{\underline{m}}(t) = t$, then $F_n \gamma_{\underline{m}} = \gamma_{\underline{m}}$ for all n.

How does one compute F_n? There is a formal "trick" for that. Let $\mathcal{G}$ be a formal group. One has formally the identity

$$V_m F_m \gamma = \sum_{i=1}^{m}{}_{\mathcal{G}} [\zeta^i]\gamma, \tag{4.32}$$

where ζ is a primitive m-th root of unity (which need not exist in K) and where $\sum_{\mathcal{G}}$ denotes summation $\underset{\mathcal{G}}{+}$ (see [**69**, 15.1]). For example, $V_m F_m \gamma_{\underline{m}} = \zeta t \underset{\widehat{\mathbb{G}}_m}{+} \cdots \underset{\widehat{\mathbb{G}}_m}{+} \zeta^m t = 1 - \prod_{i=1}^m (1 - \zeta^i t) = t^m$, which gives $F_m \gamma_{\underline{m}} = \gamma_{\underline{m}}$. More generally,

$$\begin{aligned} V_m F_m \gamma &= \left(\sum a_n \zeta^n t^n\right) \underset{\widehat{\mathbb{G}}_m}{+} \cdots \underset{\widehat{\mathbb{G}}_m}{+} \left(\sum a_n \zeta^{mn} t^n\right) \\ &= 1 - \prod_{i=1}^{m}\left(1 - \sum_n a_n \zeta^{in} t^n\right). \end{aligned} \tag{4.33}$$

DEFINITION 4.8. Let $\mathcal{V}$ be a formal variety and let $(\gamma_i)_{i\in I}$ be an indexed set of curves in $\mathcal{V}$. It is a called a *basic set of curves* if $(\mathcal{T}\gamma_i)_{i\in I}$ is a K-basis for $\mathcal{T}\mathcal{V}$.

PROPOSITION 4.9 ([**91**, III 6.1]). *Let $\mathcal{G}$ be a formal group and $(\gamma_i)_{i\in I}$ be a basic set of curves. Then any curve $\gamma \in \mathfrak{C}(\mathcal{G})$ can be written uniquely as*

$$\gamma = \sum_{\substack{m\geq 1\\ i\in I}}{}_{\mathcal{G}} V_m [x_{m,i}]\gamma_i, \quad \forall m\ (x_{m,i})_{i\in I} \in K^{(I)}. \tag{4.34}$$

Conversely, every such expression defines a curve $\gamma \in \mathfrak{C}(\mathcal{G})$.

EXAMPLE 4.10. In the formal group $\widehat{\mathbb{G}}_a$, the set $\{\gamma_{\underline{a}}\}$ (consisting of the single element $\gamma_{\underline{a}}(t) = t$) is a basic set of curves. Every curve can be written uniquely as

$$\sum_{n=1}^{\infty} a_n t^n = \sum_{n=1}^{\infty}{}_{\mathcal{G}} V_n[a_n]\gamma_{\underline{a}}. \tag{4.35}$$

EXAMPLE 4.11. In the group $\widehat{\mathbb{G}}_m$, the set $\{\gamma_{\underline{m}}\}$ (consisting of the single element $\gamma_{\underline{m}}(t) = t$) is a basic set of curves. What is $\sum_{\widehat{\mathbb{G}}_m} V_n[x_n]\gamma_{\underline{m}}$? If we write the result as $\sum a_n t^n$ then $\sum a_n t^n = \sum_{\mathcal{G}} x_n t^n$. Equivalently,

$$1 - \sum_{n=1}^{\infty} a_n t^n = \prod_{n=1}^{\infty} (1 - x_n t^n). \tag{4.36}$$

One proves that the a_i's and the x_i's are functions of each other. We may write

$$\sum V_n[x_n]\gamma_{\underline{m}} + \sum V_n[y_n]\gamma_{\underline{m}} = \sum V_n[z_n]\gamma_{\underline{m}}, \tag{4.37}$$

and the z_n are determined by the identity

$$\prod (1 - x_n t^n) \prod (1 - y_n t^n) = \prod (1 - z_n t^n). \tag{4.38}$$

We define[2]:

- $\mathbb{W}^+(K) = K^{\mathbb{N}^+}$ with the addition law $(x_n) + (y_n) = (z_n)$.
- $\widehat{\mathbb{W}}^+(K) = K^{(\mathbb{N}^+)}$ with the addition law $(x_n) + (y_n) = (z_n)$.

EXERCISE 4.12. *Calculate that* $z_1 = x_1 + y_1$, $z_2 = x_2 + y_2 - x_1 y_1$ *and* $z_3 = x_3 + y_3 - x_1^2 y_1 - x_1 y_1^2$.

REMARK 4.13. One can prove that the group $\mathbb{W}^+(K)$, canonically isomorphic to $\mathfrak{C}(\widehat{\mathbb{G}}_m)$, is isomorphic to the group of infinite Witt vectors over $\mathbb{Z}$! See [**69**, §17.1], "Lots of Witt vectors," for lots on Witt vectors. See also §6.1 below. The group $\widehat{\mathbb{W}}^+(K)$ is an infinite-dimensional formal group.

Let $\gamma_{\underline{w}}$ be the canonical curve in $\widehat{\mathbb{W}}^+$ given by $\gamma_{\underline{w}}(t) = (1, 0, 0, \dots)$ and define curves ϵ_i by $\epsilon_i(t) = (0, \dots, 0, \underset{i}{1}, 0, \dots)$. In fact, $\epsilon_i = F_i \gamma_{\underline{w}}$.

EXERCISE 4.14. *Using Equation* (4.32) *prove that* $\epsilon_i = F_i \gamma_{\underline{w}}$.

THEOREM 4.15 (Representation Theorem, [**91**, III.4]). *There exists a canonical isomorphism*

$$\mathfrak{C}(\mathcal{G}) \cong \mathrm{Hom}(\widehat{\mathbb{W}}^+, \mathcal{G}). \tag{4.39}$$

It is defined as follows: Given $\gamma \in \mathfrak{C}(\mathcal{G})$ *there exists a unique homomorphism of formal groups*

$$u_\gamma \colon \widehat{\mathbb{W}}^+ \to \mathcal{G}, \tag{4.40}$$

such that

$$\gamma = u_\gamma \circ \gamma_{\underline{w}}. \tag{4.41}$$

COROLLARY 4.16. *There is a canonical bijection*

$$\mathrm{Cart}(K) \to \mathfrak{C}(\widehat{\mathbb{W}}^+). \tag{4.42}$$

[2]Recall that $\mathbb{N}^+ = \{1, 2, 3, \dots\}$.

PROOF. Given $x \in \mathrm{Cart}(K)$ associate to it the curve $x\gamma_{\underline{w}}$ in $\mathfrak{C}(\widehat{\mathbb{W}}^+)$. For any formal group $\mathcal{G}$ and $\gamma \in \mathfrak{C}(\mathcal{G})$ we have,

$$x\gamma = x(u_\gamma \circ \gamma_{\underline{w}}) = u_\gamma \circ (x\gamma_{\underline{w}}). \tag{4.43}$$

This shows that the map

$$\mathrm{Cart}(K) \to \mathfrak{C}(\widehat{\mathbb{W}}^+), \quad x \mapsto x\gamma_{\underline{w}}, \tag{4.44}$$

is injective. To show it is surjective, we argue as follows: Given $\gamma' \in \mathfrak{C}(\widehat{\mathbb{W}}^+)$, define an operator x in $\mathrm{Cart}(K)$ by

$$x\gamma := u_\gamma \circ \gamma'. \tag{4.45}$$

We need to check that for any morphism $f\colon \mathcal{G} \to \mathcal{G}'$ we have $x(f \circ \gamma) = f \circ (x\gamma)$. Indeed $x(f\circ\gamma) = u_{f\circ\gamma}\circ\gamma' = f\circ u_\gamma\circ\gamma' = f\circ(x\gamma)$. Since this holds for any morphism, we may apply that to $m\colon \mathcal{G}\times\mathcal{G} \to \mathcal{G}$ to deduce that x is additive. Hence, indeed $x \in \mathrm{Cart}(K)$. □

We have found a bijection $\mathrm{Cart}(K) \to \mathfrak{C}(\widehat{\mathbb{W}}^+)$ by sending x to $x\gamma_{\underline{w}}$. The curves $\{\epsilon_i : i \in \mathbb{N}^+\}$ form a basic set of curves in $\widehat{\mathbb{W}}^+$ and hence by Proposition 4.9 every curve can be uniquely expressed as

$$\sum_{i,m\in\mathbb{N}^+} V_m[x_{m,i}]\epsilon_i = \sum_{i,m\in\mathbb{N}^+} V_m[x_{m,i}]F_i\gamma_{\underline{w}}, \quad \forall m, (x_{m,i})_i \in K^{(\mathbb{N}^+)}. \tag{4.46}$$

COROLLARY 4.17. *The ring* Cart(K) *has the following presentation*:

$$\mathrm{Cart}(K) = \Big\{ \sum_{i,m\in\mathbb{N}^+} V_m[x_{m,i}]F_i : \forall m \ (x_{m,i})_i \in K^{(\mathbb{N}^+)} \Big\}. \tag{4.47}$$

5. Modules over Cart(K)

We have seen that for any (commutative, smooth) formal group $\mathcal{G}$, the group $\mathfrak{C}(\mathcal{G})$ of curves is a module over the ring $\mathrm{Cart}(K)$. The modules of $\mathrm{Cart}(K)$ arising this way have certain special properties which we attempt to single out now. Our motivation is to use such modules to establish an equivalence of categories between them and formal groups. The main new ingredient we need to pay attention to is topological properties.

Let $x = \sum_{i,m\in\mathbb{N}^+} V_m[x_{m,i}]F_i$ be an element of $\mathrm{Cart}(K)$. Put

$$\mathrm{ord}(x) = \min\{m : \exists n\ x_{m,n} \neq 0\}. \tag{5.1}$$

So, for example $\mathrm{ord}(V_m) = m$ and $\mathrm{ord}(F_i) = 1$ for every m and i. One may verify ([**91**, IV.2]) that

$$\begin{gathered} \mathrm{ord}(x \pm y) \geq \min\{\mathrm{ord}(x), \mathrm{ord}(y)\}, \quad \mathrm{ord}(xy) \geq \mathrm{ord}(x), \\ \mathrm{ord}([a+b] - [a] - [b]) > 1. \end{gathered} \tag{5.2}$$

Define ideals I_n of $\mathrm{Cart}(K)$ by

$$I_n = \{x : \mathrm{ord}(x) \geq n\}. \tag{5.3}$$

Under the topology defined by these ideals $\mathrm{Cart}(K)$ becomes a complete Hausdorff topological ring.

DEFINITION 5.1. (1) A *uniform* Cart(K)-module is a topological continuous Cart(K)-module C such that for any indexed set $(x_j)_{j\in J}$ in Cart(K) converging to zero, and for any set of elements $(\gamma_j)_{j\in J}$ of C, the sum $\sum_{j\in J} x_j\gamma_j$ converges in C.

(2) For a uniform module C we let C_n be the closure of the sum of all the subgroups V_iC for $i \geq n$. Then C is a complete Hausdorff module with respect to the C_n's. Let

$$\mathbf{gr}_n C = C_n/C_{n+1}. \tag{5.4}$$

(3) A *reduced* Cart(K)-module C is a uniform module C such that the following hold:

- Its topology is the C_n topology.
- The homomorphism $\mathbf{gr}_1 C \to \mathbf{gr}_n C$ given by multiplication by V_n is bijective for every n.
- The K-module $\mathbf{gr}_1 C$ is free.

The following is the main theorem of this theory. It is due to P. Cartier.

THEOREM 5.2 (Main Theorem, [**91**, III.11]). *The category of formal groups over K with morphisms given by homomorphisms of formal groups is equivalent to the category of reduced* Cart(K)*-modules with morphisms given by continuous* Cart(K)*-linear maps. The equivalence is given by*

$$\mathcal{G} \mapsto \mathfrak{C}(\mathcal{G}). \tag{5.5}$$

Unfortunately, this theorem is not stated precisely enough in some early references, a fact that led to minor inaccuracies in the applications of the theory to deformations of abelian varieties.

It is of interest to understand, at least "qualitatively," how one associates a formal group to a reduced Cart(K)-module C. As a "simple" motivating example one may keep in mind the universal formal deformation of a supersingular elliptic curve.

Given any uniform Cart(K)-module C, one associates to it a functor Γ from the category of formal varieties to the category of abelian groups

$$\mathcal{V} \mapsto \Gamma(\mathcal{V}). \tag{5.6}$$

The definition of Γ is motivated by the following reasoning: If C comes from a formal group $\mathcal{G}$, then $\Gamma(\mathcal{V})$ should be equivalent to $\mathrm{Mor}(\mathcal{V},\mathcal{G})$. Since the functor $\mathfrak{C}$ is faithful, we have an injection

$$\mathrm{Mor}(\mathcal{V},\mathcal{G}) \hookrightarrow \mathrm{Mor}\big(\mathfrak{C}(\mathcal{V}),\mathfrak{C}(\mathcal{G})\big). \tag{5.7}$$

Since C is to be identified with $\mathfrak{C}(\mathcal{G})$, one defines $\Gamma(\mathcal{V})$ as a certain functorial subgroup of $\mathrm{Mor}(\mathfrak{C}(\mathcal{V}), C)$. It consists of all the maps of the following form:

Let $(\gamma_j)_{j\in J}$ be any indexed set of elements of C and $(\pi_j)_{j\in J}\colon \mathcal{V} \to D^{(J)}$ a morphism. Define a morphism $f\colon \mathfrak{C}(\mathcal{V}) \to C$ by

$$f(\delta) = \sum_{j\in J} \mathrm{comp}(\pi_j \circ \delta)\cdot \gamma_j, \quad \delta\colon D \to \mathcal{V}. \tag{5.8}$$

One denotes f by $\sum \pi_j \star \gamma_j$.

Let us assume now further that C is reduced. A V*-basis* for C is an indexed set $(\gamma_j)_{j\in J}$ of elements of C such that the γ_j mod C_2 are a K-basis of the free K-module $\mathbf{gr}_1 C = C/C_2$. One then proves that every element in C may be written uniquely as $\sum_{n,j\in\mathbb{N}^+} V_{n,j}[x_{n,j}]\gamma_j$ where for every n we have $(x_{n,j})_{j\in J} \in K^{(J)}$. One shows that C is in bijective correspondence with $\mathfrak{C}(D^{(J)})$ given by associating to a curve $\varphi = (\varphi_j)\colon D \to D^{(J)}$ the element $\sum_{j\in J} \operatorname{comp}(\varphi_j)\cdot\gamma_j$ of C. One further shows that there is a functorial isomorphism of topological groups

$$\operatorname{Mor}(\mathcal{V}, D^{(J)}) \cong \Gamma(\mathcal{V}), \tag{5.9}$$

obtained by associating to $(\pi_j)\colon \mathcal{V} \to D^{(J)}$ the element $\sum_{j\in J} \pi_j \star \gamma_j$.

6. The case of $\mathbb{Q}$-algebras

THEOREM 6.1[3] (The $\mathbb{Q}$ Theorem, [**91**, II.3]). *Assume that K is a $\mathbb{Q}$-algebra and $\mathcal{G}$, $\mathcal{G}'$ are two formal groups over K. Then, for any K-linear map $u\colon \mathcal{TG} \to \mathcal{TG}'$, there exists a unique formal group homomorphism $f\colon \mathcal{G} \to \mathcal{G}'$ such that $\mathcal{T}f = u$.*

Thus, the category of formal groups over K is equivalent to the category of free K-modules. The equivalence is given by the fully-faithful functor $\mathcal{T}$. In particular, there exist unique formal group isomorphisms:

$$\log_{\mathcal{G}}\colon \mathcal{G} \to (\mathcal{TG})^+, \quad \exp_{\mathcal{G}}\colon (\mathcal{TG})^+ \to \mathcal{G}, \tag{6.1}$$

which are inverses of each other and such that

$$\mathcal{T}\log_{\mathcal{G}} = \mathrm{Id}, \quad \mathcal{T}\exp_{\mathcal{G}} = \mathrm{Id}. \tag{6.2}$$

EXAMPLE 6.2. For $\widehat{\mathbb{G}}_a$ the logarithm and exponent are of course the identity maps. For $\widehat{\mathbb{G}}_m$ we have

$$\log_{\widehat{\mathbb{G}}_m}(x) = -\log(1-x), \quad \exp_{\widehat{\mathbb{G}}_m}(x) = 1-\exp(-x). \tag{6.3}$$

For $F_{\epsilon,\delta}$ see Example 3.4. For the case of $\widehat{\mathbb{W}}^+$ see §6.1. For the case of elliptic curves, see [**156,175**]. We can not resist mentioning the following beautiful theorem connecting the arithmetic and geometry of elliptic curves.

THEOREM 6.3 (Honda, [**70**]). *Let E be an elliptic curve over $\mathbb{Q}$. Let $L(E/\mathbb{Q}, s) = \sum_{n=1}^{\infty} a_n n^{-s}$ be its L-function. Put $f(x) = \sum_{n=1}^{\infty} a_n n^{-1} x^n$ and let $\mathcal{G}$ be the formal group over $\mathbb{Q}$ with logarithm $\log_{\mathcal{G}} = f(x)$. Then $\mathcal{G}$ is a formal group over $\mathbb{Z}$, i.e., the power series defining it has integer coefficients, and is isomorphic to the formal group of the Néron model of E over $\mathbb{Z}$.*

6.1. Digression on Witt vectors. We follow [**69**, §17]. Define polynomials

$$w_n(x_1,\dots,x_n) = \sum_{d|n} d x_d^{n/d} \in \mathbb{Z}[x_1,\dots,x_n]. \tag{6.4}$$

We shall write $\underline{x} = (x_1, x_2, \dots)$ and would consider w_n also as functions of $\underline{x}$. For example,

$$\begin{aligned} w_1(\underline{x}) &= x_1,\\ w_2(\underline{x}) &= x_1^2 + 2x_2,\\ w_3(\underline{x}) &= x_1^3 + 3x_3,\\ w_4(\underline{x}) &= x_1^4 + 2x_2^2 + 4x_4. \end{aligned}$$

[3]See also [**69**, §§9.7, (11.1.6), (14.2.3)] that discuss also the noncommutative case.

One proves that there exists polynomials

$$\begin{aligned} &\Sigma_n(x_1,\dots,x_n;y_1,\dots,y_n)\in\mathbb{Z}[x_1,\dots,x_n;y_1,\dots,y_n], \quad n=1,2,\dots,\\ &\Pi_n(x_1,\dots,x_n;y_1,\dots,y_n)\in\mathbb{Z}[x_1,\dots,x_n;y_1,\dots,y_n], \quad n=1,2,\dots,\\ &\iota_n(x_1,\dots,x_n)\in\mathbb{Z}[x_1,\dots,x_n], \quad n=1,2,\dots, \end{aligned} \tag{6.5}$$

such that for every n

$$\begin{aligned} w_n\big(\Sigma(\underline{x};\underline{y})\big) &= w_n(\underline{x})+w_n(\underline{y}),\\ w_n\big(\Pi(\underline{x};\underline{y})\big) &= w_n(\underline{x})\cdot w_n(\underline{y}),\\ w_n\big(\iota(\underline{x})\big) &= -w_n(\underline{x}), \end{aligned} \tag{6.6}$$

where we have put $\Sigma=(\Sigma_1,\Sigma_2,\dots),\Pi=(\Pi_1,\Pi_2,\dots)$ and $\iota=(\iota_1,\iota_2,\dots)$.

Define a ring functor $\mathbb{W}$ from the category of commutative rings to itself

$$\mathbb{W}\colon \textbf{Rings}\to\textbf{Rings},\quad A\mapsto\mathbb{W}(A), \tag{6.7}$$

by letting $\mathbb{W}(A)$ be the ring $A^{\mathbb{N}^+}$ with addition and multiplication defined by Σ,Π and i ($\mathbb{W}(A)$ is called the Witt ring of A, or the ring of infinite Witt vectors over A). Namely,

$$\begin{aligned} \underline{x}+\underline{y} &= \Sigma(\underline{x};\underline{y}),\\ \underline{x}\cdot\underline{y} &= \Pi(\underline{x};\underline{y})\\ -\underline{x} &= \iota(\underline{x}). \end{aligned} \tag{6.8}$$

Still more visibly,

$$\begin{aligned} (a_1,a_2,\dots)+(b_1,b_2,\dots) &= (\Sigma_1(a_1;b_1),\Sigma_2(a_1,a_2;b_1,b_2),\dots),\\ (a_1,a_2,\dots)\cdot(b_1,b_2,\dots) &= (\Pi_1(a_1;b_1),\Pi_2(a_1,a_2;b_1,b_2),\dots),\\ -(a_1,a_2,\dots) &= (\iota(a_1),\iota(a_1,a_2),\dots). \end{aligned}$$

One may amuse himself by calculating some of these polynomials:

$$\begin{aligned} \Sigma_1(x_1;y_1) &= x_1+y_1, & \Sigma_2(x_1,x_2;y_1,y_2) &= x_2+y_2-x_1y_1,\\ \Pi_1(x_1;y_1) &= x_1y_1, & \Pi_2(x_1,x_2;y_1,y_2) &= x_2y_1^2+y_2x_1^2+2x_2y_2,\\ \iota_1(x_1) &= -x_1, & \iota_2(x_1,x_2) &= -x_1^2-x_2. \end{aligned}$$

One can prove using Equation (6.6) that this indeed gives a ring structure. Let $\widehat{\mathbb{W}}$ be the sub-functor

$$\widehat{\mathbb{W}}(A)=A^{(\mathbb{N}^+)}. \tag{6.9}$$

Then clearly the underlying additive group of $\widehat{\mathbb{W}}$, denoted $\widehat{\mathbb{W}}^+$ is a formal group. Indeed, the formal group law is none other than

$$F(\underline{x},\underline{y})=(\Sigma_1(\underline{x};\underline{y}),\Sigma_2(\underline{x};\underline{y}),\dots). \tag{6.10}$$

The map

$$\underline{x}=(x_1,x_2,\dots)\xrightarrow{\log} w(x)=(w_1(\underline{x}),w_2(\underline{x}),\dots), \tag{6.11}$$

is the logarithm of the formal group $\widehat{\mathbb{W}}^+$.

Consider now the ring $1+tA[\![t]\!]$ endowed with a ring structure as follows: The usual multiplication of power series is decreed *addition*. To define *multiplication*, write formally

$$(6.12)\quad 1+x_1t+x_2t^2+\cdots=\prod_{i=1}^{\infty}(1-\xi_it),\quad 1+y_1t+y_2t^2+\cdots=\prod_{j=1}^{\infty}(1-\eta_jt).$$

Then, we may write using suitable polynomials P_i with integer coefficients,

$$(6.13)\qquad \prod_{i,j=1}^{\infty}(1-\xi_i\eta_jt)=1+P_1(x_1;y_1)t+P_2(x_1,x_2;y_1,y_2)t^2+\cdots.$$

Multiplication is defined by the rule

$$(6.14)\quad (1+x_1t+x_2t^2+\cdots)\star(1+y_1t+y_2t^2+\cdots)$$
$$:=1+P_1(x_1;y_1)t+P_2(x_1,x_2;y_1,y_2)t^2+\cdots.$$

The map

$$(6.15)\qquad \mathbf{E}\colon \mathbb{W}(A)\to 1+tA[\![t]\!],\quad \mathbf{E}(a_1,a_2,\ldots)=\prod_{i=1}^{\infty}(1-a_it^i),$$

is an isomorphism of rings. This justifies Remark 4.13, as well as the notation.

Fix a prime p. We may then consider only "the p-part" of $\mathbb{W}$ and $\widehat{\mathbb{W}}$. That is, consider only the indices n's that are power of p. Bearing this in mind, we change notation (as is customary) and let

$$(6.16)\qquad \phi_n(x_0,\ldots,x_n)=\sum_{i=0}^{n}p^ix_i^{p^{n-i}}=x_0^{p^n}+\cdots+p^{n-1}x_{n-1}^p+p^nx_n.$$

That is, $\phi_n=w_{p^n}$. Then, the polynomials $s_n=\Sigma_{p^n}$, $p_n=\Pi_{p^n}$ and $i_n=\iota_{p^n}$ define a ring structure on $W(A)=A^{\mathbb{N}}$ and $\widehat{W}(A)=A^{(\mathbb{N})}$. This is clear from what was said above.

A fundamental result concerning $W(K)$, where K is a finite field of characteristic p, is that $W(K)$ is the unique, up to isomorphism, discrete complete valuation ring of characteristic 0 with residue field K. The maximal ideal is

$$(6.17)\qquad \mathfrak{m}=\{(0,x_1,x_2,\cdots):x_i\in K\}.$$

The map

$$(6.18)\qquad K^\times\to W(K),\quad x\mapsto(x,0,0,\ldots),$$

is the *Teichmüller lift*. That is, the image of x is the unique root of unity reducing to x modulo the maximal ideal. The Teichmüller lift is a multiplicative map.

EXERCISE 6.4. *Prove that* $(x,0,0,\ldots)(y,0,0,\ldots)=(xy,0,0,\ldots)$, *but in general* $(x,0,0,\ldots)+(y,0,0,\ldots)\neq(x+y,0,0,\ldots)$.

Since W is a functor, one may ask what is the functorial map $W(K)\to W(K)$ corresponding to the Frobenius map $x\mapsto x^p$ on K? It is given just by

$$(6.19)\qquad (x_0,x_1,x_2,\ldots)\mapsto {}^F(x_0,x_1,x_2,\ldots):=(x_0^p,x_1^p,x_2^p,\ldots).$$

There is another map, the *Verschiebung*,

$$(6.20)\qquad V\colon W(K)\to W(K),\quad {}^V(x_0,x_1,\ldots)=(0,x_0,x_1,\ldots).$$

It is a homomorphism of the underlying additive group, but does not respect multiplication. A fundamental relation is

$$FV = VF = [p]. \tag{6.21}$$

EXERCISE 6.5. *Let A be a ring. Note that for every n the polynomial ϕ_n gives a ring homomorphism*

$$\phi_n \colon W(A) \to A. \tag{6.22}$$

Prove that there exists a unique homomorphism $x \mapsto {}^F x$ that is functorial in A and satisfies the identity:

$$\phi_n({}^F x) = \phi_{n+1}(x), \quad \forall n \in \mathbb{N}. \tag{6.23}$$

We call this isomorphism the Frobenius homomorphism of W. Note that it induces a homomorphism of A if p is zero in A. In this case, the definition is consistent with the one in Equation (6.19).

7. Formal groups in characteristic p

If one is willing to work in characteristic p only, then the constructions above can be simplified. Let us assume henceforth that K is of positive characteristic p.

One replaces the ring $\mathrm{Cart}(K)$ by the ring $\mathrm{Cart}_p(K)$. It is the subring of $\mathrm{Cart}(K)$ consisting of all the sums

$$\sum_{i,n \in \mathbb{N}} V^n [x_{n,i}] F^i, \quad \forall n \ (x_{n,i})_i \in K^{(\mathbb{N})}, \tag{7.1}$$

where

$$V^n := V_{p^n}, \quad F^i := F_{p^i}. \tag{7.2}$$

Let $V = V^1$ and $F = F^1$. Recall the relations given in Theorem 4.5:

$$\begin{aligned} F^i F^j &= F^{i+j}, \quad V^i V^j = V^{i+j}, \quad FV = p \cdot 1, \\ [a]V &= V[a^p], \quad F[a] = [a^p]F, \quad [ab] = [a][b]. \end{aligned} \tag{7.3}$$

A special feature of characteristic p is the identity:

$$VF = p \cdot 1. \tag{7.4}$$

One can prove that the map

$$\mathrm{op} \colon W(K) \to \mathrm{Cart}_p(K), \tag{7.5}$$

defined by

$$\mathrm{op}(\underline{x}) = \mathrm{op}(x_0, x_1, \dots) = \sum_{n=0}^{\infty} V^n [x_n] F^n, \tag{7.6}$$

is a ring homomorphism, and the following identities hold:

$$F^m(\mathrm{op})(\underline{x}) = \mathrm{op}({}^{F^m}\underline{x}) F^m, \quad \mathrm{op}(\underline{x}) V^m = V^m \, \mathrm{op}({}^{F^m}\underline{x}). \tag{7.7}$$

In particular, every $\mathrm{Cart}_p(K)$*-module is canonically a* $W(K)$*-module via the ring homomorphism* op.[4]

---

[4]We remark that the same holds for $\mathrm{Cart}(K)$ for a general ring K. The map $\mathrm{op}(x_1, x_2, \dots) = \sum V_n [x_n] F_n$ is an embedding of $\mathbb{W}(K)$ in $\mathrm{Cart}(K)$.

One defines in the same way the notions of uniform and reduced $\mathrm{Cart}_p(K)$-modules (but see below). An element γ in a $\mathrm{Cart}(K)$-module is called *p-typical* if $F_n\gamma = 0$ for every n prime to p. Let

$$\mathfrak{C}_p(\mathcal{G}) \subset \mathfrak{C}(\mathcal{G}) \tag{7.8}$$

be the subgroup of p-typical curves.

THEOREM 7.1 (Main Local Theorem, [**91**, IV.7, IV.8]). *Let K be a ring of characteristic p. The functor*

$$\mathcal{G} \mapsto \mathfrak{C}_p(\mathcal{G}), \tag{7.9}$$

is an equivalence of categories from the category of formal groups over K, to the category of reduced $\mathrm{Cart}_p(K)$-modules.

Further simplifications occurring in characteristic p are:

- A $\mathrm{Cart}_p(K)$-module C is uniform if and only if $\bigcap_n V^n C = \{0\}$.
- A $\mathrm{Cart}_p(K)$-module C is reduced if and only if it is uniform, V is injective and C/VC is a free K-module.
- Every $\gamma \in \mathfrak{C}(\mathcal{G})$ has a unique expression $\gamma = \sum_{(n,p)=1} V_n\gamma_n$ for appropriate $\gamma_n \in \mathfrak{C}_p(\mathcal{G})$. This implies that $\mathfrak{C}_p(\mathcal{G})/V\mathfrak{C}_p(\mathcal{G}) \cong \mathcal{T}\mathcal{G}$.

REMARK 7.2. In the definition of reduced module one requires that $\bigcap_i V^i C = \{0\}$. Why? The operation V is coming from the "Frobenius map,"

$$R[\![x_1, \dots, x_n]\!] \to R[\![x_1, \dots, x_n]\!], \tag{7.10}$$

taking x_i to x_i^p. This makes it clear it should be topologically nilpotent. But note that the corresponding action on the p-divisible group is again Frobenius. Thus, in the covariant theory that we present, the operator V on the module corresponds to the Frobenius morphism on the p-divisible group. Since the p-divisible group is connected, the nilpotence of V is reflecting the fact that on a connected p-divisible group Frobenius is nilpotent.

8. Classification of p-divisible groups in characteristic p, Newton polygons and types

Our conventions in the case of real multiplication are as in situation $\mathbf{II}_p$ (see p. 194). Let K stand for a characteristic p ring and k for an algebraically closed field of characteristic p. We use R to denote a general ring but whenever we say $\mathrm{Cart}_p(R)$-module, or consider $\mathfrak{C}_p(\mathcal{G})$ for a formal group $\mathcal{G}$ over R, it is tacitly assumed that R has characteristic p. All formal groups in this section are finite-dimensional.

DEFINITION 8.1. (1) A formal group with real multiplication by L, or simply, *formal group with RM*, is a formal group $\mathcal{G}$ over a ring R, together with an embedding of rings

$$\iota\colon \mathcal{O}_L \to \mathrm{End}_R(\mathcal{G}), \tag{8.1}$$

making $\mathcal{T}\mathcal{G}$ into a locally-free $\mathcal{O}_L \otimes R$-module of rank 1.

(2) A $\mathrm{Cart}(R)$-module (resp. $\mathrm{Cart}_p(R)$-module) with RM, is a reduced $\mathrm{Cart}(R)$-module (resp. $\mathrm{Cart}_p(R)$-module) C together with an embedding of rings

$$\iota\colon \mathcal{O}_L \to \mathrm{End}(C) \tag{8.2}$$

(endomorphisms of topological Cartier modules), making $\mathbf{gr}_1 C$ into a locally free $\mathcal{O}_L \otimes R$-module of rank 1.

REMARK 8.2. These definitions are too restrictive. abelian varieties over R with real multiplication, in the sense of having an action of $\mathcal{O}_L$, yield formal groups on which $\mathcal{O}_L$ act faithfully. It need not be the case that the tangent space to the formal group (which is the tangent space to the abelian variety at the origin) is a projective $\mathcal{O}_L \otimes R$- module. For example, if R is a characteristic p ring and p is ramified in L, there exists a $g-2$-dimensional closed subvariety of the moduli space of abelian varieties with RM and satisfying condition (DP), where the tangent space is not projective.

Even in the case where p is unramified in L, there are abelian varieties with RM such that the tangent space is not a projective $\mathcal{O}_L \otimes R$-module. Assuming again that R is a field of characteristic p, one find such abelian varieties as "generating" Moret–Bailly families. See [**3**].

Theorem 5.2 (resp. Theorem 7.1) implies that the functor

$$\mathcal{G} \mapsto \mathfrak{C}(\mathcal{G}), \quad (\text{resp. } \mathcal{G} \mapsto \mathfrak{C}_p(\mathcal{G})), \tag{8.3}$$

is an equivalence of categories between formal groups with RM and morphisms commuting with the $\mathcal{O}_L$-structure, to the category of reduced modules with morphisms of Cartier modules commuting with the $\mathcal{O}_L$-structure.

Recall that a formal group $\mathcal{G}$ over k is p-divisible, if, fixing an isomorphism $\mathcal{G} \cong D^g$, the coordinate ring $k[\![x_1, \ldots, x_g]\!]$ of D^g becomes a finite module over itself via the homomorphism $[p]^* \colon k[\![x_1, \ldots, x_g]\!] \to k[\![x_1, \ldots, x_g]\!]$ induced by multiplication by p on the formal group. Equivalently, the module $\mathfrak{C}_p(\mathcal{G})/p\mathfrak{C}_p(\mathcal{G})$ is a finite k-module. We remark that for a p-divisible group $\mathcal{G}$ one has that $\mathfrak{C}_p(\mathcal{G})$ is a free $W(k)$-module of finite rank.

Two p-divisible groups $\mathcal{G}_1$, $\mathcal{G}_2$, over k are called *isogenous* (notation: $\mathcal{G}_1 \sim \mathcal{G}_2$) if there exists a surjective homomorphism $f \colon \mathcal{G}_1 \to \mathcal{G}_2$ with finite kernel. This is an equivalence relation. Indeed, the kernel $\mathrm{Ker}(f)$ is contained in $\mathcal{G}_1[p^n]$ for large enough n. This implies that there is an isogeny $g \colon \mathcal{G}_2 \to \mathcal{G}_1$ such that the following diagram commutes:

$$\begin{array}{ccc} \mathcal{G}_1 & \xrightarrow{f} & \mathcal{G}_2 \\ & \searrow^{[p^n]} & \downarrow g \\ & & \mathcal{G}_1. \end{array} \tag{8.4}$$

On the other hand, multiplication by p^n induces an isomorphism $\mathcal{G}_1/\mathcal{G}_1[p^n] \cong \mathcal{G}_1$.

The groups $\mathcal{G}_1$ and $\mathcal{G}_2$ are isogenous if and only if $\mathfrak{C}_p(\mathcal{G}_1) \otimes \mathbb{Q}$ and $\mathfrak{C}_p(\mathcal{G}_2) \otimes \mathbb{Q}$ are isomorphic as $\mathrm{Cart}_p(k) \otimes \mathbb{Q}$ modules.

The classification of p-divisible formal groups over k up to isogeny was carried out by Dieudonné, and up to isomorphism by Manin in [**93**]. Note that the theory in [**93**] is contravariant, while we survey the covariant theory.

THEOREM 8.3 (Dieudonné, [**91**, VI.7, VI.8], [**93**, II.4]). *The category of p-divisible formal groups up to isogeny, over an algebraically closed field k of characteristic p, is semi-simple. The simple objects are precisely the p-divisible groups $\mathcal{G}_{m,n}$ for $m, n \in \mathbb{N}^+$ and* $\gcd(m,n) = 1$ *or* $(m,n) = (1,0)$.

The group $\mathcal{G}_{m,n}$ *has dimension* m *and is determined by its* $\mathrm{Cart}_p(k)$ *module* $\mathcal{C}_{m,n} = \mathfrak{C}_p(\mathcal{G}_{m,n})$ *where*

$$\mathcal{C}_{m,n} = \mathrm{Cart}_p(k)/(F^m - V^n). \tag{8.5}$$

REMARK 8.4. The group $\mathcal{G}_{1,0}$ is just $\widehat{\mathbb{G}}_m$ and is isomorphic to the formal group of an ordinary elliptic curve over k. The group $\mathcal{G}_{1,1}$ is isomorphic to the formal group of a supersingular elliptic curve over k. Sometimes one puts $\mathcal{G}_{0,1} := \underline{\mathbb{Q}_p/\mathbb{Z}_p}$, even though this is not a formal group in our sense. One notes that $\mathcal{G}^t_{m,n} = \mathcal{G}_{n,m}$.

EXERCISE★ 8.5. *Prove that* $\mathcal{C}_{m,n}$ *is isogenous to* $\mathrm{Cart}_p(k)/(p^m - V^{n+m})$. *Assume that* $k \supseteq \mathbb{F}_{p^{n+m}}$. *Show that* $\mathrm{End}(\mathcal{C}_{m,n}) \otimes \mathbb{Q}$ *is "the" central simple* $\mathbb{Q}_p$*-algebra of dimension* $(m+n)^2$ *over* $\mathbb{Q}_p$ *and invariant* $n/(n+m)$.

DEFINITION 8.6. We say that $\mathcal{G}_{m,n}$ (or $\mathcal{C}_{m,n}$) has *slope* $m/(m+n)$ of *length* $m+n$. Recall that the height of a p-divisible group $\mathcal{G}$ is the integer h such that the p-torsion subgroup scheme of $\mathcal{G}$ has rank p^h. The group $\mathcal{G}_{m,n}$ has height $m+n$. (Thus, the length of $\mathcal{C}_{m,n}$ over $W(k)$ is the height of $\mathcal{G}_{m,n}$). One gives $\mathcal{G}_{0,1}$ slope 0 of length 1.

If $\mathcal{G}$ is any p-divisible group over k and

$$\mathcal{G} \sim \bigoplus \mathcal{G}_{m_i,n_i} \tag{8.6}$$

(repetitions allowed), we associate to it a lower convex polygon in the plane. It is the polygon having increasing slopes, such that the polygon starts at $(0,0)$ and ends at $\big(\mathrm{height}(\mathcal{G}), \dim(\mathcal{G})\big)$, and has for every (m_i, n_i) a segment of slope $m_i/(m_i+n_i)$ of length m_i+n_i. We call it the *Newton polygon* of $\mathcal{G}$. If $\mathcal{G}$ is the p-divisible group of an abelian variety A/k, $\mathcal{G} = A(p)$, then the Newton polygon satisfies the following symmetry condition: For every slope λ with multiplicity r (i.e., the length of the segment of slope λ is r), the slope $1-\lambda$ appears and with the same multiplicity r.

EXERCISE 8.7. *Use the existence of a polarization* $\lambda\colon A \to A^t$ *to prove the symmetry condition.*

Thus, every Newton polygon of an abelian variety of dimension g (i) starts at $(0,0)$ and ends at $(2g, g)$; (ii) has increasing slopes; (iii) is lower convex and symmetric with respect to the transformation $(a,b) \mapsto \big(2g-a, g-(a-b)\big)$, i.e., if a slope λ appears so does $1-\lambda$ and with the same multiplicity; (iv) has integer breaking points. It thus lies between the ordinary polygon and the supersingular polygon. See Figure 8.1.

REMARK 8.8. Let $A/\mathbb{F}_q$ be a g-dimensional abelian variety (q a power of p) and let ϕ be that iterate of Frobenius satisfying $\phi(x) = x^q$. Assume that $|A[p](\overline{\mathbb{F}_q})| = 1$. Then $A(p)$ is a connected p-divisible group, or, equivalently, a formal group, of dimension g and height $2g$. Its Newton polygon satisfies the restrictions above.

Consider ϕ as an endomorphism of A, hence of $A(p)$, or of the free $W(\overline{\mathbb{F}_q})$-module $\mathfrak{C}_p\big(A(p)\big)$. On the latter it is the operator V. Viewed in this way, it has a characteristic polynomial over $W(\overline{\mathbb{F}_q})$, whose Newton polygon *is* the Newton polygon of $A(p)$ as a formal group.

Suppose we are given a reduced $\mathrm{Cart}_p(k)$-module C. How does one find the Newton polygon?

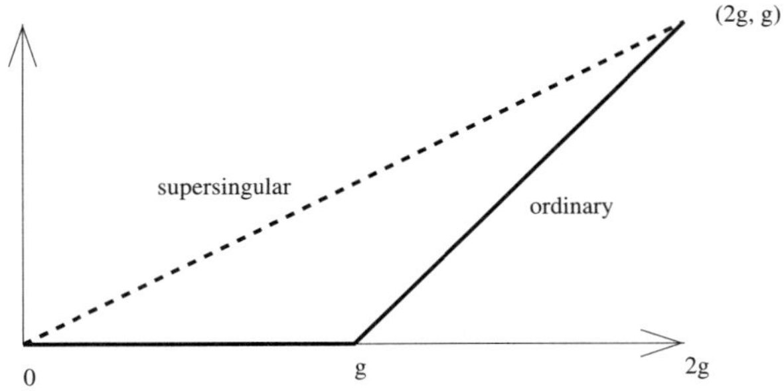

FIGURE 8.1. The ordinary and supersingular Newton polygons.

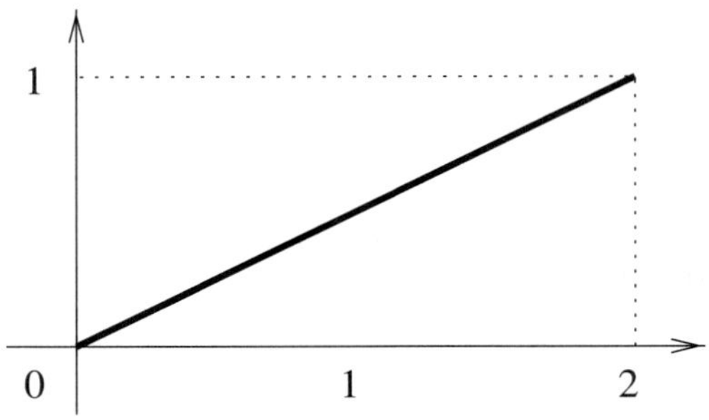

FIGURE 8.2. Newton polygon of $x^2 + p$.

In practice this is not easy to find! But consider a very lucky scenario: Suppose that with respect to some basis $(e_1, \ldots, e_n)$ of the module C (a free $W(k)$-module of rank $n = \text{height}(\mathcal{G})$), the matrix $[V]$ of the operator V has entries in $W(\mathbb{F}_{p^m})$. The operator V^m is represented by the matrix

$$[V^m] = [V][V]^{\sigma^{-1}} \ldots [V]^{\sigma^{-(m-1)}} \tag{8.7}$$

(here σ stands for Frobenius). Considering V^m as an operator on the $W(\mathbb{F}_{p^m})$ span of $(e_1, \ldots, e_n)$, we find that it is a linear operator represented by the matrix $[V^m]$. Take the Newton polygon $\mathcal{N}'$ of the characteristic polynomial of this matrix, and let $\mathcal{N}$ be the polygon obtained by dividing $\mathcal{N}'$ by m. That is, $(x, y) \in \mathcal{N}$ iff $(x, my) \in \mathcal{N}'$. Then $\mathcal{N}$ is the Newton polygon of C.

EXAMPLE 8.9 (B. Gross). Consider the field $\mathbb{F}_{p^2}$ where $p \equiv 3 \bmod 4$ and $i \in \mathbb{F}_{p^2}$ satisfies $i^2 = -1$. Suppose that V is given with respect to some basis by the matrix

$$[V] = \frac{-1}{2} \begin{pmatrix} 1-p & (p+1)i \\ (p+1)i & p-1 \end{pmatrix}. \tag{8.8}$$

The usual linear polynomial is $x^2 + p$, which would give the Newton polygon in Figure 8.2.

However, this is not the Newton polygon of the module. Indeed, if one changes basis by the matrix

$$N = \begin{pmatrix} 1 & i \\ i & 1 \end{pmatrix}, \tag{8.9}$$

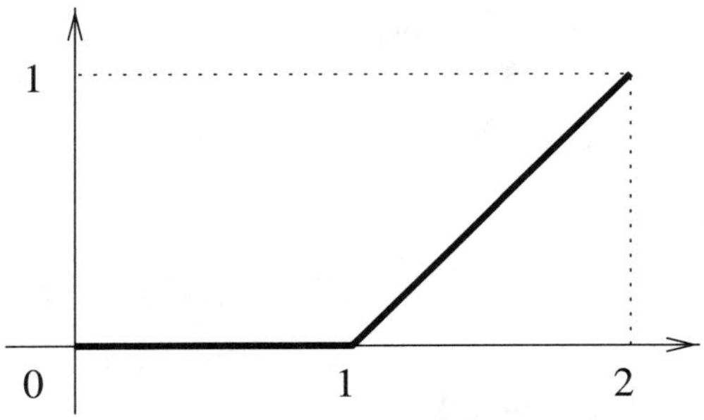

FIGURE 8.3. The correct Newton polygon of V.

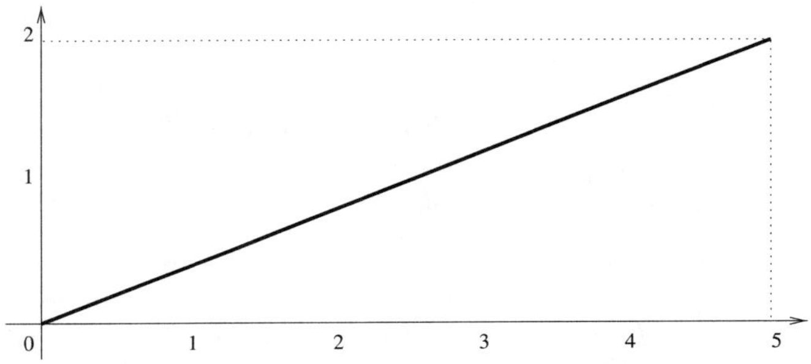

FIGURE 8.4. The Newton polygon of $x^5 - p^2$.

then the resulting matrix of V is

$$N[V](N^{-1})^{\sigma^{-1}} = \begin{pmatrix} p & 0 \\ 0 & 1 \end{pmatrix}. \tag{8.10}$$

Evidently, the Newton polygon is as in Figure 8.3.

Note also that V^2 is given by

$$\frac{1}{2}\begin{pmatrix} p^2+1 & (p^2-1)i \\ (1-p^2)i & p^2+1 \end{pmatrix}. \tag{8.11}$$

Its linear characteristic polynomial is $(x^2 - 2(p^2+1)x + p^2)/4$ and its Newton polygon is twice the Newton polygon in Figure 8.3.

EXAMPLE 8.10. The module $\mathcal{C}_{2,3}$ is generated freely by $F, 1, V, V^2, V^3$. The matrix of V is

$$\begin{pmatrix} 0 & 0 & 0 & 0 & p \\ p & 0 & 0 & 0 & 0 \\ 0 & 1 & 0 & 0 & 0 \\ 0 & 0 & 1 & 0 & 0 \\ 0 & 0 & 0 & 1 & 0 \end{pmatrix}. \tag{8.12}$$

It has characteristic polynomial $x^5 - p^2$ whose Newton polygon is given in Figure 8.4.

Another criterion of a different flavor that is very useful in many applications is the following. Let k be an algebraically closed field of characteristic $p \geq 3$. Let

C be a nonsingular curve of genus 2 over k and choose a hyperelliptic equation for C, say $y^2 = f(x)$, where $f(x)$ is a separable polynomial of degree 5. Write

$$f(x)^{(p-1)/2} = \sum c_i x^i. \tag{8.13}$$

Let

$$M = \begin{pmatrix} c_{p-1} & c_{p-2} \\ c_{2p-1} & c_{2p-2} \end{pmatrix}, \quad M^{(p)} = \begin{pmatrix} c_{p-1}^p & c_{p-2}^p \\ c_{2p-1}^p & c_{2p-2}^p \end{pmatrix}. \tag{8.14}$$

Then the Jacobian of C is:

- Ordinary iff M has rank 2.
- Supersingular iff $MM^{(p)} = 0$.
- Superspecial iff $M = 0$.

(Cf. the criterion for elliptic curves in Chapter 1, p. 31). For references and related matters see [**71**]. See also [**174**].

EXERCISE 8.11. *Let C be the curve $y^2 = x^5 + 1$. Prove that C is a nonsingular curve for $p > 5$. Prove that whether the reduction of C mod p is ordinary, supersingular, or superspecial depends only on the factorization of p in $\mathbb{Q}(\zeta_5)$. This is an example of a general result in complex multiplication. See* [**47**].

Let h and d be positive integers. Let ℓ be the graph of a convex function $f\colon [0,h] \to \mathbb{R}$ satisfying $f(h) = d$. Let $\mathcal{G}$ be a d-dimensional p-divisible formal group over R of height h. We say that the Newton polygon of $\mathcal{G}$ is *above* ℓ if for every homomorphism $\varphi\colon R \mapsto k$, from R to an algebraically closed field k, the Newton polygon of $\mathcal{G} \otimes_R k$ is above ℓ in the sense that no point of the polygon is strictly below ℓ (if (x,y) is a point of that polygon then $y \geq f(x)$). We remark that this property only depends on the point $\operatorname{Ker}(\varphi) \in \operatorname{Spec}(R)$.

THEOREM 8.12 (Grothendieck's Specialization Theorem, [**82**, 2.3.1]). *The set of points in* $\operatorname{Spec}(R)$ *for which the Newton polygon of $\mathcal{G}$ is above ℓ is Zariski closed.*

COROLLARY 8.13. *In case* $\mathbf{I}_p$ *(resp.* $\mathbf{II}_p$*), there exists for every ℓ a closed subset $\mathcal{N}_\ell$ of $\mathcal{A}_{g,n}(\mathbb{F}_p) = \mathcal{A}_{g,n} \otimes \mathbb{F}_p$ (resp. $\mathfrak{M}(\mathbb{F}, \mu_N) = \mathfrak{M}(\mu_N) \otimes \mathbb{F}$) universal for the condition that the Newton polygon is above ℓ. That is, if $(A, \lambda, \alpha)/R$ (resp. (A, ι, α)) has the property that the Newton polygon of $A(p)/R$ is above ℓ, then the unique classifying morphism* $\operatorname{Spec}(R) \to \mathcal{A}_{g,n}(\mathbb{F}_p)$ *(resp.* $\operatorname{Spec}(R) \to \mathfrak{M}(\mathbb{F}, \mu_N)$*) corresponding to $(A, \lambda, \alpha)/R$ (resp. (A, ι, α)) factors through $\mathcal{N}_\ell$.*

REMARK 8.14. Of course one can restrict to ℓ which is actually one of the Newton polygons appearing. Conjectures regarding those subschemes $\mathcal{N}_\ell$ appeared for case $\mathbf{I}_p$ in [**130**] and for case $\mathbf{II}_p$ in [**50**].

de Jong and Oort proved the following theorem:

THEOREM 8.15 (Purity Theorem, [**76**]). *Let R be a Noetherian local ring of characteristic p and $\mathcal{G}$ a p-divisible group over R. Assume that the Newton polygon of $\mathcal{G}$ is constant over* $\operatorname{Spec}(R) \setminus \{\mathfrak{m}_R\}$. *Then, either* $\dim(R) \leq 1$ *or the Newton polygon of $\mathcal{G}$ is constant over R.*

EXAMPLE 8.16. Let E/k be a supersingular elliptic curve and let $\mathcal{E}/\operatorname{Spec}(R)$ be its universal equi-characteristic local deformation, where $R = k[\![t]\!]$. The Newton polygon over $\operatorname{Spec}(R) \setminus \{\mathfrak{m}_R\}$ is of course constant and is the ordinary Newton polygon. On the other hand the Newton polygon of $\mathcal{E}$ is not constant over $\operatorname{Spec}(R)$—it is the supersingular polygon over the point $\mathfrak{m}_R$. Hence the possibility $\dim(R) \leq 1$ in the theorem.

On the other hand, let R be a two-dimensional Noetherian local ring of characteristic p. Let $\mathcal{E}$ be an elliptic curve over $\mathrm{Spec}(R)$. Assume that the Newton polygon of $\mathcal{E}$ is constant over $\mathrm{Spec}(R) \setminus \{\mathfrak{m}_R\}$. Consider the j-invariant morphism $j\colon \mathrm{Spec}(R) \to \mathbb{A}^1_R$ (see [**104**]). Since we *know* that the supersingular locus is a divisor on the coarse moduli scheme for elliptic curves $\mathbb{A}^1$, it follows that the supersingular locus on $\mathrm{Spec}(R)$ is either empty, an effective divisor or $\mathrm{Spec}(R)$. It follows that the Newton polygon of $\mathcal{E}$ is constant over $\mathrm{Spec}(R)$.

Much in the same spirit of this example, knowledge of the structure of the Newton strata on $\mathcal{A}_{g,n}$ (in particular its dimension) would imply the Purity Theorem. Conversely, using this theorem one obtains lower bounds on $\dim(\mathcal{N}_\ell)$ and upper bounds on the dimension of components of $\mathcal{N}_\ell$ containing *supersingular* points (loc. cit., Introduction). These are points such that the Newton polygon of the associated abelian variety is a straight line from $(0,0)$ to $(2g, g)$. Among them, the easiest to describe are the superspecial points. A geometric point x of $\mathcal{A}_{g,n}(\mathbb{F}_p)$, or $\mathfrak{M}(\mathbb{F}, \mu_N)$, is *superspecial* if the associated abelian variety has p-divisible group *isomorphic* to $\mathcal{G}_{1,1}^g$. By a theorem of Oort [**128**] an abelian variety is superspecial iff its a-number is maximal (equal to its dimension). Up to isomorphism, the $\mathrm{Cart}_p(k)$-module of such a p-divisible group is generated freely over $W(k)$ by generators $x_1, y_1, \dots, x_g, y_g$ on which F acts by

$$(8.15)\qquad \begin{pmatrix} 0 & -p & & & & & 0 \\ 1 & 0 & & & & & \\ & & 0 & -p & & & \\ & & 1 & 0 & & & \\ & & & & \ddots & & \\ & & & & & 0 & -p \\ & 0 & & & & 1 & 0 \end{pmatrix}.$$

A geometric point x is supersingular iff the corresponding abelian variety A_x is isogenous to a product of supersingular elliptic curves.

DEFINITION 8.17. Let $\mathcal{G}$ be a formal group with RM over k. We define the *type* of $\mathcal{G}$ to be the isomorphism class of the representation of $\mathcal{O}_L \otimes k$ on $\mathbb{D}_\alpha(\mathcal{G})$ where

$$(8.16)\qquad \mathbb{D}_\alpha(\mathcal{G}) = \mathfrak{C}_p(\mathcal{G})/\big(V\mathfrak{C}_p(\mathcal{G}) + F\mathfrak{C}_p(\mathcal{G})\big).$$

Note that $\mathbb{D}_\alpha(\mathcal{G}) = \mathcal{TG}/F\mathcal{TG}$. We remark that if $\mathcal{G}$ is the formal group of the connected part of the p-divisible group of an abelian variety A/k, $\mathcal{G} = A(p)^0$, then $\mathbb{D}_\alpha(\mathcal{G})$ is the covariant Dieudonné module of $\alpha(A)$—the alpha group of A (See Appendix A, §6) equal to $\mathrm{Ker}(\mathrm{Fr}) \cap \mathrm{Ker}(\mathrm{Ver})$. Since $\alpha(A)$ is self dual, $\mathbb{D}_\alpha(\mathcal{G})$ is also the contravariant Dieudonné module of $\alpha(A)$.

The isomorphism class of the representation on $\mathbb{D}_\alpha(\mathcal{G})$ can be written as a formal sum $\sum_{(j,i)} \epsilon_{(j,i)} \sigma_{(j,i)}$, where $\epsilon_{(j,i)}$ is zero or one, or also with a vector

$$(8.17)\qquad \tau(\mathcal{G}) = \big(\tau_1(\mathcal{G}), \dots, \tau_r(\mathcal{G})\big),$$

where $\epsilon_{(j,i)} = 1$ iff $(j,i) \in \tau_j(\mathcal{G})$ (thus, $\tau_j(\mathcal{G})$ is a subset of $\{(j,i) : 1 \leq i \leq f_j\}$).

DEFINITION 8.18. Given a formal group $\mathcal{G}$ over k and given any $\tau = (\tau_1, \dots, \tau_r)$ with τ_j a subset of $\{(j,i) : 1 \leq i \leq f_j\}$, we say that $\tau(\mathcal{G}) \geq \tau$, if for every j we have $\tau_j(\mathcal{G}) \supseteq \tau_j$.

Given a formal group $\mathcal{G}$ over an $\mathbb{F}$-algebra R, we say that $\tau(\mathcal{G}) \geq \tau$ if for every homomorphism $\varphi\colon R \to k$ of R to an algebraically closed field k, we have $\tau(\mathcal{G} \times_R k) \supseteq \tau$. (This depends only on the point $\mathrm{Ker}(\varphi) \in \mathrm{Spec}(R)$).

THEOREM 8.19 (Specialization Theorem). *Given a type τ and a formal group $\mathcal{G}$ over R, the set of points of* $\mathrm{Spec}(R)$ *where $\tau(\mathcal{G}) \geq \tau$ is Zariski closed.*

PROOF. Consider the linearization $F^\sharp$ of F,

$$F^\sharp\colon \mathcal{TG} \otimes_{R,\sigma} R \to \mathcal{TG}, \tag{8.18}$$

where $\sigma\colon R \to R$ is the Frobenius and $F^\sharp(x \otimes \lambda) = \lambda F(x)$. Note that $F^\sharp(x \otimes \lambda^p) = \lambda^p F(x) = F(\lambda x) = F(\lambda x \otimes 1)$. Hence, $F^\sharp$ is well-defined.

We could have formulated the notion of type using $\mathcal{TG}/\mathrm{Im}(F^\sharp)$. First, for $m \in \mathcal{O}_L$ we have $F^\sharp(mx \otimes \lambda) = \lambda F(mx) = \lambda m F(x) = mF^\sharp(x \otimes \lambda)$. Thus, $F^\sharp$ is an $\mathcal{O}_L \otimes R$-linear map. If we decompose

$$\mathcal{TG} = \bigoplus_{j=1}^{r} \bigoplus_{i=1}^{f_j} R_{(j,i)}, \tag{8.19}$$

according to the decomposition

$$\mathcal{O}_L \otimes R = \bigoplus_{j=1}^{r} \bigoplus_{i=1}^{f_j} R_{(j,i)}, \tag{8.20}$$

(where every $R_{(j,i)} = R$) then for every (j,i) we have

$$F^\sharp\colon R_{(j,i-1)} \otimes_{R,F} R \to R_{(j,i)}. \tag{8.21}$$

In fact this map is given by multiplication by an element $a_{(j,i)} \in R$. Then the subset of $\mathrm{Spec}(R)$ such that (j,i) does not belong to the type is precisely the basic open set $B_{a_{(j,i)}}$ (i.e., where $a_{(j,i)}$ is invertible and there "$R = a_{(j,i)}R$"). □

Given an abelian scheme with RM (A, λ, ι) over an $\mathbb{F}$-scheme S, we define $\tau(A) = \tau(A(p)^0)$.

COROLLARY 8.20. *For every τ there exists a Zariski closed subset W_τ of the moduli space $\mathfrak{M}(\mu_N, \mathbb{F})$ which is universal for the property $\tau(A, \lambda, \iota) \supseteq \tau$.*

9. Mid-way summary

We fix the following notation and assumptions

k	an algebraically closed field of positive characteristic p.
$\mathcal{C}_{k,p}$	category of local artinian k-algebras with residue field k.
L	a totally real number field of degree g over $\mathbb{Q}$ in which p is inert.
$\mathbb{F}$	a fixed field of p^g elements.
$\mathfrak{M}$	moduli space of abelian varieties with RM by $\mathcal{O}_L$.
x	a k-rational point of $\mathfrak{M}$ parameterizing the object (A, ι).

There is an implicit μ_N-level ($N \geq 4$ structure and prime to p) whenever we deal with $\mathfrak{M}$ or x. Assume that the p-divisible group $A(p)$ is connected and let $\mathcal{G}_k$ be the corresponding p-divisible formal group. Then the completion of the local ring of x in $\mathfrak{M}$ can be identified with certain universal equi-characteristic deformation rings:

$$\widehat{\mathcal{O}}_{\mathfrak{M}\otimes\mathbb{F},x} \cong R^{\mathrm{univ}}_{(A,\iota)} \cong R^{\mathrm{univ}}_{(A(p),\iota(p))} \cong R^{\mathrm{univ}}_{(\mathcal{G}_k,\iota_k)}. \tag{9.1}$$

Let us call the last ring simply $\mathbf{R}^{\mathbf{U}}$. In the above isomorphisms, the universal rings are with respect to deformations over objects of $\mathcal{C}_{k,p}$ (and are just the reduction mod p of the universal ring of deformations over $\mathcal{C}_k$).

It follows that there exists a $\mathrm{Cart}_p(\mathbf{R}^{\mathbf{U}})$-module $\mathbf{C}^{\mathbf{U}}$ such that for every ring R in $\mathcal{C}_{k,p}$ and a deformation $(\mathcal{G}_R, \iota_R)$ of $(\mathcal{G}_k, \iota_k)$ over R, there exists a unique ring homomorphism $\varphi\colon \mathbf{R}^{\mathbf{U}} \to R$ such that $\mathbf{C}^{\mathbf{U}} \otimes_{W(\mathbf{R}^{\mathbf{U}})} W(R) \cong \mathfrak{C}_p(\mathcal{G}_R)$ as $\mathcal{O}_L \otimes \mathrm{Cart}_p(R)$-modules.

We remark that since p is inert, $A(p)$ is connected if and only if A is not ordinary. This follows from observing that $A[p](k)$ is a module over the field of p^g elements $\mathcal{O}_L/(p)$.

Our next goal is to describe a method to construct effectively the ring $\mathbf{R}^{\mathbf{U}}$ and the module $\mathbf{C}^{\mathbf{U}}$. Our motivation comes from the fact that much of the geometry of $\mathfrak{M} \otimes \mathbb{F}$ (as well as arithmetic of modular forms, L-functions and geometric zeta functions, i.e., the zeta functions of the abelian varieties themselves [**85**]) is exhibited by the variation of the Frobenius morphism of the abelian variety as one "moves" in the moduli space $\mathfrak{M} \otimes \mathbb{F}$. The local aspect of this variation is reflected in the module $\mathbf{C}^{\mathbf{U}}$ over $\mathrm{Cart}_p(\mathbf{R}^{\mathbf{U}})$.

10. Displays

Displays are a machinery developed to simplify and normalize the presentation of a reduced $\mathrm{Cart}_p(R)$-module, where R is a simple enough ring; e.g., a complete Noetherian local ring. The need in finding a simplified presentation arises when one wants to study explicitly the local deformations of abelian varieties or p-divisible groups, for example, to study how the Newton polygon or the type varies locally.

Let k be an algebraically closed field of characteristic $p > 0$, and let $\mathcal{G}$ be a formal group over k. There is no doubt that the Cartier–Dieudonné module of $\mathcal{G}$ is "the right thing." One of the reasons the theory works so well is the following: Let R be a perfect ring of characteristic p. Let $W(R)[F,V]$ be the ring consisting of all the expressions

$$\left\{ a_0 + \sum_{i=1}^{n} a_i F^i + \sum_{i=1}^{m} b_i V^i : a_i, b_i \in W(R) \right\}. \tag{10.1}$$

Make it into a ring by the obvious addition and multiplication determined by the relations:

$$FV = VF = p, \quad Fa = {}^F a F, \quad V^F a = aV. \tag{10.2}$$

Then:

- this makes $W(R)[F,V]$ into a ring! (perfectness is needed), and
- the image of the natural map $W(R)[F,V] \to \mathrm{Cart}_p(R)$ is dense.

This phenomenon is responsible for a considerable simplification in the study of $\mathrm{Cart}_p(R)$-modules. The question is how to *extend* this notion to the case of, say, local artinian ring of characteristic p with residue field k (we do narrow our ambitions here, but that, in fact, suffices for the applications later). One way is, of course, to use $\mathfrak{C}_p$ as described above, but it turns out that such modules are not "nice enough." A modification was found recently by Thomas Zink [**178**], following ideas of Mumford, Norman and Oort.

10.1. Basics. Let p be a prime. Let R be a characteristic p ring. To be consistent with the notation in [**178**] we denote the Frobenius and Verschiebung homomorphisms of $W(R)$ by

$$x \mapsto {}^F x, \quad x \mapsto {}^V x. \tag{10.3}$$

These are the same maps as in Equations (6.19), (6.20), (6.23). Let

$$I_R = {}^V W(R) = \{(0, r_1, r_2, \dots) : r_i \in R\}. \tag{10.4}$$

Note that $R = W(R)/I_R$.

DEFINITION 10.1. Let M and N be $W(R)$-modules. An additive map

$$\alpha \colon M \to N \tag{10.5}$$

is called a ^{F}*-linear map* if it satisfies

$$\alpha(\lambda \cdot m) = {}^F\lambda \cdot \alpha(m), \quad \forall \lambda \in W(R), m \in M. \tag{10.6}$$

Let $\alpha^\sharp$ be the $W(R)$-linear map,

$$\alpha^\sharp \colon W(R) \otimes_{W(R),F} M \to N, \quad \alpha^\sharp(\lambda \otimes m) = \lambda \cdot \alpha(m). \tag{10.7}$$

We say that α is an *epimorphism* (resp. *monomorphism, isomorphism*) if $\alpha^\sharp$ is an epimorphism (resp. monomorphism, isomorphism).

DEFINITION 10.2. A 3n-display[5] over R is a quadruple (P, Q, F, V^{-1}) such that:

- P is a finitely generated projective $W(R)$-module.
- $Q \subset P$ is a $W(R)$-submodule.
- $F \colon P \to P$ and $V^{-1} \colon Q \to P$ are ^{F}-linear maps.

The following holds:

(1) $I_R P \subset Q \subset P$ and there is a decomposition $P = L \oplus T$ as $W(R)$-module such that $Q = L \oplus I_R T$.
(2) $V^{-1} \colon Q \to P$ is an ^{F}-linear epimorphism.
(3) $\forall x \in P, w \in W(R)$, we have

$$V^{-1}({}^V w \cdot x) = w \cdot Fx. \tag{10.8}$$

REMARK 10.3. Note that "there is no V map." Note also the identity

$$Fx = V^{-1}({}^V 1 \cdot x) \tag{10.9}$$

$(= p \cdot V^{-1}x$ if $x \in Q)$. Thus, F is determined by V^{-1}.

Displays form an additive category. A morphism

$$\varphi \colon (P_1, Q_1, F_1, V_1^{-1}) \to (P_2, Q_2, F_2, V_2^{-1}) \tag{10.10}$$

is a morphism $\varphi \colon P_1 \to P_2$ of $W(R)$-modules such that:

$$\varphi(Q_1) \subset Q_2, \quad \varphi F_1 = F_2 \varphi, \quad \varphi V_1^{-1} = V_2^{-1} \varphi. \tag{10.11}$$

Assume that R is a Noetherian local ring of characteristic p. Then the p-divisible formal groups over R correspond to reduced $\mathrm{Cart}_p(R)$-modules C such that C is a finitely generated projective $W(R)$-module. If R is also *perfect*, this means that C/pC is a finitely generated R-module, because in this case $I_R =$

---

[5] 3n stands for **n**ot **n**ecessarily **n**ilpotent.

$pW(R)$. Such modules can then equivalently be described as free $W(R)$-modules M of finite rank, endowed with additive maps $F, V\colon M \to M$ such that

$$F(w \cdot x) = {}^F w \cdot x, \quad V({}^F w \cdot x) = w \cdot Vx, \quad FV = VF = p. \tag{10.12}$$

Furthermore, M/VM and VM/pM are free R-modules and there exists an n such that $V^n M \subset pM$. The last condition is implicit for every reduced $\mathrm{Cart}_p(R)$-module M, because $\cap V^i C = \{0\}$. The equivalence rests on the fact that the natural map $W(R)[F, V] \to \mathrm{Cart}_p(R)$ is injective with dense image.

We recall the notion of a Dieudonné module. If R is a perfect ring, one denotes by $W(R)[F, V]$ the noncommutative polynomial ring in the variables F and V, subject to the relations

$$FV = VF = p, \quad Fa = {}^F aF, \quad V^F a = aV. \tag{10.13}$$

One calls a $W(R)[F, V]$-module M which is a finitely generated projective $W(R)$ module, a *Dieudonné module* if M/VM and VM/pM are projective R-modules.

Now let R be *any* ring of characteristic p. Let (P, Q, F, V^{-1}) be a display over R with a decomposition $P = L \oplus T$ as in Definition 10.2. After localizing on R we may assume that P, Q, L and T are all free $W(R)$-modules. Choose bases $e_1, \ldots, e_d$ of T and $e_{d+1}, \ldots, e_h$ of L (sic!). There exist scalars $\alpha_{ij} \in W(R)$ such that

$$\begin{aligned} Fe_j &= \sum_{i=1}^{h} \alpha_{ij} e_i, \qquad j = 1, \ldots, d, \\ V^{-1} e_j &= \sum_{i=1}^{h} \alpha_{ij} e_i, \qquad j = d+1, \ldots, h. \end{aligned} \tag{10.14}$$

Note that this determines $V^{-1}|_L$ and $V^{-1}|_{I_R T}$ because $V^{-1}({}^V w \cdot x) = w \cdot Fx$. Thus (10.14) determines V^{-1} and hence also F. Moreover, the matrix (α_{ij}) is invertible. This follows from

LEMMA 10.4 ([**178**]). *The map*

$$V^{-1} \oplus F\colon L \oplus T \to P, \tag{10.15}$$

given by $(x, y) \mapsto V^{-1}x + Fy$, *is a* ^{F}*-linear isomorphism.*

Conversely, given Equations (10.14) for (α_{ij}) invertible in $M_h(W(R))$, we can define a 3n-display over R. Indeed, let T be the free $W(R)$-module on $e_1, \ldots, e_d$, and let L be the free $W(R)$-module on $e_{d+1}, \ldots, e_h$. Put

$$P = L \oplus T, \quad Q = L \oplus I_R T, \tag{10.16}$$

and define F and V^{-1} by the additional relations:

$$\begin{aligned} Fe_j &= \sum_{i=1}^{h} p\alpha_{ij} e_i, \qquad j = d+1, \ldots, h, \\ V^{-1}({}^V w e_j) &= \sum_{i=1}^{h} w\alpha_{ij} e_i, \qquad j = 1, \ldots, d. \end{aligned} \tag{10.17}$$

REMARK 10.5. The name "displays" is articulating the fact that in Equations (10.14) and (10.17) the maps F and V^{-1} are "displayed."

In general a V operator does not exist. The following definition attempts to define the nilpotency of V "were it to exist." We remark that a more natural (and complicated) definition can be given, without using the choice of display, which actually describes the nilpotence of some operator that always exists. See [**178**].

DEFINITION 10.6. Let (P, Q, F, V^{-1}) be a 3n-display and (α_{ij}) a displaying matrix as in Equation (10.14). Let

$$(\beta_{kl}) = (\alpha_{ij})^{-1}, \quad B = (\overline{\beta_{kl}})_{d+1 \leq k,l \leq h}, \tag{10.18}$$

where $\overline{\beta}$ denotes β mod I_R. Let $B^{(p^i)}$ denote the matrix obtained from B by raising each coefficient to the p^i-power. We say (P, Q, F, V^{-1}) is a *display* if for some N,

$$B^{(p^N)} \cdots B^{(p)} B = 0. \tag{10.19}$$

We also say then that V^{-1} satisfies the *nilpotence condition.*

REMARK 10.7. If R is perfect then V^{-1} is described by the matrix $\left(\begin{smallmatrix} A & B \\ C & D \end{smallmatrix}\right)$ with respect to the basis $pe_1, \ldots, pe_d, e_{d+1}, \ldots, e_h$. In this case we may also define an operator V, given with respect to the same basis by

$$\left((\alpha_{ij})^{-1}\right)^{F^{-1}} = (\beta_{ij})^{F^{-1}}. \tag{10.20}$$

To know whether the operator V is nilpotent on the whole module P with respect to the p-adic topology (as is required for a Dieudonné module), we need to care only about

$$(\beta_{ij})^{F^{-1}}_{d+1 \leq i,j \leq h} =: B^{F^{-1}} \mod p. \tag{10.21}$$

The nilpotence is just that for some N

$$B^{F^{-1}} B^{F^{-2}} \cdots B^{F^{-(N+1)}} = 0. \tag{10.22}$$

Operating by F^{N+1} we get precisely the condition above.

The proof of the following proposition is straight-forward.

PROPOSITION 10.8 ([**178**]). *The category of 3n-displays over a perfect ring R is equivalent to the category of Dieudonné modules over R. Moreover, the displays correspond exactly to the Dieudonné modules for which V is topologically nilpotent for the p-adic topology; i.e., to those which extend to a reduced p-divisible* $\mathrm{Cart}_p(R)$*-module.*

It follows that the category of displays over a perfect ring R is equivalent to the category of p-divisible formal groups over R.

DEFINITION 10.9. A *3n-display with RM* is a 3n-display (P, Q, F, V^{-1}) together with an embedding of rings, $\iota\colon \mathcal{O}_L \to \mathrm{End}\big((P, Q, F, V^{-1})\big)$, that makes P/Q into a locally-free (on R) $\mathcal{O}_L \otimes R$-module of rank 1. A *display with RM* is a 3n-display with RM which is a display.

Note that $I_R P \subset Q$ so the definition makes sense and in fact P/Q is isomorphic to $T/I_R T$.

10.2. Examples.

(1) *The multiplicative display* $\mathcal{P}_{\mathrm{m}}$. Let $\mathcal{P}_{\mathrm{m}} = (P, Q, F, V^{-1})$ be defined as follows: $P = W(R)$, $Q = I_R$, $Fw = {}^F w$ and $V^{-1}({}^V w \cdot x) = w \cdot F(x)$. One obtains a decomposition $P = L \oplus T$ by putting $L = \{0\}$. Thus $d = h = 1$ and the nilpotency condition holds vacuously. We remark that we have seen that $\mathfrak{C}_p(\widehat{\mathbb{G}}_m) = \widehat{W}(R)^+$ and that justifies the name.

Let $\mathcal{P}_{\mathrm{m}}^g$ be the g-fold product of $\mathcal{P}$. Write $\mathcal{P}_{\mathrm{m}}^g = (P^g, Q^g, F, V^{-1})$. Say R is in $\mathbb{F}$-**Alg**. Then

$$P^g = W(R)^g \underset{\varphi}{\cong} W(R) \otimes \mathcal{O}_L. \tag{10.23}$$

This gives, via φ, an action of $\mathcal{O}_L$ on P^g, which makes it into a display with RM. Such displays are obtained from ordinary abelian varieties with RM.

(2) *The superspecial display* $\mathcal{P}_{\mathrm{sp}}$. Let $\mathcal{P}_{\mathrm{sp}} = (P, Q, F, V^{-1})$ where:

$$P = W(R)e_2 \oplus W(R)e_1 = L \oplus T, \tag{10.24}$$

and $Q = L \oplus I_R T$. Let the matrix $\left(\begin{smallmatrix} 0 & -1 \\ 1 & 0 \end{smallmatrix}\right)$ define the display. One checks easily that the nilpotence condition holds ($B = 0$).

We define a superspecial display to be one isomorphic to $\mathcal{P}_{\mathrm{sp}}^n$ for some n, and a superspecial display with RM is a display with RM isomorphic to $\mathcal{P}_{\mathrm{sp}} \otimes \mathcal{O}_L$. Such displays are obtained from superspecial abelian varieties with RM.

The *standard superspecial display* is defined as follows: Let $\mathcal{P} = (P, Q, F, V^{-1})$ with the action of F given by the $2g \times 2g$ matrix

$$\begin{pmatrix} 0_2 & & & & \cdots & \begin{smallmatrix} 0 & -p \\ 1 & 0 \end{smallmatrix} \\ \begin{smallmatrix} 0 & -p \\ 1 & 0 \end{smallmatrix} & 0_2 & & & & \\ & \begin{smallmatrix} 0 & -p \\ 1 & 0 \end{smallmatrix} & & & & \\ & \vdots & & 0_2 & & \\ \cdots & \cdots & \cdots & & \begin{smallmatrix} 0 & -p \\ 1 & 0 \end{smallmatrix} & 0_2 \end{pmatrix}. \tag{10.25}$$

(We let 0_n stand for the $n \times n$ zero matrix). The action of $\mathcal{O}_L$ is given by

$$a \mapsto \operatorname{diag}(\sigma_1(a), \sigma_1(a), \ldots, \sigma_g(a), \sigma_g(a)), \tag{10.26}$$

One can prove this defines a superspecial display with RM. In fact, it is canonically isomorphic to $\mathcal{P}_{\mathrm{sp}} \otimes_{W(\mathbb{F}_p)} \mathcal{O}_L$: Let e_1, e_2 be as in Equation (10.24) and let $u_1, \ldots, u_g$ be the idempotents of $W(\mathbb{F}) \otimes \mathcal{O}_L$ then the basis $e_1 \otimes u_1, e_2 \otimes u_1, \ldots, e_1 \otimes u_g, e_2 \otimes u_g$ provides a basis in which Frobenius is given by the matrix in Equation (10.25). Note though that to obtain a display we have to re-order the basis vectors.

We remark that if k is an algebraically closed field of characteristic p then there are finitely many isomorphism classes of superspecial abelian varieties with RM over k (in fact also in the situation $\mathbf{I}_p$). One can prove that over such a field every superspecial display with RM is isomorphic to the standard superspecial display. However, there is usually more than one isomorphism class of superspecial abelian varieties.

The reason superspecial abelian varieties (or displays) are of such importance is the following

THEOREM 10.10. *Let $\tau \subset \{1, \dots, g\}$. Every component of W_τ contains a superspecial point.*

For the proof see [**50**]. One may think of the superspecial points as "extreme points" of $\mathfrak{M} \otimes \mathbb{F}$, which often play a role similar to the cusps.

10.3. The main result. We present a somewhat simplified form of Zink's theorem. Its main advantage over Proposition 10.8 is that it deals with non-perfect rings. This is crucial for the theory since non-perfect rings like $k[\![t_1, \dots, t_g]\!]$ appear as universal local deformation rings.

THEOREM 10.11 (Zink [**178**]). *Let R be a ring of characteristic p. Assume that R is a complete local ring, or a ring such that R is an algebra of finite type over a field k.*

There exists a functor $\mathcal{BT}$ from the category of displays over R to the category of p-divisible formal groups over R, which is an equivalence of categories.

See [**178**] for the definition of the functor $\mathcal{BT}$. It would suffice to know that the $\mathrm{Cart}_p(R)$ module of $\mathcal{BT}(\mathcal{P})$ is

$$\mathrm{Cart}_p(R) \otimes_{W(R)} \mathcal{P}/\langle F \otimes x - 1 \otimes Fx, V \otimes V^{-1}y - 1 \otimes y\rangle_{x \in \mathcal{P}, y \in Q} \tag{10.27}$$

where the brackets denote the $\mathrm{Cart}_p(R)$-module generated by the specified generators.

Thus, heuristically speaking, the display $\mathcal{P}$ singles out the essential part of the Cartier–Dieudonné module.

11. The universal display and applications to abelian varieties

11.1. Base change and deformations. Let $\phi \colon S \to R$ be a ring homomorphism and let $\mathcal{P} = (P, Q, F, V^{-1})$ be a 3n-display over S. Define the *base change to R* of $\mathcal{P}$, denoted $\mathcal{P}_R = (P_R, Q_R, F_R, V_R^{-1})$, as follows:

- $P_R = W(R) \otimes_{W(S)} P$.
- $Q_R = \mathrm{Ker}\big(W(R) \otimes_{W(S)} P \to R \otimes_S (P/Q)\big)$.
- $F_R = {}^F \otimes F$. That is, $F_R(\lambda \otimes x) = {}^F\lambda \otimes F(x)$.
- V_R^{-1} is defined as the unique $W(R)$- ^{F}-linear homomorphism satisfying:

$$V_R^{-1}(w \otimes y) = {}^F w \otimes V^{-1} y, \quad w \in W(R), y \in Q; \tag{11.1}$$

$$V_R^{-1}({}^V w \otimes x) = w \otimes Fx, \qquad x \in P. \tag{11.2}$$

We remark that if $P = L \oplus T$, then $Q_R = W(R) \otimes_{W(S)} L \oplus I_R \otimes_{W(S)} T$.

It is easy to verify that if $\mathcal{P}$ is a display (not merely a 3n-display) then $\mathcal{P}_R$ is a display as well. If every element of $\mathrm{Ker}(\phi)$ is nilpotent, the converse is true.

DEFINITION 11.1. Let $\mathcal{P}_0$ be a display over R. Let $\phi \colon S \to R$ be a homomorphism of rings. A *deformation* of $\mathcal{P}_0$ over S is a display $\mathcal{P}$ over S together with an isomorphism $\mathcal{P}_R \cong \mathcal{P}_0$.

11.2. Universal display. Let k be a field of characteristic p. Let $\mathcal{C}_{k,p}$ be the category of local artinian k-algebras with residue field k. We will describe the universal display over $\mathcal{C}_{k,p}$.

Let Λ be a topological ring such that the topology on Λ is given by ideals

$$\Lambda = \mathfrak{a}_0 \supset \mathfrak{a}_1 \supset \cdots \supset \mathfrak{a}_n \supset \cdots, \tag{11.3}$$

such that $\mathfrak{a}_i\mathfrak{a}_j \subset \mathfrak{a}_{i+j}$. Assume that Λ is complete and separated with respect to this topology and is of characteristic p. A 3n-display over Λ is called a display if its base change to $\Lambda/\mathfrak{a}_i$ is a display, in the sense of Definition 10.2, for every i.

We remark that this modification is especially tailored to suit rings like the ring $k[\![t_1, \dots, t_g]\!]$ with $\mathfrak{a}_i$ equal to the power series in monomials of degree greater or equal to i.

DEFINITION 11.2. Let $\mathcal{P} = (P, Q, F, V^{-1})$ be a display over k. Define the *functor of deformations*

$$\mathbf{Def}_{\mathcal{P}} \colon \mathcal{C}_{k,p} \to \mathbf{Sets}, \tag{11.4}$$

sending each ring R in $\mathcal{C}_{k,p}$ to the isomorphism classes of pairs $(\widetilde{\mathcal{P}}, j)$ consisting of a display $\widetilde{\mathcal{P}}$ over R and an isomorphism $j \colon \widetilde{\mathcal{P}}_k \to \mathcal{P}$.

One can prove that this functor is pro-representable by a ring

$$\mathbf{R}^{\mathbf{U}} = k[\![t_{k\ell} : 1 \le k \le d, d+1 \le \ell \le h]\!], \tag{11.5}$$

where $t_{k\ell}$ are free parameters and d and h are as in Equation (10.14) for $\mathcal{P}$. Let $T_{k\ell}$ be the Teichmüller lifts (or any other lifts) of $t_{k\ell}$ to $W(\mathbf{R}^{\mathbf{U}})$. Let $T = (T_{k\ell})$.

Let (α_{ij}) be the displaying $h \times h$ matrix for $\mathcal{P}$. Note that $(\alpha_{ij}) \in \mathrm{GL}_h(W(k))$ and can thus be considered in $\mathrm{GL}_h(W(\mathbf{R}^{\mathbf{U}}))$. Consider the matrix

$$(\alpha^{\mathbf{u}}_{ij}) = \begin{pmatrix} \mathrm{I}_d & T \\ 0 & \mathrm{I}_{h-d} \end{pmatrix} (\alpha_{ij}). \tag{11.6}$$

The symbol I_r stands for the $r \times r$ identity matrix. Write

$$(\alpha_{ij}) = \begin{pmatrix} A & B \\ C & D \end{pmatrix}, \tag{11.7}$$

where A is of size $d \times d$, B is of size $d \times (h-d)$, etc. Then, we may write

$$(\alpha^{\mathbf{u}}_{ij}) = \begin{pmatrix} A + TC & B + TD \\ C & D \end{pmatrix}. \tag{11.8}$$

Let $\mathcal{P}^{\mathbf{U}}$ denote the display over $\mathbf{R}^{\mathbf{U}}$ (meaning, modules over $W(\mathbf{R}^{\mathbf{U}})$ etc.) defined by the matrix $(\alpha^{\mathbf{U}}_{ij})$ in (11.8)*), then $\mathcal{P}^{\mathbf{U}}$ is the universal display* (*see* [**178**]).

It may be beneficial to recall what that means! It says that for every ring R in $\mathcal{C}_{k,p}$ and a deformation $(\widetilde{\mathcal{P}}, j)$ over R with an isomorphism $j \colon \widetilde{\mathcal{P}}_k \to \mathcal{P}$, there exists a unique homomorphism $\phi \colon \mathbf{R}^{\mathbf{U}} \to R$ such that $(\mathcal{P}^{\mathbf{U}}_{\phi,R}, j^{\mathbf{U}}_{\phi,R})$ is isomorphic to $(\widetilde{\mathcal{P}}, j)$.

REMARK 11.3. Using Mumford's theory of bi-extensions [**106**], Norman and Oort [**117, 118**] proved that in situation $\mathbf{I}_p$, if we choose a symplectic basis for $\mathcal{P}$, the condition for keeping the polarization is Riemann's relations on T; i.e., that T is symmetric (cf. Chapter 1, Remark 6.37). This is also explained in [**178**].

11.3. Endomorphisms conditions. We would like to examine now the universal display in the case of RM. We are in situation $\mathbf{II}_p$; the totally real field L of degree g is fixed, and *we assume for simplicity that p is inert in L*. We consider a display $\mathcal{P} = (P, Q, F, V^{-1})$ with RM defined over an algebraically closed field k of characteristic p, such that the free $W(k)$ module P is a free $\mathcal{O}_L \otimes W(k)$ module of rank 2. These are the displays that actually come from abelian varieties with

RM satisfying Rapoport condition (R)—the abelian varieties with RM that are parameterized by the moduli space. We have

$$\mathcal{O}_L \otimes_{W(\mathbb{F}_p)} W(k) \cong \bigoplus_{i=1}^{g} W(k). \tag{11.9}$$

One may thus decompose P as

$$P = \bigoplus_{i=1}^{g} P_i, \tag{11.10}$$

where each P_i is a free $W(k)$-module of rank 2, and $\mathcal{O}_L$ acts on P_i via its i-th embedding in $W(k)$. See also 1.7. Since the maps F and V^{-1} are ^{F}-linear, it follows that for every i

$$F(P_i) \subset P_{i+1}, \quad V^{-1}(P_i) \subset P_{i+1}. \tag{11.11}$$

In the case at hand, both L and T of the normal decomposition are of rank g because T/I_kT is of rank g over k (where $I_k \lhd W(k)$ is the maximal ideal). It is important to note that the normal decomposition is not canonical in any sense. In fact, since k is perfect V exists, and we may find basis x_i, y_i to P_i such that y_i is in the image of V and the image of V (i.e., Q) is spanned by $px_1, y_1, \ldots, px_g, y_g$. Then L can be taken to be the span of the y_i's and T of the x_i's. We consider the basis

$$x_1, \ldots, x_g, y_1, \ldots, y_g. \tag{11.12}$$

The matrix (α_{ij}) giving the display of $\mathcal{P}$ with respect to *this* basis is of the form

$$(\alpha_{ij}) = \begin{pmatrix} \mathfrak{d}_2(a_2, \ldots, a_g, a_1) & \mathfrak{d}_2(b_2, \ldots, b_g, b_1) \\ \mathfrak{d}_2(c_2, \ldots, c_g, c_1) & \mathfrak{d}_2(d_2, \ldots, d_g, d_1) \end{pmatrix}, \tag{11.13}$$

where the notation $\mathfrak{d}_2$ stands for a sub-diagonal matrix. That is:

$$\mathfrak{d}_2(s_2, \ldots s_g, s_1) = \begin{pmatrix} 0 & & & & s_1 \\ s_2 & 0 & & & \\ & s_3 & 0 & & \\ & & \ddots & \ddots & \\ & & & s_g & 0 \end{pmatrix}. \tag{11.14}$$

Follow now the recipe for the universal display $\mathcal{P}^{\mathbf{U}} = (P^{\mathbf{U}}, Q^{\mathbf{U}}, F^{\mathbf{U}}, V^{-1,\mathbf{U}})$ using this displaying matrix. Consider also the universal display with RM, $\mathcal{P}^{\mathbf{U},\mathbf{L}}$, which is a base change of $\mathcal{P}^{\mathbf{U}}$, obtained by dividing $\mathbf{R}^{\mathbf{U}}$ by an ideal.

To compute this ideal we argue as follows: assuming (as we do) that one knows $\mathfrak{M} \otimes \mathbb{F}$ to be nonsingular of dimension g, it is enough to find *some* extension of the $\mathcal{O}_L$-action to $\mathcal{P}^{\mathbf{U}}$, and *some* ideal $\mathfrak{a}$ such that the following hold:

- Considered mod $\mathfrak{a}$, i.e., on $\mathcal{P}^{\mathbf{U}}_{\mathbf{R}^{\mathbf{U}}/\mathfrak{a}}$, this extended action of $\mathcal{O}_L$ defines a display with RM.
- The formal variety $\mathrm{Spf}(\mathbf{R}^{\mathbf{U}}/\mathfrak{a})$ is a nonsingular g-dimensional formal subscheme of $\mathrm{Spf}(\mathbf{R}^{\mathbf{U}})$.

Thus, first write

$$\mathcal{P}^{\mathbf{U}} = \bigoplus_{i=1}^{g} \mathcal{P}^{\mathbf{U}}_i, \tag{11.15}$$

where each $\mathcal{P}_i^{\mathbf{U}}$ is a free $W(\mathbf{R^U})$-module of rank 2, extending the decomposition in (11.10).

Second, choose $\mathfrak{a}$ to be the ideal $(t_{ij} : i \neq j)$. The condition that one needs to verify is that mod $\mathfrak{a}$ we have an equality:

$$\begin{pmatrix} \Sigma(a) & 0 \\ 0 & \Sigma(a) \end{pmatrix} \begin{pmatrix} A + TC & p(B+TD) \\ C & pD \end{pmatrix} = \begin{pmatrix} A + TC & p(B+TD) \\ C & pD \end{pmatrix} \begin{pmatrix} \Sigma(a) & 0 \\ 0 & \Sigma(a) \end{pmatrix}^F . \tag{11.16}$$

This is immediate to verify. Therefore, we have proven

THEOREM 11.4 ([**50**]). *The universal ring for deformations of a display $\mathcal{P}$ with RM over k to displays with RM over rings of $\mathcal{C}_{k,p}$ is the ring $\mathbf{R^{U,L}} = k[\![t_1, \dots, t_g]\!]$. If*

$$\begin{pmatrix} \mathfrak{d}_2(a_2, \dots, a_g, a_1) & \mathfrak{d}_2(b_2, \dots, b_g, b_1) \\ \mathfrak{d}_2(c_2, \dots, c_g, c_1) & \mathfrak{d}_2(d_2, \dots, d_g, d_1) \end{pmatrix} \tag{11.17}$$

is the displaying matrix for $\mathcal{P}$, then the displaying matrix for the universal display with RM, $\mathcal{P}^{\mathbf{U,L}}$, can be take to be

(11.18)
$$\begin{pmatrix} \mathfrak{d}_2(a_2+T_2c_2, \dots, a_g+T_gc_g, a_1+T_1c_1) & \mathfrak{d}_2(b_2+T_2d_2, \dots, b_g+T_gd_g, b_1+T_1d_1) \\ \mathfrak{d}_2(c_2, \dots, c_g, c_1) & \mathfrak{d}_2(d_2, \dots, d_g, d_1) \end{pmatrix},$$

where T_i is the Teichmüller lift of t_i to $W(k[\![t_1, \dots, t_g]\!])$.

11.4. The local structure of W_τ and Hasse invariants. We are now in position to determine the local structure of the varieties W_τ and to gain better understanding of the Hasse invariants.

Let $k \in \mathbb{F}$-**Alg** and let (A, ι) be an abelian variety with RM over k (with μ_N-level, $N \geq 4$, that we suppress from the notation), such that $A(p)$ is connected (equivalently, since p is inert, A is not ordinary). Let $\mathcal{P}$ be the corresponding display defined by a displaying matrix as in (11.17). We first notice the following

LEMMA 11.5. *We have $i \in \tau(A)$ if and only if $a_i = 0 \bmod p$.*

PROOF. Identify $\mathcal{P}/p\mathcal{P}$ with the Dieudonné module of $A[p]$, and $\mathcal{P}/(V\mathcal{P}+F\mathcal{P})$ with the Dieudonné module of $\alpha(A)$. Then $i \in \tau(A)$ iff $\mathcal{P}_i/(V\mathcal{P}_{i+1} + F\mathcal{P}_{i-1})$ is not zero, where the subscript i denotes, here and below, the i-th isotypical component for the action of $\mathcal{O}_L$. Since $\mathcal{P}_i/p\mathcal{P}_i$ is two-dimensional over k and both $V\mathcal{P}_{i+1}/p\mathcal{P}_i$, $F\mathcal{P}_{i-1}/p\mathcal{P}_i$ are one-dimensional, we deduce that $i \in \tau(A)$ iff $V\mathcal{P}_{i+1} = F\mathcal{P}_{i-1}$. Equivalently, iff Fx_{i-1} and Fy_{i-1} belong to $V\mathcal{P}_{i+1}$. (The basis $x_1, \dots, x_g, y_1, \dots, y_g$ is as in (11.12).)

Now, since $y_{i-1} \in V\mathcal{P}$, the condition imposed on it is vacuous. We conclude that $i \in \tau(A)$ iff $Fx_{i-1} = a_ix_i + c_iy_i \in (V\mathcal{P})_i$. Since $(V\mathcal{P})_i/p\mathcal{P}_i$ is spanned by y_i we find that $i \in \tau(A)$ iff $a_i \equiv 0 \bmod p$. □

Now, if $a_i \neq 0 \bmod p$ (i.e., $i \notin \tau(A)$) then $a_i + t_ic_i \pmod p$ is an invertible element of $k[\![t_1 \dots, t_g]\!]$ and hence $a_i + T_ic_i$ is invertible in $W(k[\![t_1, \dots, t_g]\!])$. Therefore, every specialization of the universal display would have type not containing i. That is, if A has type τ every "generification" of A has type contained in τ. Moreover, it is evident that for every $\rho \subset \tau$ the variety W_ρ is defined locally at x (where x is the moduli point of (A, ι)) by the ideal $\mathfrak{a} = (t_i : i \in \rho)$. One obtains

THEOREM 11.6 (Goren–Oort [**50**]). *The subvarieties W_τ are nonsingular varieties of pure dimension $g - |\tau|$. Furthermore,*

$$W_\tau \cap W_\sigma = W_{\tau\cup\sigma}. \tag{11.19}$$

We now discuss the Hasse invariants. Recall first the definition (see Chapter 5, §3.1). Given A/R with RM, where R is an $\mathbb{F}$-algebra, and a nonvanishing differential $\omega \in \mathfrak{t}^*_{A/R}$, we let $e_1, \ldots, e_g$ be the orthogonal idempotents of $\mathcal{O}_L \otimes_{\mathbb{F}} R$, and we obtain a basis $(e_1\omega, \ldots, e_g\omega)$ to $\mathfrak{t}^*_{A/R}$. We take a dual basis (with respect to some $\mathcal{O}_L$-linear polarization) to $\mathfrak{t}_{A^t/R}$, say $(\eta_1, \ldots, \eta_g)$ and put

$$h_i(A, \iota, \omega) = F\eta_{i-1}/\eta_i. \tag{11.20}$$

We could have equally taken a basis $(\eta'_1, \ldots, \eta'_g)$ for $\mathfrak{t}_{A/R}$ and let $h_i(A, \iota, \omega) = F\eta'_{i-1}/\eta'_i$.

Now, $\mathfrak{t}_{A/R} \cong P/Q$ where $\mathcal{P} = (P, Q, F, V^{-1})$ is the display associated to A. The isomorphism is chosen to respect the $\mathcal{O}_L$-action and therefore (in the notation of §11.3) the element η'_i is a multiple by an invertible element of $W(k)$ of x_i, at least mod p. The same arguments may now be applied to the universal deformation of $\mathcal{P}$ given by Equation (11.18). Thus, if $\mathbf{R}^{\mathbf{U},\mathbf{L}} = k[\![t_1, \ldots, t_g]\!]$, we get

LEMMA 11.7. *There exists $c \in W(\mathbf{R}^{\mathbf{U},\mathbf{L}})^\times$ such that*

$$h_i(A^{\mathbf{U}}, \iota^{\mathbf{U}}, \omega) = c(a_i + T_i c_i) \bmod p. \tag{11.21}$$

COROLLARY 11.8. *We have an equality of divisors:*

$$(h_i) = W_i. \tag{11.22}$$

In particular, the divisor of the partial Hasse invariant h_i is reduced. Let $H = h_1 \cdots h_g$ be the total Hasse invariant, then the divisor of H is a reduced normal crossing divisor, equal to the complement of the ordinary locus.

The following corollary may give some "intuitive feeling" to why every component of W_τ must contain a superspecial point. We recall a theorem of Bailly saying that if $\det \underline{\omega}$ is the sheaf of Hilbert modular forms of parallel weight 1, then there exists an r such that $\det \underline{\omega}^{\otimes r}$ is an ample line bundle.

COROLLARY 11.9 (Raynaud's trick). *The divisor (H) is ample. Thus, every family of ordinary abelian varieties with RM over a complete (irreducible) variety $S \to \operatorname{Spec}(\mathbb{F})$, say $\mathcal{A} \to S$, is isotrivial.*

PROOF. Indeed, for large enough k, we have that $rk \cdot (H)$ is the divisor of the section H^{rk} of the very ample line bundle $\det \underline{\omega}^{\otimes rk(p-1)}$, hence a very ample divisor. Thus, (H) is ample.

Now, the sheaf $\det \underline{\omega}^{\otimes r}$ extends to an ample line bundle over a compactification $\mathfrak{M}^*(\mathbb{F})$, and therefore the complement of (H) in $\mathfrak{M}^*(\mathbb{F})$, which contains the ordinary locus, is affine. Thus, a complete subvariety of $\mathfrak{M}^*(\mathbb{F})$ that is disjoint with (H) must be zero-dimensional.

Given the family $\mathcal{A} \to S$, one may find a complete (connected) variety T and a finite morphism $\pi\colon T \to S$ such that $\pi^*\mathcal{A} \to T$ is endowed with a μ_N-level structure ($N \geq 4$). Let $\mathfrak{M}(\mu_N, \mathbb{F})$ stand for the moduli space of abelian varieties with RM and this level structure. Then, there exists a morphism $\psi\colon T \to \mathfrak{M}(\mu_N, \mathbb{F})$ such that $\pi^*\mathcal{A}$ is the pull-back by ψ of the universal family over $\mathfrak{M}(\mu_N, \mathbb{F})$. But, by the above, the image of ψ is connected and zero-dimensional. That is, $\pi^*\mathcal{A}$ is constant, that is to say, $\mathcal{A} \to S$ is isotrivial. □

11.5. Polarization conditions. Let k be a perfect field of characteristic p. The notion of a principal polarization λ on a g-dimensional abelian variety X/k translates into a symmetric isomorphism $\lambda(p)\colon X(p) \to X(p)^t$ or, equivalently, into a perfect symmetric pairing

$$\langle\ ,\ \rangle_\lambda \colon X(p) \times X(p) \to \widehat{\mathbb{G}}_m. \tag{11.23}$$

By the Serre–Tate Theorem deforming the data (X, λ) is equivalent to deforming the data $\big(X(p), \lambda(p)\big)$. Letting $\mathcal{P}$ be the associated display, one obtains a perfect alternating pairing

$$\langle\ ,\ \rangle_\lambda \colon \mathcal{P} \times \mathcal{P} \to \mathcal{P}_{\mathrm{m}}, \tag{11.24}$$

satisfying $\langle V^{-1}(x), V^{-1}(y)\rangle^V = \langle x, y\rangle$, where $\mathcal{P}_{\mathrm{m}}$ is the multiplicative display of Example (1) on p. 225. Deformations of $\big(X(p), \lambda(p)\big)$ correspond to deformations of the display $(\mathcal{P}, \langle\ ,\ \rangle_\lambda)$ into displays with a perfect alternating pairing valued in $\mathcal{P}_{\mathrm{m}}$ satisfying $\langle V^{-1}(x), V^{-1}(y)\rangle^V = \langle x, y\rangle$ ([**118, 178**]). One can prove (loc. cit.) that if one chooses the basis $e_1, \dots, e_{2g}$ for the display to be symplectic, the universal equi-characteristic deformation as a principally polarized abelian variety (over the universal equi-characteristic local deformation space $k[\![t_{ij}]\!]$) is given by the closed subscheme $t_{ij} = t_{ji}$, $\forall i, j$. This is a very important result and it is useful to record it:

Let (X, λ) be a principally polarized g-dimensional abelian variety over a perfect characteristic p field k; let $\mathcal{P}$ be the corresponding display with the induced perfect pairing $\langle\ ,\ \rangle_\lambda$; let $e_1, \dots, e_{2g}$ be a basis for the display $\mathcal{P}$ satisfying the usual condition ($e_1, \dots, e_g$ span L, etc.) which is moreover a symplectic basis: $\langle e_i, e_j\rangle = \delta_{i,j-g}$ for every $1 \le i \le g$ and $g+1 \le j \le 2g$; let $\left(\begin{smallmatrix} A & B \\ C & D\end{smallmatrix}\right)$ be the display. Then the universal principally polarized display is given over $k[\![t_{ij}]\!]/(t_{ij} - t_{ji})$ by

$$\begin{pmatrix} A + TC & B + TD \\ C & D \end{pmatrix}, \tag{11.25}$$

where $T = (T_{ij})$ and T_{ij} is a lift of t_{ij} to $W\big(k[\![t_{ij}]\!]/(t_{ij} - t_{ji})\big)$.

The symmetry condition is the characteristic p counterpart of Riemann's relations. See Chapter 1, Remark 6.37.

Here is an example of how one calculates with this. Assume that k is algebraically closed. Let $x \in \mathcal{A}_{2,n}$ be a k-rational point corresponding to a principally polarized superspecial abelian surface (with some rigid level n structure, $(n, p) = 1$). One can show that the display $\mathcal{P}$ of this data has a symplectic basis $e_1, \dots, e_4$ such that the displaying matrix is

$$\begin{pmatrix} 0 & 0 & -1 & 0 \\ 0 & 0 & 0 & -1 \\ 1 & 0 & 0 & 0 \\ 0 & 1 & 0 & 0 \end{pmatrix} \tag{11.26}$$

(cf. Equation (8.15)). The universal is display is thus

$$\begin{pmatrix} T_{11} & T_{12} & -1 & 0 \\ T_{12} & T_{22} & 0 & -1 \\ 1 & 0 & 0 & 0 \\ 0 & 1 & 0 & 0 \end{pmatrix}, \tag{11.27}$$

and the action of Frobenius modulo p is therefore given by the matrix

$$\begin{pmatrix} t_{11} & t_{12} & 0 & 0 \\ t_{12} & t_{22} & 0 & 0 \\ 1 & 0 & 0 & 0 \\ 0 & 1 & 0 & 0 \end{pmatrix}. \tag{11.28}$$

The nonordinary locus is given by the condition that the determinant of the Hasse–Witt matrix, namely $\det\left(\begin{smallmatrix} t_{11} & t_{12} \\ t_{12} & t_{22} \end{smallmatrix}\right)$, is zero. That is, locally (in the formal sense) around the point x the nonordinary locus is given by the cone

$$t_{11}t_{22} - t_{12}^2 = 0. \tag{11.29}$$

The supersingular locus is characterized by the property that Fr^2 is equal to zero modulo p, namely, by the condition

$$\begin{pmatrix} t_{11} & t_{12} \\ t_{12} & t_{22} \end{pmatrix} \begin{pmatrix} t_{11}^p & t_{12}^p \\ t_{12}^p & t_{22}^p \end{pmatrix} = \begin{pmatrix} t_{11}^{p+1} + t_{12}^{p+1} & t_{11}t_{12}^p + t_{12}t_{22}^p \\ t_{12}t_{11}^p + t_{22}t_{12}^p & t_{12}^{p+1} + t_{22}^{p+1} \end{pmatrix} = 0. \tag{11.30}$$

Let ζ be a $p+1$-st root of -1. Using the upper left entry we get that a branch of this locus is defined by $t_{12} = \zeta t_{11}$. Since the supersingular locus is contained in the nonordinary locus, we obtain that $t_{22} = \zeta^2 t_{11}$. We conclude that the supersingular locus has $p+1$ branches at x and is defined locally, in the formal neighborhood of x, by the equations

$$\begin{cases} t_{11}t_{22} - t_{12}^2 = 0, \\ t_{11}^{p+1} + t_{12}^{p+1} = 0. \end{cases} \tag{11.31}$$

Finally, the superspecial locus is identified by the property $\mathrm{Fr} = 0$ modulo p. We see that this has no deformations.

This example was one of the first ever calculated. See Koblitz [**85**, §IV–11]. The interest in it was spurred by the wish to understand the variation of the zeta function of abelian varieties (loc. cit.).

As an another example, we consider the structure of the moduli scheme $\mathcal{A}_{2,n}$ at a supersingular point with a-number equal 1 (Appendix A, §6). Some of the considerations can be applied in the case $a = 1$ in higher dimensions.

Let $x \in \mathcal{A}_{2,n}$ be a k-rational point such that the corresponding abelian surface has a-number 1. It follows that the corresponding Dieudonné module D is cyclic over $W(k)[F,V]$ (because $D/(FD + VD)$ is one-dimensional over $W(k)/pW(k)$). Let x be a generator. Then one can show that for a suitable x, the elements x, Fx, V^2x, Vx form a symplectic basis. Note that F must be nilpotent so F^2x is equal to $aFx + bV^2x + cVx$ for suitable $a, b, c \in W(k)$ with $p \mid a$. We may display the module by the matrix

$$\begin{pmatrix} 0 & 0 & 1 & 0 \\ 1 & a & 0 & 0 \\ 0 & b & 0 & 1 \\ 0 & c & 0 & 0 \end{pmatrix}. \tag{11.32}$$

The universal display as a principally polarized abelian surface is given by

$$\begin{pmatrix} 0 & bT_{11} + cT_{12} & 1 & T_{11} \\ 1 & a + bT_{12} + cT_{22} & 0 & T_{12} \\ 0 & b & 0 & 1 \\ 0 & c & 0 & 0 \end{pmatrix}. \tag{11.33}$$

The nonordinary locus is thus given locally formally by the linear equation

$$bt_{11} + ct_{12} = 0. \tag{11.34}$$

The supersingular locus is given locally formally by the equation

$$\begin{pmatrix} 0 & bt_{11} + ct_{12} \\ 1 & bt_{12} + ct_{22} \end{pmatrix} \begin{pmatrix} 0 & b^p t_{11}^p + c^p t_{12}^p \\ 1 & b^p t_{12}^p + c^p t_{22}^p \end{pmatrix} \tag{11.35}$$

$$= \begin{pmatrix} bt_{11} + ct_{12} & (bt_{11} + ct_{12})(bt_{12} + ct_{22})^p \\ bt_{12} + ct_{22} & (bt_{11} + ct_{12})^p + (bt_{12} + ct_{22})(bt_{12} + ct_{22})^p \end{pmatrix} = 0.$$

That is, the supersingular locus is defined locally, at the formal neighborhood of x, by the system of equations

$$\begin{cases} bt_{11} + ct_{12} = 0, \\ bt_{12} + ct_{22} = 0. \end{cases} \tag{11.36}$$

APPENDIX A

Group Schemes

The problem we face in this appendix is to stay within certain length constraints, while, at the same time, to give the reader the feeling that he (or she) knows what group schemes are about, make him aware of the main examples, and allow him to follow the arguments concerning them that are spread all over this book. Since extensive treatises and several survey papers do exist (e.g., [**28–30, 126, 163, 168**]), we shall offer a rather peculiar choice of topics, assuming that the interested reader would consult the above references for a more complete picture. For example, we will not prove the main theorems, but will prove certain exotic statements that are relevant to the issues discussed in the book.

We shall assume that the reader speaks, though not necessarily fluently, the language of schemes. This is imperative since many of the group schemes that interest us are not reduced, and hence are beyond the scope of classical algebraic geometry. Perhaps the inevitability of learning the subject is clear when one learns that the group scheme of the p-torsion of an elliptic curve over a field of characteristic p has either p geometric points (ordinary case) or one geometric point (supersingular case). The only way to retain the harmony of "p-torsion being of order p^2" is to consider the group scheme of p-torsion, which is indeed of rank p^2, as a group scheme.

1. Some definitions

Throughout this appendix let S stand for a noetherian scheme. Let $\pi\colon G \to S$ be a scheme over S. One says G is a group scheme if there exist S-morphisms

(1.1) $$\mathrm{inv}\colon G \to G, \quad m\colon G \times_S G \to G,$$

and a section

(1.2) $$e\colon S \to G,$$

such that the following diagrams commute (compare Chapter 1, (1.3)):

(1.3) $$\begin{array}{ccc} G\times_S G\times_S G & \xrightarrow{m\times \mathrm{id}} & G\times_S G \\ {\scriptstyle \mathrm{id}\times m}\downarrow & & \downarrow{\scriptstyle m,} \\ G\times_S G & \xrightarrow[m]{} & G \end{array} \qquad \begin{array}{ccc} G & \xrightarrow{(e\circ\pi,\mathrm{id})} & G\times_S G \\ {\scriptstyle (\mathrm{id},e\circ\pi)}\downarrow & \overset{\mathrm{id}}{\searrow} & \downarrow{\scriptstyle m,} \\ G\times_S G & \xrightarrow[m]{} & G \end{array} \qquad \begin{array}{ccc} G & \xrightarrow{(\mathrm{inv},\mathrm{id})} & G\times_S G \\ {\scriptstyle (\mathrm{id},\mathrm{inv})}\downarrow & \overset{e\circ\pi}{\searrow} & \downarrow{\scriptstyle m.} \\ G\times_S G & \xrightarrow[m]{} & G \end{array}$$

As schemes are completely determined by their functor of points [**111**, II.6], an equivalent definition is: The functor of points h_G of G gives a functor

(1.4) $$h_G\colon \mathbf{Sch}_S \to \mathbf{Groups}\,.$$

It is often this last definition that is easily verified.

A group scheme $\pi\colon G \to S$ is called *affine* if the morphism π is affine. It means that for every open affine $U = \mathrm{Spec}(R)$ of S we have $G|_U := G \times_S U$ is affine, say equal to $\mathrm{Spec}(R[x_1, \dots, x_{r(U)}]/I_U)$.[1] In this case the group law is determined by morphisms that are *R-algebra homomorphisms* (compare Chapter 1 (1.4)):

$$\widetilde{m}\colon R[x_1, \dots, x_{r(U)}]/I_U \to R[x_1, \dots, x_{r(U)}]/I_U \otimes_R R[x_1, \dots, x_{r(U)}]/I_U, \tag{1.5}$$

$$\widetilde{\mathrm{inv}}\colon R[x_1, \dots, x_{r(U)}]/I_U \to R[x_1, \dots, x_{r(U)}]/I_U, \tag{1.6}$$

$$\tilde{e}\colon R[x_1, \dots, x_{r(U)}]/I_U \to R. \tag{1.7}$$

The kernel of $\tilde{e}$ is called the *augmentation ideal.*

A group scheme $\pi\colon G \to S$ is called *flat* if the morphism π is flat. That means that one can cover S by open affine sets $U = \mathrm{Spec}(R)$ such that over each U there exists a covering of G by open affine sets $V = \mathrm{Spec}(R')$ such that R' is a flat R-algebra. Recall that this means that the functor $M \mapsto M \otimes_R R'$, from R-modules to R'-modules is exact. Though probably not clear from the definition, this means that G varies continuously over S. See [**111**, Chapter III.10].

A group scheme $\pi\colon G \to S$ is called *finite* if π is a finite morphism. Thus, locally on S, one can write G as $\mathrm{Spec}(R[x_1, \dots, x_{r(U)}]/I_U) \to \mathrm{Spec}(R)$ and the module $R[x_1, \dots, x_{r(U)}]/I_U$ is a finite R-module. If G is also flat, then this R module is locally free of a certain rank. This rank is constant if S is connected. In general, given a finite flat group scheme G over S, we assume that the rank is constant and call it the *rank of* G.

A *closed subgroup scheme* H of a group scheme $\pi\colon G \to S$ is a closed subscheme that is a group scheme under the morphisms induced from those of G. Suppose that $\pi\colon G \to S$ is an affine group scheme and that $S = \mathrm{Spec}(R)$ is affine. Write $G = \mathrm{Spec}(R_G)$ for a suitable R algebra R_G, and let I_G be the augmentation ideal of G. Let H be a closed subgroup scheme, then $H = \mathrm{Spec}(R_G/J)$ for a some ideal J. The properties forced on J are that $J \subset I_G$, that $\widetilde{m}(J) \subset J \otimes R_G + R_G \otimes J$ and that $\widetilde{\mathrm{inv}}(J) \subset J$. Conversely, every ideal J with such properties defines a closed subgroup scheme.

Let $G' = \mathrm{Spec}(R_{G'})$ be another group scheme affine over S. Giving a homomorphism $f\colon G \to G'$ of S-group schemes is equivalent to giving a homomorphism $\phi\colon R_{G'} \to R_G$ that commutes with the maps $\widetilde{m}$, $\widetilde{\mathrm{inv}}$ and $\tilde{e}$. The *kernel* of f is defined as the fiber product

$$\begin{array}{ccc} \mathrm{Ker}(f) & \longrightarrow & G \\ \downarrow & & \downarrow{\scriptstyle f} \\ S & \xrightarrow{\ e\ } & G' \end{array} \tag{1.8}$$

It is always a subgroup scheme—one does not need G and G' to be affine for this to hold—that represents the functor on $\mathbf{Sch}_{/S}$, taking T to $\mathrm{Ker}\big(G(T) \to G'(T)\big)$. In the affine case, the ideal J of R_G defining $\mathrm{Ker}(f)$ is $\phi(I_{G'})R_G$.

The notion of a quotient group scheme is much harder, even if one attempts to divide by a normal subgroup scheme. We remark that the quotient by a finite flat group scheme very often exists. For example, if G/S is a quasi-projective S-group scheme and H is a closed subgroup scheme, which is proper and flat over S then

[1] We do allow infinitely many variables here. A more precise notation is $R[x_\alpha : \alpha \in J(U)]/I_U$, where $J(U)$ is an index set depending of U.

G/H exists. See [**152**, §2], [**107**, §12] and [**163**, §3.4] for results, discussion and additional references.

2. Digression on Frobenius and Verschiebung

Let S be a scheme of characteristic p. That is, $p = 0$ in the structure sheaf $\mathcal{O}_S$ of S. There is then a morphism

$$\mathrm{Fr}^{\mathrm{abs}} \colon S \to S, \tag{2.1}$$

the *absolute Frobenius*, that is given as the identity map on the underlying topological space of S and as the map of raising to the p-power on the structure sheaf $\mathcal{O}_S$.

Let $f \colon X \to S$ be a scheme over S. Define $X^{(p)}$ as a fiber product in the following cartesian diagram

$$\begin{array}{ccc} X^{(p)} & \longrightarrow & X \\ \downarrow & & \downarrow f \\ S & \xrightarrow{\mathrm{Fr}^{\mathrm{abs}}} & S. \end{array} \tag{2.2}$$

The operation $X \mapsto X^{(p)}$ is a covariant *functor* from the category of S schemes to itself.

EXERCISE 2.1. *Determine how to define the Frobenius of a morphism and verify that we indeed get a functor. Is it a faithful functor? fully-faithful?*

The commutative diagram

$$\begin{array}{ccc} X & \xrightarrow{\mathrm{Fr}^{\mathrm{abs}}} & X \\ \downarrow f & & \downarrow f \\ S & \xrightarrow{\mathrm{Fr}^{\mathrm{abs}}} & S. \end{array} \tag{2.3}$$

produces by the universal property of fiber product a morphism $\mathrm{Fr} = \mathrm{Fr}_X$, called the *Frobenius morphism*,

$$\mathrm{Fr} \colon X \to X^{(p)}. \tag{2.4}$$

Note that it is a *morphism of S schemes.*

For example, let k be a perfect field of characteristic p, $S = \mathrm{Spec}(k)$, and let X be the scheme

$$X = \mathrm{Spec}\big(k[x_1, \dots, x_n]/(f_1, \dots, f_m)\big). \tag{2.5}$$

Then the scheme $X^{(p)}$ is given by

$$X^{(p)} = \mathrm{Spec}\big(k[x_1, \dots, x_n]/(g_1, \dots, g_m)\big), \tag{2.6}$$

where g_i is obtained from f_i by raising each coefficient of f_i to the p power. (In particular, if the polynomials f_i are all defined over $\mathbb{F}_p$ then $X^{(p)} \cong X$). The morphism $\mathrm{Fr} \colon X \to X^{(p)}$ is given by the homomorphism of k-algebras

$$k[x_1, \dots, x_n]/(g_1, \dots, g_m) \to k[x_1, \dots, x_n]/(f_1, \dots, f_m), \tag{2.7}$$

determined uniquely by $x_i \mapsto x_i^p$ for $i = 1, \dots, n$. In terms of the functor of points, the morphism $\mathrm{Fr} \colon X \to X^{(p)}$ corresponds to

$$h_X \mapsto h_{X^{(p)}} \tag{2.8}$$

given for any k-algebra R by

$$(2.9)\quad h_X(R) = \{(r_1, \ldots, r_n) : f_i(r_1, \ldots, r_n) = 0 \ \forall i\} \mapsto h_{X^{(p)}}(R) = \{(r_1, \ldots, r_n) : g_i(r_1, \ldots, r_n) = 0 \ \forall i\}$$

by

$$(2.10)\qquad (r_1, \ldots, r_n) \mapsto (r_1^p, \ldots, r_n^p).$$

Returning to the general case, the functoriality of the construction easily gives that if $G \to S$ is a group scheme then so is $G^{(p)} \to S$ and $\mathrm{Fr}\colon G \to G^{(p)}$ is a group homomorphism.

Assume that G is a finite flat commutative group scheme or an abelian scheme (see below) over S. Duality theory provides one with a morphism $\mathrm{Ver}\colon G^{(p)} \to G$, called the *Verschiebung morphism*, as follows: One considers the morphism

$$(2.11)\qquad \mathrm{Fr}\colon G^t \to (G^t)^{(p)}.$$

Here G^t is the dual group scheme (see below) if G is a finite flat group scheme, or the dual abelian scheme if G is an abelian variety. Upon dualizing we obtain a morphism $\mathrm{Ver} = \mathrm{Ver}_{G^{(p)}}$

$$(2.12)\qquad \mathrm{Ver} := \mathrm{Fr}^t\colon G^{(p)} \to G.$$

The main property of Ver is that it is a group homomorphism satisfying

$$(2.13)\qquad \mathrm{Fr}_G \circ \mathrm{Ver}_{G^{(p)}} = [p]_{G^{(p)}}, \quad \mathrm{Ver}_{G^{(p)}} \circ \mathrm{Fr}_G = [p]_G.$$

3. Important examples

3.1. The multiplicative group $\mathbb{G}_m$. For every ring R let

$$(3.1)\qquad \mathbb{G}_{m/R} := \mathrm{Spec}(R[x, x^{-1}]),$$

with

$$(3.2)\qquad \widetilde{m}(x) = x \otimes x, \quad \widetilde{\mathrm{inv}}(x) = x^{-1}, \quad \tilde{e}(x) = 1.$$

It is the group scheme associating to any R-algebra T the group $T^\times$ of invertible elements in T. The augmentation ideal is generated by $x - 1$. For every R-algebra R' we have $\mathbb{G}_{m/R'} = \mathbb{G}_{m/R} \times_{\mathrm{Spec}(R)} \mathrm{Spec}(R')$ – a feature of many of the examples below.

3.2. The roots of unity μ_N. Let N be a positive integer. For every ring R we define the group of N-th roots of unity by

$$(3.3)\qquad \mu_N = \mathrm{Spec}\big(R[x]/(x^N - 1)\big),$$

with

$$(3.4)\qquad \widetilde{m}(x) = x \otimes x, \quad \widetilde{\mathrm{inv}}(x) = x^{-1}, \quad \tilde{e}(x) = 1.$$

It is the group scheme associating to any R-algebra R' the multiplicative group $\{\zeta \in R' : \zeta^N = 1\}$ (the N-th roots of unity in R'; there might be more, or less, than N of them). The group μ_N is the kernel of the homomorphism $\mathbb{G}_m \to \mathbb{G}_m$ given by $x \mapsto x^N$, or, in terms of points, for every R-algebra R'

$$(3.5)\qquad \mu_N(R') \to \mu_N(R'), \quad \zeta \mapsto \zeta^N.$$

Note that by our recipe for kernels μ_N is to be defined by the image of $I_{\mathbb{G}_m}$, which is indeed the case. We further remark that if R has characteristic p then μ_p is the kernel of $\mathrm{Fr}\colon \mathbb{G}_{m/R} \to \mathbb{G}_{m/R}$. More generally, $\mu_{p^n} = \mathrm{Ker}(\mathrm{Fr}^n)$.

EXERCISE 3.1. *Let N be prime and let R be an integral domain. Prove that $\mu_{N/R}$ has no proper subgroup schemes. Note that for many R' the abstract group $\mu_N(R')$ is not simple.*

3.3. The additive group $\mathbb{G}_a$. For every ring R define the group scheme $\mathbb{G}_a$ as

$$\mathbb{G}_a := \mathrm{Spec}(R[x]), \tag{3.6}$$

with

$$\tilde{m}(x) = x \otimes 1 + 1 \otimes x, \quad \widetilde{\mathrm{inv}}(x) = -x, \quad \tilde{e}(x) = 0. \tag{3.7}$$

It is the group scheme assigning to every R-algebra R' the underlying additive group of R'. In general the endomorphisms of $\mathbb{G}_a$ are just the maps induced by $x \mapsto rx$ for $r \in R$ that include the multiplication by n maps $x \mapsto nx$. Assume however that R has characteristic p—a prime number. Then the map of raising-to-the-p-power, given on the coordinate ring by $x \mapsto x^p$, is a homomorphism of groups. It is in fact the Frobenius morphism defined above.

Thus, if R if a perfect ring of characteristic p, $\mathrm{End}(\mathbb{G}_a/R) \supset R\{\tau\}$—the non-commutative ring of polynomials in the variable τ. Every element f of $R\{\tau\}$ has a unique expression of the form $f = r_0 + r_1\tau + \cdots + r_n\tau^n$ and $\tau r = r^p\tau$. The action of f on $\mathbb{G}_a/R$ is given by

$$x \mapsto r_0 x + r_1 x^p + \cdots + r_n x^{p^n}. \tag{3.8}$$

This structure is fundamental to the theory of Drinfeld modules. See [**35**].

3.4. The group α_{p^r}. Let R be a ring of characteristic p. We define the group scheme $\alpha_{p^r/R}$ as the kernel of $\mathrm{Fr}^r\colon \mathbb{G}_a \to \mathbb{G}_a$. Thus, by our recipe for kernels,

$$\alpha_{p^r/R} = \mathrm{Spec}\big(R[x]/(x^{p^r})\big), \tag{3.9}$$

with

$$\tilde{m}(x) = x \otimes 1 + 1 \otimes x, \quad \widetilde{\mathrm{inv}}(x) = -x, \quad \tilde{e}(x) = 0. \tag{3.10}$$

It associates to every R-algebra R' the additive group of nilpotent elements of order p^r of R'. That is, $\{a \in R' : a^{p^r} = 0\}$.

EXERCISE 3.2. *Let R be a field of characteristic p. Prove that* $\mathrm{End}(\alpha_{p/R}) = R$.

3.5. The group GL_n. For notational simplicity we just define the group scheme GL_2. For every R we let

$$\mathrm{GL}_{2/R} := \mathrm{Spec}(R[a, b, c, d, (ad - bc)^{-1}]), \tag{3.11}$$

with $\tilde{m}$ given by

$$\begin{aligned} \tilde{m}(a) &= a \otimes a + b \otimes c, & \tilde{m}(b) &= a \otimes b + b \otimes d, \\ \tilde{m}(c) &= c \otimes a + d \otimes c, & \tilde{m}(d) &= c \otimes b + d \otimes d, \end{aligned} \tag{3.12}$$

with $\widetilde{\mathrm{inv}}$ given by

$$\begin{aligned} \widetilde{\mathrm{inv}}(a) &= d(ad - bc)^{-1}, & \widetilde{\mathrm{inv}}(b) &= -b(ad - bc)^{-1}, \\ \widetilde{\mathrm{inv}}(c) &= -c(ad - bc)^{-1}, & \widetilde{\mathrm{inv}}(d) &= a(ad - bc)^{-1}, \end{aligned} \tag{3.13}$$

and with $\tilde{e}$ given by

$$\tilde{e}(a) = 1, \quad \tilde{e}(b) = 0, \quad \tilde{e}(c) = 0, \quad \tilde{e}(d) = 1. \tag{3.14}$$

It is the group scheme associating to each R-algebra R' the group of 2×2 invertible matrices with entries in R.

The reader is well acquainted with the group homomorphism

$$\det\colon \mathrm{GL}_2 \to \mathbb{G}_m \tag{3.15}$$

given by the determinant.

EXERCISE 3.3. *Write this homomorphism in terms of the coordinate rings. The kernel is a the group scheme* SL_2. *What is the ideal defining it?*

3.6. A non-commutative group scheme of rank p^2. Let R be a ring of characteristic p. We define a subgroup scheme of GL_2 in functorial terms. For every R-algebra R' it is given by the matrices

$$\begin{pmatrix} \zeta & \alpha \\ 0 & 1 \end{pmatrix}, \quad \zeta \in \mu_p(R'), \alpha \in \alpha_p(R'). \tag{3.16}$$

EXERCISE 3.4. *Check that this a group scheme and write it as an affine group scheme. Prove it is of rank* p^2 *and non-commutative.*

EXERCISE★ 3.5. *Prove that this group scheme is isomorphic to* $\mu_p \ltimes \alpha_p$. *For that you have to first make sense of the last expression.*

3.7. The constant group scheme $\underline{\Gamma}$. Let Γ be a finite abelian group in the usual sense of group theory. Let R be a ring and $S = \mathrm{Spec}(R)$. We define the constant group ring $\underline{\Gamma}$ defined by Γ as

$$\underline{\Gamma} = \coprod_{\gamma \in \Gamma} \left(\mathrm{Spec}(R)\right)_\gamma = \mathrm{Spec} \bigoplus_{\gamma \in \Gamma} R_\gamma = \mathrm{Spec}\, R^\Gamma. \tag{3.17}$$

This defines an S-scheme $\pi\colon \underline{\Gamma} \to S$. Suppose that S is connected and T is a connected S-scheme. Then $\underline{\Gamma}(T) = \Gamma$. This explains the name "constant."

We may identify R^Γ with $R[\Gamma] := \{\sum a_\gamma \cdot \gamma \colon a_\gamma \in R\}$. This identification sends $\gamma \in \Gamma \subseteq R[\Gamma]$ to the delta function at γ—an element of R^Γ. Therefore, the multiplication law induced on $R[\Gamma]$ is *not* the usual one of the group ring, but rather

$$\left(\sum a_\gamma \cdot \gamma\right)\left(\sum b_\gamma \cdot \gamma\right) = \sum a_\gamma b_\gamma \cdot \gamma. \tag{3.18}$$

To emphasize that we shall write $R[\underline{\Gamma}]$. So far what we said holds for every finite set. The group scheme structure on $\underline{\Gamma}$ comes from the group structure on Γ. One finds that the comorphisms are

$$\widetilde{m}(\gamma) = \sum_{\delta \in \Gamma} \gamma\delta \otimes \delta^{-1}, \quad \widetilde{\mathrm{inv}}(\gamma) = \gamma^{-1}, \quad \tilde{e}(\gamma) = 0. \tag{3.19}$$

Thus, co-multiplication is convolution of functions

EXERCISE 3.6. *Show that if T is a connected S-scheme then* $\underline{\Gamma}(T) = \Gamma$ *as groups, where the group structure on the left-hand side is the one induced from Equation* (3.19).

EXERCISE 3.7. *Given a finite flat commutative group scheme $\pi\colon G \to S$, one can define a dual group scheme. The construction being local on the base, we may restrict to $S = \operatorname{Spec}(R)$. Then G is given by an R-algebra, say T. It comes equipped with co-multiplication, co-inverse and augmentation maps $\widetilde{m}\colon T \to T \otimes_R T$, $\widetilde{\operatorname{inv}}\colon T \to T$ and $\tilde{e}\colon T \to R$. Let us denote the multiplication by $\mu\colon T \otimes T \to T$ and the structure map by $\epsilon\colon R \to T$.*

Consider now the finite R-module $T^* := \operatorname{Hom}_R(T, R)$. Show that the map $\widetilde{m}$ induces multiplication on T^* and that T^* becomes an R-algebra with structure map induced from $\tilde{e}$. Show that there is a natural group structure on $\operatorname{Spec}(T^*)$ for which co-multiplication is induced by μ, co-inverse is induced by $\widetilde{\operatorname{inv}}$ and augmentation (co-unit) is induced by ϵ. The group scheme $\operatorname{Spec}(T^*)$ is called the dual group scheme to G and is usually denoted G^*, G^t, $\widehat{G}$ or $G^\vee$.

The adjective "dual" is justified in that there is a canonical perfect pairing

$$G \times G^t \to \mathbb{G}_m, \tag{3.20}$$

and $(G^t)^t$ is naturally isomorphic to G. See [**152**, §4], [**163**, §§(2.9), (2.10), (3.8)], [**168**, §2.4]

EXERCISE 3.8. *Let $H = \operatorname{Spec}(A)$ be a group scheme. Show that to give a homomorphism $H \to \mathbb{G}_{m/A}$ is equivalent to giving an element $x \in A$ such that $\widetilde{m}(x) = x \otimes x$. Such elements are called "group-like elements".*

EXERCISE★ 3.9. *Prove that for an R-algebra R' we have*

$$\begin{aligned} G^t(R') &\cong \{\text{group-like elements in } T \otimes_R R'\} \\ &\cong \operatorname{Hom}_{\mathbf{Groups}/R'}(G \otimes_R R', \mathbb{G}_{m/R'}). \end{aligned} \tag{3.21}$$

Thus G^t represents the functor on R-algebras

$$R' \mapsto \operatorname{Hom}_{\mathbf{Groups}/R'}(G \otimes_R R', \mathbb{G}_{m/R'}). \tag{3.22}$$

In particular, there exists a natural pairing

$$G(R') \times G^t(R') \to \mathbb{G}_m(R'), \tag{3.23}$$

under which $(a, f) \mapsto f_{R'}(a)$ ($f \in \operatorname{Hom}_{\mathbf{Groups}/R'}(G \otimes_R R', \mathbb{G}_{m/R'})$ is a collection of homomorphisms $f_{R''}$ for R'-algebras R''). This pairing need not be perfect on a particular R'.

EXERCISE★ 3.10. *Assume that T is free over R and let $e_1, \ldots, e_d$ be a basis for T over R. Let $e_1^t, \ldots, e_d^t$ be the dual basis for T^*. Show that the pairing in (3.20) corresponds to the function $\sum_{i=1}^d e_i \otimes e_i^t$.*

EXERCISE 3.11. *Show that the dual group scheme of $\underline{\Gamma}$ is the diagonalizable group scheme, discussed extensively in Chapter 1, §1 (but we allow p-torsion now), whose coordinate ring is $R[\Gamma]$. Note that this is the same set as the coordinate ring $R[\underline{\Gamma}]$ but now multiplication is the "usual" multiplication in a group ring:*

$$\Big(\sum_{\gamma\in\Gamma} a_\gamma \cdot \gamma\Big)\Big(\sum_{\gamma\in\Gamma} b_\gamma \cdot \gamma\Big) = \sum_{\gamma\in\Gamma}\sum_{\delta\in\Gamma} a_{\gamma\delta} b_{\delta^{-1}} \cdot \gamma, \tag{3.24}$$

while co-multiplication is given by $\gamma \mapsto \gamma \otimes \gamma$.

EXERCISE 3.12. *Show that μ_N and $\underline{\mathbb{Z}/N\mathbb{Z}}$ are dual to one another.*

3.8. Étale group schemes. Recall that a morphism of schemes $\pi: T \to S$ is called *étale* if it is locally of finite type, flat and unramified. Equivalently, locally of finite type, flat and $\Omega^1_{T/S} = 0$. For example a morphism that locally Zariski on S is of the form $\mathrm{Spec}\big(R[x_1, \ldots, x_n]/(f_1, \ldots, f_n)\big) \to \mathrm{Spec}(R)$ with $\det(\partial f_i/\partial x_j)$ a unit. Heuristically, this concept puts together the notion of a topological covering map and a separable field extension. See [**99**] for more on étale morphisms. A group scheme $\pi: G \to S$ is called étale if the morphism π is étale. The main fact one employs about étale group schemes is that after a suitable base change they become constant group schemes. The group μ_N is an étale group scheme if and only if the characteristic of every geometric point of S is prime to N.

EXERCISE 3.13. *Let R be a field of characteristic p and let ℓ be a prime. Prove that the group scheme $\mu_{\ell/R}$ is étale over R if and only if $p \neq \ell$.*

Suppose that R is a field, $S = \mathrm{Spec}(R)$ and K is an algebraic closure of R. Then the category of finite étale group schemes is equivalent to the category of finite groups with a $\mathrm{Gal}(K/R)$ action. See [**163**] or [**99**].

3.9. The p-torsion group scheme $A[p]$. Let $\pi: A \to S$ be an abelian scheme. That is, $\pi: A \to S$ is a group scheme, and the morphism π is proper, smooth, with geometrically connected fibers. An abelian scheme is always commutative. It should be thought of as a continuously varying family of abelian varieties (possibly over fields of different characteristics). The simplest example is given by a Weierstrass equation $y^2z + a_1xyz + a_3yz^2 = x^3 + a_2x^2z + a_4xz^2 + a_6z^3$ in $\mathbb{P}^2$ with $a_1, \ldots, a_6$ integers. Assume that the discriminant Δ of this equation in non-zero. Then this equation defines an abelian scheme of relative dimension 1 over $\mathbb{Z}[\Delta^{-1}]$. A much deeper and typical example is the relative Jacobian. If $C \to S$ is a family of curves, then one can put their Jacobian varieties together to one abelian scheme $\pi: A \to S$ whose fibers are the Jacobian varieties of the corresponding curves. See also Chapter 3, Section 1.3.

Let $\pi: A \to S$ be an abelian scheme of relative dimension g. For every integer n we denote by $[n]$ the *multiplication by n map*. It is a proper flat morphism (see [**113**]) and its kernel is a finite flat group scheme of order n^{2g} denoted by $A[n]$. Let S_0 be the open subscheme of S where the primes dividing n are invertible. Then $A[n]$ is étale over S_0, and S_0 is maximal with such property. That is, if k is a field of characteristic p and A/k is an abelian variety then $A[p]$ is never étale. In fact, its largest étale quotient is of order $\leq p^g$. If equality exists, one says that A is *ordinary*.

We provide some examples of the structure of p-torsion of a g-dimensional abelian variety A over an algebraically closed field of characteristic p.

EXAMPLE 3.14 ($g = 1$). Every elliptic curve is automatically principally polarized. This implies that $A[p]$ is self-dual. See Chapter 3, §1.1. There are two possibilities:

- A is an ordinary elliptic curve:

$$A[p] \cong \mu_p \oplus \underline{\mathbb{Z}/p\mathbb{Z}}. \tag{3.25}$$

Here μ_p is the kernel of Frobenius and $\underline{\mathbb{Z}/p\mathbb{Z}}$ is the kernel of Verschiebung. We remark here that for any abelian variety of dimension g in characteristic p Frobenius has degree p^g.

• A is not an ordinary elliptic curve. Then A is called a supersingular elliptic curve. One has a non-split exact sequence

$$0 \to \alpha_p \to A[p] \to \alpha_p \to 0. \tag{3.26}$$

The embedded α_p is unique and is both the kernel of Frobenius and Verschiebung. For every two supersingular elliptic curves A_1, A_2 over k (algebraically closed!) we have $A_1[p] \cong A_2[p]$. One may choose a model over $\mathbb{F}_p$ with $\mathrm{Fr} = -\mathrm{Ver}$. We shall denote this group scheme by M.

EXAMPLE 3.15 ($g = 2$). We assume that A is principally polarized, hence $A[p]$ is self-dual. There are four possibilities:

• A is ordinary:

$$A[p] \cong \mu_p^2 \oplus \underline{\mathbb{Z}/p\mathbb{Z}}^2. \tag{3.27}$$

Here μ_p^2 is the kernel of Frobenius and $\underline{\mathbb{Z}/p\mathbb{Z}}^2$ is the kernel of Verschiebung.

• A has étale part of order p. In this case, as in fact forced by self-duality, we have

$$A[p] \cong M \oplus \mu_p \oplus \underline{\mathbb{Z}/p\mathbb{Z}}, \tag{3.28}$$

where M is the p-torsion of a supersingular elliptic curve. Thus $A[p]$ contains a unique α_p. In this case, the kernel of Frobenius is $\alpha_p \oplus \mu_p$ and the kernel of Verschiebung is $\alpha_p \oplus \underline{\mathbb{Z}/p\mathbb{Z}}$.

• $A[p]$ has no étale part. In this case A is supersingular (but be careful: for $g \geq 3$ having no physical p-torsion (i.e., trivial étale quotient) does not imply supersingularity, though supersingularity implies no physical p-torsion). There are two possibilities:

(i) A is superspecial. That is, A is isomorphic to a product of supersingular elliptic curves. In this case

$$A[p] \cong M^2. \tag{3.29}$$

Note that $\alpha_p \oplus \alpha_p$ embeds in $A[p]$ and is in fact the kernel of both the Frobenius and Verschiebung morphisms.

(ii) A is not superspecial. We remark that A is always isogenous to a product of two supersingular elliptic curves. In this case one has a filtration

$$H \subset G \subset A[p], \tag{3.30}$$

where $H \cong \alpha_p$, where $G/H \cong \alpha_p \oplus \alpha_p$, and where $A[p]/G \cong \alpha_p$. The kernel of Frobenius G_1 and the kernel of Verschiebung G_2 are contained in G and we have an exact sequence

$$0 \to H \to G_1 \oplus G_2 \to G \to 0. \tag{3.31}$$

We remark that neither G_1 nor G_2 is isomorphic to M. The group scheme G_1 is killed by Fr and Ver^2, the group scheme G_2 is killed by Fr^2 and Ver and G_1 is dual to G_2.

EXERCISE 3.16. *What possibilities are there when one does not assume principal polarization?*

EXERCISE 3.17. *If A is a principally polarized abelian variety of dimension three over an algebraically closed field k of characteristic p, there are precisely eight possibilities for $A[p]$. Find eight self-dual non-isomorphic commutative group schemes of order p^6 over k that are killed by p.*

For the general theory see [**131**] *that discusses also the resulting stratification of the moduli space of principally polarized abelian varieties.*

4. The basic exact sequence

Let $(R, \mathfrak{m})$ be a Henselian local ring. Recall that this means that Hensel's lemma holds in R. That is, if $f(x) \in R[x]$ is a monic polynomial and $f(x) \equiv g_0(x)h_0(x) \mod \mathfrak{m}$, a product of monic polynomials then $f(x) = g(x)h(x)$, for suitable monic polynomials, such that $g(x) \equiv g_0(x) \mod \mathfrak{m}$, and $h(x) \equiv h_0(x) \mod \mathfrak{m}$. Examples include fields and complete local rings.

Let $S = \operatorname{Spec}(R)$ and let $\pi \colon G \to S$ be a finite flat group scheme. Then there exists a canonical exact sequence

$$0 \to G^0 \to G \to G^{\text{ét}} \to 0. \tag{4.1}$$

In this sequence G^0 is the connected component of the identity. The group $G^{\text{ét}}$ is étale and is in fact the largest étale quotient of G. If R is a perfect field then the sequence splits and G is a semi-direct product $G^0 \rtimes G^{\text{ét}}$. For proofs see [**163**]. If R is a perfect field and G is commutative then we may decompose further G^0 and $G^{\text{ét}}$. The procedure being similar, we explicate only the case of G^0. One considers the dual group scheme $H = (G^0)^t$. It can be decomposed as $H = H^0 \times H^{\text{ét}}$. Dualizing, we get $G^0 = G^{0\text{-}0} \times G^{0\text{-ét}}$ where $G^{0\text{-}0}$ is connected with connected dual, $G^{0\text{-ét}}$ is connected with étale dual. Similarly, $G^{\text{ét}}$ decomposes into a direct sum of an étale group with connected dual $G^{\text{ét-}0}$ and an étale group with étale dual $G^{\text{ét-ét}}$. All together

$$G \cong G^{0\text{-}0} \times G^{0\text{-ét}} \times G^{\text{ét-}0} \times G^{\text{ét-ét}}. \tag{4.2}$$

Using this decomposition, the category of finite flat commutative group schemes over a perfect field R decomposes into a direct sum of four categories: connected groups with connected dual, connected groups with étale dual, étale groups with connected dual, and étale groups with étale dual.

If R is a field of characteristic zero then all the categories except $G^{\text{ét-ét}}$ consist of the zero group alone. The category $G^{\text{ét-ét}}$ is equivalent to the category of finite commutative groups with $\operatorname{Gal}(R^{\text{alg}}/R)$ action. If R is a field of characteristic p then representatives of the categories are given by

$$\alpha_p, \quad \mu_p, \quad \underline{\mathbb{Z}/p\mathbb{Z}}, \quad \underline{\mathbb{Z}/\ell\mathbb{Z}} \tag{4.3}$$

(where ℓ is any prime not equal to p), respectively. In fact, if R is algebraically closed then the only simple objects of these abelian categories are given by (4.3).

We provide some further remarks about connected group schemes. Let R be a perfect field and G a finite commutative connected group scheme over R. Then the underlying topological space of G consists of only one point, equivalently G is the spectrum of a local artinian ring.

Assume now that R is a perfect field of characteristic p. Then one can "effectively" construct the connected component of the identity in G. Let $G = \operatorname{Spec}(T)$ for an R-algebra T, and let I_G be the augmentation ideal. Let $I_G^{(p^b)}$ be the ideal generated by all p^b powers of elements of I (usually strictly included in I^{p^b}). Let $I_G^\infty = \cap_b I_G^{(p^b)}$. Then

$$G^0 = \operatorname{Spec}(T/I_G^\infty). \tag{4.4}$$

Equivalently, let $G[\mathrm{Fr}^b]$ denote the kernel of the Frobenius morphism iterated b times, $\mathrm{Fr}\colon G \to G^{(p^b)}$. Then

$$G^0 = \bigcup_b G[\mathrm{Fr}^b]. \tag{4.5}$$

5. Group schemes over a perfect field of characteristic p

In this section we make the following standing assumptions: k is a perfect field of characteristic p; $\pi\colon G \to \operatorname{Spec}(k)$ is a finite commutative group scheme of order p^g for some g. We let $W(k)$ be the ring of infinite Witt vectors over k and we let $Q(k) = W(k) \otimes \mathbb{Q}$ be its quotient ring.

The main tool in studying a group scheme like G is its Dieudonné module. Here we give the recipe of covariant Dieudonné modules. It differs from the theory exposed in [**28, 120**] by taking duals.

Consider the non-commutative ring $A = W(k)[F, V]$ with the Frobenius automorphism $\sigma\colon W(k) \to W(k)$, where F and V are variables and

$$FV = VF = p, \quad F\lambda = \lambda^\sigma F, \quad \lambda V = V\lambda^\sigma, \quad \forall \lambda \in W(k). \tag{5.1}$$

There is an equivalence of categories between finite commutative group schemes G over k of p-power order and finite length left A-modules that we denote by

$$G \mapsto \mathcal{D}(G). \tag{5.2}$$

It has the following properties:

(1) It commutes with base change. In particular

$$\mathcal{D}(G^{(p)}) = \mathcal{D}(G) \times_{A,\sigma} A. \tag{5.3}$$

(2) Under this correspondence the map induced by $\mathrm{Fr}\colon G \to G^{(p)}$ is the $\sigma^{(-1)}$-linear map $V\colon \mathcal{D}(G) \to \mathcal{D}(G)$. Similarly, the map $\mathrm{Ver}\colon G \to G^{(1/p)}$ induces the σ-linear map $\mathrm{Fr}\colon \mathcal{D}(G) \to \mathcal{D}(G)$.

(3) There is duality:

$$\mathcal{D}(G^t) = \operatorname{Hom}_{W(k)}\big(\mathcal{D}(G), Q(k)/W(k)\big). \tag{5.4}$$

Here $\operatorname{Hom}_{W(k)}\big(\mathcal{D}(G), Q(k)/W(k)\big)$ is given a canonical A-module structure by

$$(F\varphi)(x) = \varphi(Vx)^\sigma, \quad (V\varphi)(x) = \varphi(Fx)^{\sigma^{-1}}, \tag{5.5}$$

for $\varphi \in \operatorname{Hom}_{W(k)}\big(\mathcal{D}(G), Q(k)/W(k)\big)$ and $x \in \mathcal{D}(G)$.

In particular, G is local-local (resp. local-étale, resp. étale-local) if and only if both F and V are nilpotent on $\mathcal{D}(G)$ (resp. V nilpotent and F is an isomorphism on $\mathcal{D}(G)$; resp. F nilpotent and V is an isomorphism on $\mathcal{D}(G)$).

(4) The rank of G is p^r where r is the length of $\mathcal{D}(G)$ as a $W(k)$-module.

EXAMPLE 5.1. The group α_p has the Dieudonné module k, where F, V and p act as zero, i.e., $A/(F, V)$, where (F, V) denote the left ideal $AF + AV$ generated by F and V.

EXAMPLE 5.2. The group μ_p has the Dieudonné module k, where p and V act as zero and F acts as the Frobenius automorphism of k (we remark again that we take the covariant Dieudonné module), i.e., $A/(V, 1-F)$.

EXAMPLE 5.3. The group scheme $\underline{\mathbb{Z}/p\mathbb{Z}}$ has Dieudonné module k with p and F acting as zero and V acting as the inverse of the Frobenius automorphism of k, i.e., $A/(F, 1-V)$.

EXAMPLE 5.4. The group scheme M of Example 3.14 has Dieudonné module $A/(F^2, V^2, F+V)$, while the group schemes G_1 and G_2 appearing in Example 3.15 are dual to one another and have Dieudonné modules $A/(V, F^2)$ and $A/(F, V^2)$, respectively. Furthermore, in that example the p-torsion subgroup of order p^4 containing them has the Dieudonné module $A/(F^3, V^3, F^2 - V^2)$.

Dieudonné modules are a very powerful tool in studying p-power finite commutative group schemes. For example, the k-forms of a group G over k^{sep} that can be defined over k are given by $H^1\big(\text{Gal}(k^{\text{sep}}/k), \text{Aut}\big(\mathcal{D}(G)\big)\big)$, where here and below the cohomology is always continuous cohomology. This is often readily computable.

EXAMPLE 5.5. Let $G = \alpha_p$ over k^{sep}. Then $\text{End}\big(\mathcal{D}(\alpha_p)\big) = k^{\text{sep}}$. By Hilbert's 90 we conclude that α_p has no non-trivial forms!

EXAMPLE 5.6. Let $G = \mu_p$. Then $\text{End}\big(\mathcal{D}(\mu_p)\big) = \{\lambda \in k^{\text{sep}} : \lambda^p = \lambda\}$. That is, $\text{End}\big(\mathcal{D}(\mu_p)\big) = \mathbb{F}_p$. Now, $H^1(\text{Gal}(k^{\text{sep}}/k), \mathbb{F}_p^\times) \cong k^\times/(k^\times)^{p-1}$ by Kummer theory and therefore the k-forms of μ_p are classified by $k^\times/(k^\times)^{p-1}$.

EXERCISE★ 5.7. *Do all forms of μ_p appear in the p-torsion of a suitable elliptic curve over k?*

EXAMPLE 5.8. Consider the group scheme M. Its Dieudonné module is given by $k^{\text{sep}}e_1 \oplus k^{\text{sep}}e_2$ with F acting by $Fe_1 = e_2$, $Fe_2 = 0$ and $Ve_1 = -e_2$, $Ve_2 = 0$. Note that this module is cyclic with generator e_1. An endomorphism f is thus completely determined by $f(e_1) = ae_1 + be_2$. In fact $f(e_2) = f(Fe_1) = Ff(e_1) = a^\sigma e_2$. The conditions on f being a map of Dieudonné modules is that f commutes with V and F. That is $-a^{\sigma^{-1}}e_2 = Vf(e_1) = f(Ve_1) = f(-e_2) = -a^\sigma e_2$. That is, $a \in \mathbb{F}_{p^2}$.

If we identify f with the couple (a, b), then

$$\text{End}(M) = \{(a,b) : a \in \mathbb{F}_{p^2}, b \in k^{\text{sep}}\} \tag{5.6}$$

with component-wise addition and multiplication given by

$$(\alpha, \beta)(a, b) = (\alpha a, \beta a + b\alpha^\sigma). \tag{5.7}$$

The identity is $(1, 0)$ and hence

$$\text{Aut}(M) = \{(a,b) : a \in \mathbb{F}_{p^2}^\times, b \in k^{\text{sep}}\}. \tag{5.8}$$

One has an exact sequence of $\text{Gal}(k^{\text{sep}}/k)$ modules:

$$1 \to (k^{\text{sep}}, +) \to \text{Aut}(M) \to \mathbb{F}_{p^2}^\times \to 1. \tag{5.9}$$

The maps are $\beta \mapsto (1, \beta)$ and $(\alpha, \beta) \mapsto \alpha$. In fact it splits, as Galois modules, by $\alpha \mapsto (\alpha, 0)$. That is, $\text{Aut}(M) = k^{\text{sep}} \rtimes \mathbb{F}_{p^2}^\times$.

Let us assume that $\mathbb{F}_{p^2} \subset k$. Since $H^1(\text{Gal}(k^{\text{sep}}/k), k^{\text{sep}}) = 0$ ($(k^{\text{sep}}, +)$ is cohomologically trivial), we get an injection

$$H^1\big(\text{Gal}(k^{\text{sep}}/k), \text{Aut}(M)\big) \hookrightarrow H^1\big(\text{Gal}(k^{\text{sep}}/k), \mathbb{F}_{p^2}^\times\big) = k^\times/(k^\times)^{p^2-1}, \tag{5.10}$$

by Kummer theory.

It is easy to see that $H^1\big(\mathrm{Gal}(k^{\mathrm{sep}}/k), \mathrm{Aut}(M)\big) \hookrightarrow H^1\big(\mathrm{Gal}(k^{\mathrm{sep}}/k), \mathbb{F}_{p^2}^\times\big)$ is also surjective. That would be obvious if we have dealt with usual (abelian) cohomology because $H^2(\mathrm{Gal}(k^{\mathrm{sep}}/k), k) = 0$. But, directly: Given a cocycle $\beta \in H^1(\mathrm{Gal}(k^{\mathrm{sep}}/k), \mathbb{F}_{p^2}^\times)$, $\sigma \mapsto \alpha(\sigma)$ one lifts it by $\sigma \mapsto (\alpha(\sigma), 0)$. We therefore conclude that the forms of M are classified by $k^\times/(k^\times)^{p^2-1}$.

We now look again at the case of an abelian variety A/k of dimension g. Since k is a perfect field the Verschiebung morphism $A \to A^{(1/p)}$ is well defined. Let $A[\mathrm{Fr}]$ and $A[\mathrm{Ver}]$ be the kernel of Frobenius and the kernel of Verschiebung, respectively. These are subgroup schemes of A of order p^g. We have an exact sequence:

$$0 \to A[\mathrm{Ver}] \to A[p] \xrightarrow{\mathrm{Ver}} A[\mathrm{Fr}] \to 0. \tag{5.11}$$

We apply the covariant Dieudonné functor and obtain:

$$0 \to \mathcal{D}[F] \to \mathcal{D} \xrightarrow{F} \mathcal{D}[V] \to 0, \tag{5.12}$$

where we let $\mathcal{D}$ denote the Dieudonné module $\mathcal{D}(A[p])$. Assuming that A has a polarization of degree prime to p, there is an isomorphism of $k[F,V]$-modules of $\mathcal{D}$ with $H^1_{\mathrm{dR}}(A)$ (where the action on the latter comes from the isomorphism $H^1_{\mathrm{dR}}(A) \cong H^1_{\mathrm{Crys}}(A/W(k))/pH^1_{\mathrm{Crys}}(A/W(k))$).[2] The vector space H^1_{dR} has a canonical filtration

$$0 \to H^0(A, \Omega^1_A) \to H^1_{\mathrm{dR}}(A) \to H^1(A, \mathcal{O}_A) \to 0. \tag{5.13}$$

The arrows are defined by a spectral sequence (Chapter 5, p. 176). The sequences (5.11) and (5.13) are closely related. Indeed, since Fr acts as zero on differential forms, the identification $\mathcal{D} = H^1_{\mathrm{dR}}(A)$ implies that $\mathcal{D}[F] = H^0(A, \Omega^1_A)$. We may further identify $H^1(A, \mathcal{O}_A)$ with $\mathrm{Fr}\,\mathcal{D}$. On the other hand, the map $\mathrm{Fr}\colon \mathcal{D} \to \mathcal{D}$ identifies $\mathcal{D}/\mathcal{D}[F]$ with $\mathcal{D}[V]$, only that its not linear: $\mathrm{Fr}(\lambda a) = \lambda^p\,\mathrm{Fr}(v)$. We obtain: $H^1(A, O_A) \cong \mathcal{D}[V] \otimes_{k,\mathrm{Fr}} k$. Thus,

$$\begin{array}{ccccccccc} 0 & \longrightarrow & H^0(A,\Omega^1_A) & \longrightarrow & H^1_{\mathrm{dR}}(A) & \longrightarrow & H^1(A,\mathcal{O}_A) & \longrightarrow & 0 \\ & & \downarrow\cong & & \downarrow\cong & & \downarrow\cong & & \\ 0 & \longrightarrow & \mathcal{D}(A[\mathrm{Ver}]) & \longrightarrow & \mathcal{D}(A[p]) & \xrightarrow{\mathrm{Fr}} & \mathcal{D}(A[\mathrm{Fr}]) \otimes_{k,\mathrm{Fr}} k & \longrightarrow & 0. \end{array} \tag{5.14}$$

We remark again that we are working with the covariant Dieudonné module. Oda [**119**] works with the contravariant Dieudonné theory $\mathbb{D}$ and obtains in loc. cit., Corollary 5.11, an identification

$$\begin{array}{ccccccccc} 0 & \longrightarrow & H^0(A,\Omega^1_A) & \longrightarrow & H^1_{\mathrm{dR}}(A) & \longrightarrow & H^1(A,\mathcal{O}_A) & \longrightarrow & 0 \\ & & \downarrow\cong & & \downarrow\cong & & \downarrow\cong & & \\ 0 & \longrightarrow & (k, \mathrm{Fr}^{-1}) \otimes_k \mathbb{D}(A[\mathrm{Fr}]) & \longrightarrow & \mathbb{D}(A[p]) & \xrightarrow{\mathrm{Fr}} & \mathbb{D}(A[\mathrm{Ver}]) & \longrightarrow & 0. \end{array} \tag{5.15}$$

We offer some words of explanation on that. In the contravariant theory there is a Dieudonné module $\mathbb{D}(G)$ associated to a finite group scheme G ($\mathcal{D}(G)$ is the dual of $\mathbb{D}(G)$). A morphism $G \to G'$ corresponds to a morphism of Dieudonné

[2] The canonical isomorphism is of the contravariant Dieudonné module $\mathbb{D}(A[p]) = \mathcal{D}(A[p])^t$ with $H^1_{\mathrm{dR}}(A)$. However, $\mathcal{D}(A[p])^t = \mathcal{D}(A^t[p])$, which may be identified with $\mathcal{D}(A[p])$ as Dieudonné modules if we have a polarization of degree prime to p.

modules $\mathbb{D}(G') \to \mathbb{D}(G)$. The Frobenius morphism $\mathrm{Fr}\colon G \to G^{(p)}$ induces the morphism of Dieudonné modules $\mathrm{Fr}^* : \mathbb{D}(G^{(p)}) \to \mathbb{D}(G)$ which is k-linear. Under the identification $\mathbb{D}(G^{(p)}) = \mathbb{D}(G) \otimes_k (k, \mathrm{Fr})$ we get the σ-linear map $F\colon \mathbb{D}(G) \to \mathbb{D}(G)$. Similarly, $\mathrm{Ver}\colon G^{(p)} \to G$ induces the σ^{-1}-linear map $V : \mathbb{D}(G) \to \mathbb{D}(G)$.

An endomorphism $f\colon G \to G'$ has a kernel $\mathrm{Ker}(f)$ with Dieudonné module $\mathbb{D}(G)/f^*\mathbb{D}(G')$. Consider, in particular, the morphism $\mathrm{Ver}\colon A[p] \to A^{(1/p)}[p]$. We have

$$\begin{aligned} \mathbb{D}(A[\mathrm{Ver}]) &= \mathbb{D}(A[p])/\,\mathrm{Ver}^*\,\mathbb{D}(A^{(1/p)}[p]) \\ &= \mathbb{D}(A[p])/V\mathbb{D}(A[p]) \\ &= \mathbb{D}(A[p])/\mathbb{D}(A[p])[F] \\ &= H^1_{\mathrm{dR}}(A)/H^0(A, \Omega^1_{A/k}) \\ &= H^1(A, \mathcal{O}_A). \end{aligned} \tag{5.16}$$

6. The α-group

In this section we define the alpha group of an abelian variety $A \to S$ over a scheme S of characteristic p. A caveat is that there is no good way to define this subgroup scheme as a group scheme of the abelian variety itself. We therefore define it as a subgroup scheme of the base change $A^{(p)} \to S$.

DEFINITION 6.1. Let $A \to S$ be an abelian scheme. Let

$$\alpha(A) = \mathrm{Ker}(\mathrm{Ver}\colon A^{(p)} \to A) \cap \mathrm{Ker}(\mathrm{Fr}\colon A^{(p)} \to A^{(p^2)}). \tag{6.1}$$

We call it the *alpha group* of A.

Some remarks are in order: First, note that $\alpha(A)$ is a group scheme over S. Second, the construction of $\alpha(A)$ is stable under base change: That is, for every morphism $T \to S$ we have

$$\alpha(A/S) \times_S T = \alpha(A \times_S T). \tag{6.2}$$

This is nothing more than the behaviour of Ver and Fr under base change and that Ker(Ver) and Ker(Fr) represent the functors "the kernel of Verschiebung" and "the kernel of Frobenius," respectively. Third, let S be a perfect scheme, namely, the absolute Frobenius morphism $\mathrm{Fr}^{\mathrm{abs}} : S \to S$ is an isomorphism (e.g., S is the spectrum of an algebraically closed field). We may therefore write $A = B^{(p)}$ for some abelian scheme $B \to S$. Then

$$\alpha(B)^{(p)} = \alpha(A). \tag{6.3}$$

That is, the subgroup scheme $\alpha(A) \subset A^{(p)}$ descends to a subgroup scheme of A.

DEFINITION 6.2. We say that $A \to S$ has a-number greater or equal to a and write

$$a(A) \geq a, \tag{6.4}$$

if the rank of $\alpha(A)$ is greater or equal to p^a.

We make a few remarks. First, for A over an algebraically closed field k, the a number of A in the sense above, and the a-number of A in the sense of

$$\dim_k \mathrm{Hom}(\alpha_p, A), \tag{6.5}$$

are the same. Indeed,

$$\dim_k \operatorname{Hom}(\alpha_p, A) = \dim_k \operatorname{Hom}(\alpha_p^{(p)}, A^{(p)}) = \dim_k \operatorname{Hom}\big(\alpha_p, \alpha(A)\big). \tag{6.6}$$

Second, a commutative finite flat group scheme $G \to S$ over a perfect scheme S, killed by Frobenius and Verschiebung (in the sense that $\mathrm{Ver}\colon G^{(p)} \to G$ is the zero morphism), is locally on S isomorphic to α_p^r/S.

Third, over k the embeddings

$$\alpha_p \to \alpha_p^r \tag{6.7}$$

are parameterized by surjective maps of Hopf k-algebras

$$k[x_1, \ldots, x_r]/(x_1^p, \ldots, x_r^p) \to k[t]/(t^p). \tag{6.8}$$

Consider first that case $r = 1$. Then x_1 is mapped to $f(t) = a_0 + a_1 t + \cdots + a_{p-1} t^{(p-1)}$. Then $x_1 \otimes 1 + 1 \otimes x_1$ is mapped to $f(t) \otimes 1 + 1 \otimes f(t)$ which should be equal to $m^* f(t)$, where $m\colon \alpha_p \times \alpha_p \to \alpha_p$ is the multiplication morphism. But

$$m^* f(t) = a_0 \otimes 1 + a_1(t \otimes 1 + 1 \otimes t) + \ldots a_{p-1}(t \otimes 1 + 1 \otimes t)^{p-1}. \tag{6.9}$$

Equating coefficients we get $f(t) = at$ for some $a \in k$. Thus, coming back to the general case, giving a morphism $\alpha_p \to \alpha_p^r$ is equivalent to giving a vector $(a_1, \ldots, a_r) \in k^r$. The correspondence being given by associating to the vector $(a_1, \ldots, a_r)$ the unique morphism $k[x_1, \ldots, x_r]/(x_1^p, \ldots, x_r^p) \to k[t]/(t^p)$ taking x_i to $a_i t$. The morphism of groups is injective if and only if the morphism of algebras is surjective, if and only if $(a_1, \ldots, a_r)$ is not the zero vector. Furthermore, taking the case $r = 1$, we see that the automorphisms of α_p are in natural bijection with $k^\times$, and the natural action of the automorphism associated to $a \in k^\times$ on the embeddings $\alpha_p \to \alpha_p^r$ is given by $(a_1, \ldots, a_r) \mapsto (aa_1, \ldots, aa_r)$. That is, the homomorphisms of α_p to α_p^r are naturally isomorphic to $\mathbb{A}_k^r$, and the subgroups of α_p^r that are isomorphic to α_p are in natural bijection with $\mathbb{P}_k^{r-1}$.

APPENDIX B

Calculating with Cusps

The purpose of this appendix is to explain how one calculates with the cusps of modular curves. We more or less ignore completely the theory with all its depth and subtlety, restricting ourselves to mere computational aspects. The reader interested in better understanding this important topic will do wisely to study [**26, 84**]. The author also used his notes from a course given by B. Mazur at Harvard University in 1996/7 and [**34**, §9].

The cusps of a moduli problem of elliptic curves with a certain level N structure (e.g., $\Gamma(N)$, $\Gamma_0(N)$, $\Gamma_1(N)$, μ_N) correspond to isomorphism classes of N-gons together with additional data. By that we mean (roughly), the algebraic group $\mathbb{G}_m \times \mathbb{Z}/N\mathbb{Z}$ with suitable level data modulo suitable isomorphisms.

More precisely, an *N-gon C_N* over an algebraically closed field k is a cyclic configuration of N projective lines $\mathbb{P}^1_k$, in which the components are indexed cyclically by $\mathbb{Z}/N\mathbb{Z}$ and the point ∞ on the i-th copy is glued transversely to the point 0 on the $i+1$-th copy (See Figure 1). The non-singular part C_N^{ns} is canonically isomorphic to the group scheme $\mathbb{G}_{m/k} \times \mathbb{Z}/N\mathbb{Z}$ and there is a natural action $C_N^{\mathrm{ns}} \times C_N \to C_N$.

A *generalized elliptic curve* $\mathcal{E} \to S$ of level N over a base scheme S is a relative curve (proper, flat, with geometrically connected fibers) such that $\mathcal{E}^{\mathrm{ns}}$ is a group scheme over S and there is an action (extending the group law) $\mathcal{E}^{\mathrm{ns}} \times \mathcal{E} \to \mathcal{E}$, and such that every geometric fiber is either an elliptic curve or an M-gon, where $M \mid N$.

The modular curves involving level N (e.g., $X(N)$, $X_0(N)$, $X_1(N)$, $\mathfrak{M}(\mu_N)$) parameterize generalized elliptic curves of level N with the appropriate level structure. One insists that this level structure, be it a point of order N, a subgroup of order N or a full level N structure, meets every component of every M-gon appearing as a degenerate fibre of the generalized elliptic curve. This is imposed to guarantee polarization and to have a universal object. However, for the purpose of combinatorics of the cusps alone one may as well assume that all the degenerate fibres are N-gons, not impose the condition that the level structure meets every component, and divide by the action of an appropriate automorphism group. Instead of developing a general theory we provide examples making clear what needs to be done in general.

First, the group law of $\mathbb{G}_m \times \mathbb{Z}/N\mathbb{Z}$ is given by

$$(\alpha, i)(\beta, j) = (\alpha\beta, i + j). \tag{0.1}$$

The endomorphisms of $\mathbb{G}_m \times \mathbb{Z}/N\mathbb{Z}$ are given symbolically by the matrices

$$\begin{pmatrix} \mathbb{Z} & \mu_N \\ 0 & \mathbb{Z}/N\mathbb{Z} \end{pmatrix}. \tag{0.2}$$

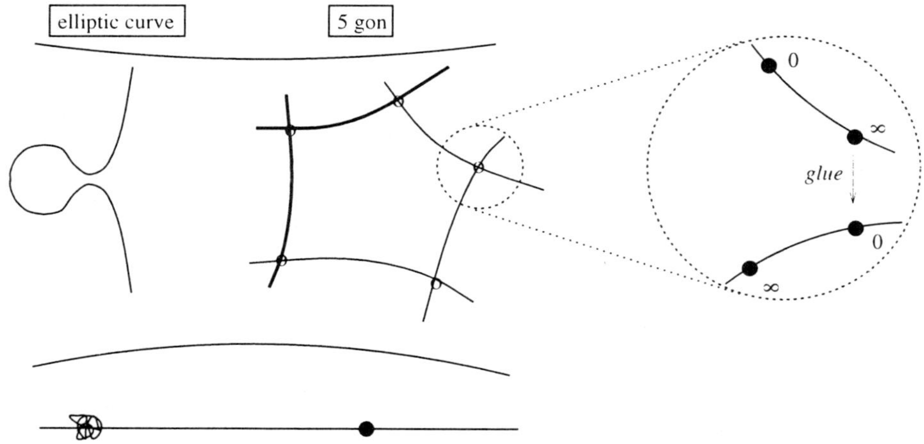

FIGURE 1. A 5-gon.

By that we mean

(0.3) $$\begin{pmatrix} a & \zeta \\ 0 & b \end{pmatrix} \begin{pmatrix} \alpha \\ i \end{pmatrix} = \begin{pmatrix} \alpha^a \cdot \zeta^i \\ bi \end{pmatrix}.$$

Therefore, the automorphisms are given by

(0.4) $$\begin{pmatrix} \pm 1 & \mu_N \\ 0 & (\mathbb{Z}/N\mathbb{Z})^\times \end{pmatrix}.$$

We define

(0.5) $$\mathrm{Aut}_1(\mathbb{G}_m \times \mathbb{Z}/N\mathbb{Z}) = \left\{ \pm \begin{pmatrix} 1 & \zeta \\ 0 & 1 \end{pmatrix} : \zeta^N = 1 \right\}.$$

The reason for that is the following: The N-gon $\mathbb{G}_m \times \mathbb{Z}/N\mathbb{Z}$ appears as a degeneration of an elliptic curve E with level N structure (as the non-singular locus of the special fiber of a regular model of E. See Figure 1).

The elliptic curve carries the canonical Weil pairing on $E[N]$ that induces such a pairing on $(\mathbb{G}_m \times \mathbb{Z}/N\mathbb{Z})[N] = \mu_N \times \mathbb{Z}/N\mathbb{Z}$. It is only natural to demand that the induced pairing on $\mu_N \times \mathbb{Z}/N\mathbb{Z}$ is the canonical one and to take the automorphisms that preserve that pairing, thus arriving at $\mathrm{Aut}_1(\mathbb{G}_m \times \mathbb{Z}/N\mathbb{Z})$.

Let ζ be a primitive N-th root of 1. Note that $\mathrm{Aut}_1(\mathbb{G}_m \times \mathbb{Z}/N\mathbb{Z})$ is generated by $[-1]$ and $[\zeta]$ acting by

(0.6) $$(\alpha, j) \overset{[-1]}{\longmapsto} (\alpha^{-1}, -j), \quad (\alpha, j) \overset{[\zeta]}{\longmapsto} (\alpha\zeta^j, j).$$

1. $\Gamma_1(N)$ -level

The cusps corresponding to $\Gamma_1(N)$-level are the points of order N in the subgroup $\mu_N \times \mathbb{Z}/N\mathbb{Z}$ of $\mathbb{G}_m \times \mathbb{Z}/N\mathbb{Z}$, taken modulo the action of $\mathrm{Aut}_1(\mathbb{G}_m \times \mathbb{Z}/N\mathbb{Z})$.

EXAMPLE 1.1. Let $N = 4$. There are 12 points of order 4:

$$\begin{array}{cccccc} (i,0) & (i,1) & (i,2) & (i,3) & (1,1) & (-1,1) \\ {\scriptstyle -1}\updownarrow & {\scriptstyle -1}\updownarrow & {\scriptstyle -1}\updownarrow & {\scriptstyle -1}\updownarrow & {\scriptstyle -1}\updownarrow & {\scriptstyle -1}\updownarrow \\ (-i,0) & (-i,3) & (-i,2) & (-i,1) & (1,3) & (-1,3). \end{array} \tag{1.1}$$

The automorphism $[i]$ acts by $(i,0) \mapsto (i,0)$, $(i,1) \mapsto (-1,1) \mapsto (-i,1) \mapsto (1,1) \mapsto (i,1)$, etc. One concludes that the cusps correspond to the orbits

$$\{(i,0),(-i,0)\},$$
$$\{(i,2),(-i,2)\},$$
$$\{(i,1),(i,3),(1,1),(-1,1),(-i,3),(-i,1),(1,3),(-1,3)\}.$$

The ramification of the cusps of $X_1(N)$ with respect to $X_1(1)$ is always the length of the orbit divided by 2 (reminiscent that $[-1]$ is an automorphism of every elliptic curve). Thus, in this example there are three cusps; two are unramified and one has ramification of order 4. The two unramified cusps have an automorphism of order two (Cf. Chapter 4, §6.3).[1]

EXERCISE 1.2. *Use the Cauchy–Frobenius formula for the number of orbits of a finite group acting on a finite set* [**140**] *to prove that the number of cusps on* $X_1(N)$ *is* $\frac{1}{2}\sum_{d|N}\varphi(d)\varphi(N/d)$ *for* $N > 4$.

The description of the cusps we gave is really of the moduli problem of an elliptic curve E with a point of order N. That is, with an embedding $\mathbb{Z}/N\mathbb{Z} \hookrightarrow E$. The absolute Galois group of $\mathbb{Q}$ acts both on the moduli problem and on the description we gave for the cusps, and the action is compatible. Thus, we easily calculate that for $N = 4$ all the cusps are defined over $\mathbb{Q}$.

EXERCISE 1.3. *Show that there are* 4 *cusps on* $X_1(5)$*; two are unramified and are defined over* $\mathbb{Q}(\sqrt{5})$*, and two have ramification of order* 5 *and are defined over* $\mathbb{Q}$.

2. $\Gamma_0(N)$-level

The cusps of level $\Gamma_0(N)$ correspond to orbits of the group $\mathrm{Aut}_1(\mathbb{G}_m \times \mathbb{Z}/N\mathbb{Z})$ in the set of subgroups of order N of $\mathbb{G}_m \times \mathbb{Z}/N\mathbb{Z}$.

EXAMPLE 2.1. Let $N = 5$. Let ζ be a primitive root of unity of order 5. There are 6 subgroups of order 5. They are:

$$H_1 = \{(1,0),(1,1),(1,2),(1,3),(1,4)\},$$
$$H_2 = \{(1,0),(\zeta,0),(\zeta^2,0),(\zeta^3,0),(\zeta^4,0)\},$$
$$H_3 = \{(1,0),(\zeta,1),(\zeta^2,2),(\zeta^3,3),(\zeta^4,4)\},$$
$$H_4 = \{(1,0),(\zeta,2),(\zeta^2,4),(\zeta^3,1),(\zeta^4,3)\},$$
$$H_5 = \{(1,0),(\zeta,3),(\zeta^2,1),(\zeta^3,4),(\zeta^4,2)\},$$
$$H_6 = \{(1,0),(\zeta,4),(\zeta^2,3),(\zeta^3,2),(\zeta^4,1)\}.$$

[1] Only the automorphism $[-1]$, if it preserves the cusp, is considered an automorphism.

The subgroups are fixed under the action of the automorphism $[-1]$. The automorphism $[\zeta]$ has the effect $H_1 \mapsto H_3 \mapsto H_5 \mapsto H_4 \mapsto H_6 \mapsto H_1$. There are therefore two cusps. One is unramified and the other has ramification 5. Both are defined over $\mathbb{Q}$.

EXERCISE 2.2. *Let p be a prime. Prove that there are two cusps on $X_0(p)$. Both are defined over $\mathbb{Q}$; one is unramified and the other ramified of order p.*

EXERCISE 2.3. *Calculate the number of cusps of $X_0(N)$ (Cf. Exercise 1.2).*

3. $\Gamma(N)$-level

The cusps of level $\Gamma(N)$ correspond to orbits of the group $\mathrm{Aut}_1(\mathbb{G}_m \times \mathbb{Z}/N\mathbb{Z})$ in the set of symplectic automorphisms of $\mu_N \times \mathbb{Z}/N\mathbb{Z}$, and if $N \geq 3$ none are ramified. That is, the cusps are in one-to-one correspondence with the coset space $\mathrm{SL}_2(\mathbb{Z}/N\mathbb{Z})/\mathrm{Aut}_1(\mathbb{G}_m \times \mathbb{Z}/N\mathbb{Z})$. In this expression we write the automorphism as the matrices of determinant 1 in

$$\begin{pmatrix} \mathbb{Z}/N\mathbb{Z} & \mathrm{Hom}(\mathbb{Z}/N\mathbb{Z}, \mu_N) \\ \mathrm{Hom}(\mu_N, \mathbb{Z}/N\mathbb{Z}) & \mathbb{Z}/N\mathbb{Z} \end{pmatrix}. \tag{3.1}$$

(Note the natural identification $\mathrm{Hom}(\mathbb{Z}/N\mathbb{Z}, \mu_N) = \mu_N$ obtained by sending f to $f(1)$). We leave the calculation of the cardinality of this coset space to the reader. We illustrate by giving an example.

EXAMPLE 3.1. Let $N = 3$. There are 24 different level 3 structures given by the matrices

$$\begin{pmatrix} a & f \\ g & d \end{pmatrix} \tag{3.2}$$

such that $ad - g \circ f(1) = 1$. The cardinality of $\mathrm{Aut}_1(\mathbb{G}_m \times \mathbb{Z}/3\mathbb{Z})$ is 6 and there are thus 4 cusps.

One can determine the field of definition of a cusp. A cusp, corresponding to a matrix $\gamma = \left(\begin{smallmatrix} a & f \\ g & d \end{smallmatrix}\right)$ is fixed by σ iff $\gamma^{-1}\gamma^\sigma \in \mathrm{Aut}_1(\mathbb{G}_m \times \mathbb{Z}/N\mathbb{Z})$. To illustrate, for $N = 3$ if we make some obvious identifications, the cusps are parameterized by $\mathrm{SL}_2(\mathbb{Z}/3\mathbb{Z})$ where the Galois group factors through $\mathrm{Gal}(\mathbb{Q}(\zeta_3)/\mathbb{Q})$ and the non-trivial automorphism σ acts by

$$\begin{pmatrix} a & b \\ c & d \end{pmatrix} \mapsto \begin{pmatrix} a & -b \\ -c & d \end{pmatrix}. \tag{3.3}$$

The cusp given by $\gamma = \left(\begin{smallmatrix} a & b \\ c & d \end{smallmatrix}\right)$ is defined over $\mathbb{Q}$ iff

$$\gamma^{-1}\gamma^\sigma = \begin{pmatrix} d & -b \\ -c & a \end{pmatrix}\begin{pmatrix} a & -b \\ -c & d \end{pmatrix} = \begin{pmatrix} ad+bc & -2bd \\ -2ac & ad+bc \end{pmatrix} \tag{3.4}$$

is in $\mathrm{Aut}_1(\mathbb{G}_m \times \mathbb{Z}/3\mathbb{Z})$. That is, iff $ac = 0$. Thus, there are two cusps that are rational and two that are interchanged by $\mathrm{Gal}(\mathbb{Q}(\zeta_3)/\mathbb{Q})$.

EXERCISE★ 3.2. *Work out the case of μ_N-level structure, especially the Galois action.*

Bibliography

1. A. Ash and G. Stevens, *Cohomology of arithmetic groups and congruences between systems of Hecke eigenvalues*, J. Reine Angew. Math. **365** (1986), 192–220.
2. ———, *Modular forms in characteristic l and special values of their L-functions*, Duke Math. J. **53** (1986), no. 3, 849–868.
3. E. Bachmat and E. Z. Goren, *On the non-ordinary locus in Hilbert–Blumenthal surfaces*, Math. Ann. **313** (1999), no. 3, 475–506.
4. P. Berthelot, L. Breen, and W. Messing, *Théorie de Dieudonné cristalline.* II, Lecture Notes in Math., vol. 930, Springer, Berlin, 1982.
5. A. Borel, *Linear algebraic groups*, 2nd edition, Grad. Texts in Math., vol. 126, Springer, New York, 1991.
6. S. Bosch, W. Lütkebohmert, and M. Raynaud, *Néron models*, Ergeb. Math. Grenzgeb. (3), vol. 21, Springer, Berlin, 1990.
7. C. Breuil, B. Conrad, F. Diamond, and R. Taylor, *On the modularity of elliptic curves over* $\mathbb{Q}$ (preprint).
8. P. Cassou-Noguès, *Valeurs aux entiers négatifs des fonctions zêta et fonctions zêta p-adiques*, Invent. Math. **51** (1979), no. 1, 29–59.
9. C.-L. Chai, *Arithmetic minimal compactification of the Hilbert–Blumenthal moduli spaces*, Ann. of Math. (2) **131** (1990), no. 3, 541–554.
10. ———, *Every ordinary symplectic isogeny class in positive characteristic is dense in the moduli*, Invent. Math. **121** (1995), no. 3, 439–479.
11. C.-L. Chai and P. Norman, *Bad reduction of the Siegel moduli scheme of genus two with $\Gamma_0(p)$-level structure*, Amer. J. Math. **112** (1990), no. 6, 1003–1071.
12. C. Chevalley, *Une démonstration d'un théorème sur les groupes algébriques*, J. Math. Pures Appl. (9) **39** (1960), 307–317.
13. H. Cohen, *Variations sur un thème de Siegel–Hecke*, Séminaire Delange–Pisot–Poitou (15 année: 1973/74), Théorie des nombres, Fasc. 1, Exp. No. 14, Secrétariat Mathématique, Paris, 1975.
14. R. F. Coleman, *Classical and overconvergent modular forms*, Invent. Math. **124** (1996), no. 1-3, 215–241.
15. R. F. Coleman, F. Q. Gouvêa, and N. Jochnowitz, *E_2, Θ and overconvergence*, Internat. Math. Res. Notices **1995**, no. 1, 23–41.
16. B. Conrad, *A modern proof of Chevalley's theorem on algebraic groups* (manuscript).
17. B. Conrad, F. Diamond, and R. Taylor, *Modularity of certain potentially Barsotti–Tate Galois representations*, J. Amer. Math. Soc. **12** (1999), no. 2, 521–567.
18. G. Cornell and J. H. Silverman (eds.), *Arithmetic geometry* (Storrs, CT, 1984), Springer, New York, 1986.
19. G. Cornell, J. H. Silverman, and G. Stevens (eds.), *Modular forms and Fermat's last theorem* (Boston, MA, 1995), Springer, New York, 1997.
20. H. Darmon, F. Diamond, and R. Taylor, *Fermat's last theorem*, Current Developments in Mathematics (Cambridge, MA, 1995), Internat. Press, Cambridge, MA, 1994, pp. 1–154.
21. P. Deligne, *Variétés abéliennes ordinaires sur un corps fini*, Invent. Math. **8** (1969), 238–243.
22. ———, *La conjecture de Weil.* I, Inst. Hautes Études Sci. Publ. Math. **43** (1974), 273–307.
23. ———, *La conjecture de Weil.* II, Inst. Hautes Études Sci. Publ. Math. **52** (1980), 137–252.
24. P. Deligne and D. Mumford, *The irreducibility of the space of curves of given genus*, Inst. Hautes Études Sci. Publ. Math. **36** (1969), 75–109.
25. P. Deligne and G. Pappas, *Singularités des espaces de modules de Hilbert, en les caractéristiques divisant le discriminant*, Compositio Math. **90** (1994), no. 1, 59–79.

26. P. Deligne and M. Rapoport, *Les schémas de modules de courbes elliptiques*, Modular Functions of One Variable. II (Antwerp, 1972), Lecture Notes in Math., vol. 349, Springer, Berlin, 1973, pp. 143–316.
27. P. Deligne and K. A. Ribet, *Values of abelian L-functions at negative integers over totally real fields*, Invent. Math. **59** (1980), no. 3, 227–286.
28. M. Demazure, *Lectures on p-divisible groups*, Lecture Notes in Math., vol. 302, Springer, Berlin–New York, 1972.
29. M. Demazure and P. Gabriel, *Groupes algébriques. I. Géométrie algébrique, généralités, groupes commutatifs* (with an appendix *Corps de classes local* by M. Hazewinkel), Masson & Cie, Paris; North-Holland, Amsterdam, 1970.
30. M. Demazure and A. Grothendieck. (eds.), *Schémas en groupes*. I, Séminaire de Géométrie Algébrique du Bois Marie 1962/64 (SGA 3), Lecture Notes in Math., vol. 151, Springer, Berlin–New York, 1962/1964; II, idem, vol. 152; III, idem, vol. 153.
31. C. Deninger and J. Murre, *Motivic decomposition of abelian schemes and the Fourier transform*, J. Reine Angew. Math. **422** (1991), 201–219.
32. E. de Shalit, *Iwasawa theory of elliptic curves with complex multiplication*, Perspect. Math., vol. 3, Academic Press, Boston, MA, 1987.
33. F. Diamond, *On deformation rings and Hecke rings*, Ann. of Math. (2) **144** (1996), no. 1, 137–166.
34. F. Diamond and J. Im, *Modular forms and modular curves*, Seminar on Fermat's Last Theorem (Toronto, ON, 1993–1994) CMS Conf. Proc., vol. 17, Amer. Math. Soc., Providence, RI, 1995, pp. 39–133.
35. V. G. Drinfeld, *Elliptic modules*, Math. USSR-Sb. **23** (1974), no. 4, 561–592.
36. B. Dwork, *The U_p operator of Atkin on modular functions of level* 2 *with growth conditions*, Modular Functions of One Variable. III (Antwerp, 1972), Lecture Notes in Math., vol. 350, Springer, Berlin, 1973, pp. 57–67.
37. B. Edixhoven, *Rational elliptic curves are modular* (*after Breuil, Conrad, Diamond and Taylor*), Séminaire Bourbaki, exp. no. 871, 2000 (preprint).
38. J. S. Ellenberg, *Hilbert modular forms and the Galois representations associated to Hilbert–Blumenthal varieties*, Ph.D. thesis, Harvard University, 1998.
39. G. Faltings and C.-L. Chai, *Degeneration of abelian varieties* (with an appendix by D. Mumford), Ergeb. Math. Grenzgeb. (3), vol. 22, Springer, Berlin, 1990.
40. G. Faltings, G. Wüstholz, F. Grunewald, N. Schappacher, and U. Stuhler, *Rational points* (Bonn/Wuppertal, 1983/1984), 3rd edition (with an appendix by G. Wüstholz), Aspects Math., vol. E6, Vieweg, Braunschweig, 1992.
41. H. M. Farkas and I. Kra, *Riemann surfaces*, 2nd edition, Grad. Texts in Math., vol. 71, Springer, New York, 1992.
42. E. Freitag, *Hilbert modular forms*, Springer, Berlin, 1990.
43. A. Fröhlich and M. J. Taylor, *Algebraic number theory*, Cambridge Stud. Adv. Math., vol. 27, Cambridge Univ. Press, Cambridge, 1993.
44. G. van der Geer, *Hilbert modular surfaces*, Ergeb. Math. Grenzgeb. (3), vol. 16, Springer, Berlin, 1988.
45. ______, *Cycles on the moduli space of abelian varieties*, Moduli of Curves and Abelian Varieties, Aspects Math., vol. E33, Vieweg, Braunschweig, 1999, pp. 65–89.
46. S. I. Gelfand and Yu. I. Manin, *Algebra. V. Homological algebra* (A. I. Kostrikin and I. R. Shafarevich, eds.), Encyclopaedia Math. Sci., vol. 38, Springer, Berlin, 1994.
47. E. Z. Goren, *On certain reduction problems concerning abelian surfaces*, Manuscripta Math. **94** (1997), no. 1, 33–43.
48. ______, *Hasse invariants for Hilbert modular varieties*, Israel J. Math. **122** (2001), 157–174.
49. ______, *Hilbert modular forms modulo p^m—the unramified case*, J. Number Theory (to appear).
50. E. Z. Goren and F. Oort, *Stratifications of Hilbert modular varieties*, J. Algebraic Geom. **9** (2000), 111–154.
51. F. Q. Gouvêa, *Arithmetic of p-adic modular forms*, Lecture Notes in Math., vol. 1304, Springer, Berlin, 1988.
52. ______, *Deforming Galois representations: controlling the conductor*, J. Number Theory **34** (1990), no. 1, 95–113.

53. ———, *Deforming Galois representations: a survey*, Seminar on Fermat's Last Theorem (Toronto, ON, 1993–1994), CMS Conf. Proc., vol. 17, Amer. Math. Soc., Providence, RI, 1995, pp. 179–207.
54. ———, *Deformations of Galois representations* (lecture notes on deformations of Galois representations given at the Park City Mathematics Institute in the summer of 1999).
55. F. Q. Gouvêa and B. Mazur, *On the density of modular representations*, Computational Perspectives on Number Theory (Chicago, IL, 1995), AMS/IP Stud. Adv. Math., vol. 7, Amer. Math. Soc., Providence, RI, 1998, pp. 127–142.
56. P. Griffiths and J. Harris, *Principles of algebraic geometry*, Pure Appl. Math., John Wiley & Sons, New York, 1978.
57. B. H. Gross, *Heights and the special values of L-series*, Number Theory (Montreal, QC, 1985), CMS Conf. Proc., vol. 7, Amer. Math. Soc., Providence, RI, 1987, pp. 115–187.
58. ———, *A tameness criterion for Galois representations associated to modular forms (mod p)*, Duke Math. J. **61** (1990), no. 2, 445–517.
59. B. H. Gross and D. B. Zagier, *On singular moduli*, J. Reine Angew. Math. **355** (1985), 191–220.
60. A. Grothendieck, *Eléments de géométrie algébrique. III. Étude cohomologique des faisceaux cohérents.* I, Inst. Hautes Études Sci. Publ. Math. **11** (1961).
61. ———, *Eléments de géométrie algébrique. III. Étude cohomologique des faisceaux cohérents.* II, Inst. Hautes Études Sci. Publ. Math. **17** (1963).
62. ———, *On the de Rham cohomology of algebraic varieties*, Inst. Hautes Études Sci. Publ. Math. **29** (1966), 95–103.
63. ———, *Crystals and the de Rham cohomology of schemes*, Dix exposés sur la cohomologie des schémas, North-Holland, Amsterdam; Masson, Paris, 1968, pp. 306–358.
64. ———, *Géométrie formelle et géométrie algébrique*, Séminaire Bourbaki, vol. 5, exp. no. 182, Soc. Math. France, Paris, 1995, pp. 193–220; Errata, p. 390.
65. ———, *Un théorème sur les homomorphismes de schémas abéliens*, Invent. Math. **2** (1966), 59–78.
66. A. Grothendieck (ed.), *Groupes de monodromie en géométrie algébrique.* I, Séminaire de Géométrie Algébrique du Bois-Marie 1967–1969 (SGA 7 I) (with the collaboration of M. Raynaud and D. S. Rim), Lecture Notes in Math., vol. 288, Springer, Berlin-New York, 1972.
67. R. Hartshorne, *Residues and duality* (with an appendix by P. Deligne), Lecture Notes in Math., vol. 20., Springer, Berlin–New York, 1966.
68. ———, *Algebraic geometry*, Grad. Texts in Math., vol. 52. Springer, New York–Heidelberg, 1977.
69. M. Hazewinkel, *Formal groups and applications*, Pure Applied Math., vol. 78. Academic Press, New York–London, 1978.
70. T. Honda, *On the theory of commutative formal groups*, J. Math. Soc. Japan **22** (1970), 213–246.
71. T. Ibukiyama, T. Katsura, and F. Oort, *Supersingular curves of genus two and class numbers*, Compositio Math. **57** (1986), no. 2, 127–152.
72. J.-I. Igusa, *Class number of a definite quaternion with prime discriminant*, Proc. Nat. Acad. Sci. U.S.A. **44** (1958), 312–314.
73. K. F. Ireland and I. M. Rosen, *A classical introduction to modern number theory*, Grad. Texts in Math. vol. 84. Springer, New York–Berlin, 1982.
74. K. Iwasawa, *Lectures on p-adic L-functions*, Ann. of Math. Stud., vol. 74, Princeton Univ. Press, Princeton, NJ; University of Tokyo Press, Tokyo, 1972.
75. A. J. de Jong, *Homomorphisms of Barsotti–Tate groups and crystals in positive characteristic*, Invent. Math. **134** (1998), no. 2, 301–333; Erratum, *idem* **138** (1999), no. 1, 225.
76. A. J. de Jong and F. Oort, *Purity of the stratification by Newton polygons*, J. Amer. Math. Soc. **13** (2000), no. 1, 209–241.
77. N. Jochnowitz, *Congruences between systems of eigenvalues of modular forms*, Trans. Amer. Math. Soc. **270** (1982), no. 1, 269–285.
78. P. L. Kassaei, *p-adic modular forms over Shimura curves over* $\mathbb{Q}$, Ph.D. thesis, MIT, 1999.
79. N. M. Katz, *p-adic properties of modular schemes and modular forms*, Modular Functions of One Variable. III (Antwerp, 1972), Lecture Notes in Math., vol. 350, Springer, Berlin, 1973, pp. 69–190.
80. ———, *Higher congruences between modular forms*, Ann. of Math. (2) **101** (1975), 332–367.

81. ______, *p-adic L-functions for CM fields*, Invent. Math. **49** (1978), no. 3, 199–297.
82. ______, *Slope filtration of F-crystals*, Journées de Géométrie Algébrique de Rennes (Rennes, 1978), vol. I, Astérisque, vol. 63, Soc. Math. France, Paris, 1979, pp. 113–163.
83. ______, *Serre–Tate local moduli*, Algebraic Surfaces (Orsay, 1976–78), Lecture Notes in Math., vol. 868, Springer, Berlin–New York, 1981, pp. 138–202.
84. N. M. Katz and B. Mazur, *Arithmetic moduli of elliptic curves*, Ann. of Math. Stud., vol. 108, Princeton Univ. Press, Princeton, NJ, 1985.
85. N. Koblitz, *p-adic variation of the zeta-function over families of varieties defined over finite fields*, Compositio Math. **31** (1975), no. 2, 119–218.
86. K. Kodaira and D. C. Spencer, *On deformations of complex analytic structures*. I, II, Ann. of Math. (2) **67** (1958), 328–466.
87. M. Kolster, *K-theory and arithmetic* (mini-course at McGill and Concordia universities, April 13–19, 2000).
88. M. Kolster and T. Nguyen Quang Do, *Syntomic regulators and special values of p-adic L-functions*, Invent. Math. **133** (1998), no. 2, 417–447.
89. S. Lang, *Elliptic functions* (with an appendix by J. Tate), 2nd edition, Grad. Texts in Math., vol. 112., Springer, New York, 1987.
90. H. Lange and Ch. Birkenhake, *Complex abelian varieties*, Grundlehren Math. Wiss., vol. 302, Springer, Berlin, 1992.
91. M. Lazard, *Commutative formal groups*, Lecture Notes in Math., vol. 443, Springer, Berlin–New York, 1975.
92. S. Lichtenbaum, *Values of zeta-functions, étale cohomology, and algebraic K-theory*, Algebraic K-Theory, II. "Classical" Algebraic K-Theory and Connections with Arithmetic (Seattle, WA, 1972), Lecture Notes in Math. vol. 342, Springer, Berlin, 1973, pp. 489–501.
93. Yu. I. Manin, *The theory of commutative formal groups over fields of finite characteristic*, Russian Math. Surveys **18** (1963), 1–80.
94. B. Mazur, *Modular curves and the Eisenstein ideal*, Inst. Hautes Études Sci. Publ. Math. **47** (1977), 33–186.
95. ______, *Deforming Galois representations*, Galois Groups over $\mathbb{Q}$ (Berkeley, CA, 1987), Math. Sci. Res. Inst. Publ., vol. 16, Springer, New York, 1989, pp. 385–437.
96. W. Messing, *The crystals associated to Barsotti–Tate groups: with applications to abelian schemes*, Lecture Notes in Math., vol. 264. Springer, Berlin-New York, 1972.
97. J.-F. Mestre, *La méthode des graphes. Exemples et applications*, Proceedings of the International Conference on Class Numbers and Fundamental Units of Algebraic Number Fields (Katata, 1986), Nagoya Univ., Nagoya, 1986, pp. 217–242.
98. J. S. Milne, *Points on Shimura varieties mod p*, Automorphic Forms, Representations and L-Functions, Part 2, Proc. Sympos. Pure Math., vol. 33, Amer. Math. Soc., Providence, RI, 1979, pp. 165–184.
99. ______, *Étale cohomology*, Princeton Math. Ser., vol. 33. Princeton Univ. Press, Princeton, NJ, 1980.
100. ______, *Abelian varieties*, Arithmetic Geometry. (Storrs, CT, 1984) (G. Cornell and J. H. Silverman, eds.) Springer, New York, 1986, pp. 103–150.
101. ______, *Jacobian varieties*, Arithmetic Geometry. (Storrs, CT, 1984) (G. Cornell and J. H. Silverman, eds.) Springer, New York, 1986, pp. 167–212.
102. T. Miyake, *Modular forms*, Springer, Berlin, 1989.
103. S. Mukai, *Duality between $D(X)$ and $D(\widehat{X})$ with its application to Picard sheaves*, Nagoya Math. J. **81** (1981), 153–175.
104. D. Mumford, *Picard groups of moduli problems*, Arithmetic Algebraic Geometry, (Purdue Univ., 1963) (O. Schilling, ed.), Harper & Row, New York, 1965, pp. 33–81.
105. ______, *On the equations defining abelian varieties*. I, Invent. Math. **1** (1966), 287–354.
106. ______, *Bi-extensions of formal groups*, Algebraic Geometry (Bombay, 1968), Oxford Univ. Press, London, 1969, pp. 307–322.
107. ______, *Abelian varieties*, Tata Inst. Fund. Res. Stud. Math., vol. 5, Oxford University Press, London, 1970.
108. ______, *Towards an enumerative geometry of the moduli space of curves*, Arithmetic and Geometry, vol. II, Progr. Math., vol. 36, Birkhäuser, Boston, MA, 1983, pp. 271–328.
109. ______, *Tata lectures on theta*. I (with the assistance of C. Musili, M. Nori, E. Previato and M. Stillman), Progr. Math., vol. 28, Birkhäuser, Boston, MA, 1983.

110. ______, *Tata lectures on theta. II. Jacobian theta functions and differential equations* (with the collaboration of C. Musili, M. Nori, E. Previato, M. Stillman, and H. Umemura), Progr. Math., vol. 43, Birkhäuser, Boston, MA, 1984.
111. ______, *The red book of varieties and schemes*, Lecture Notes in Math., vol. 1358, Springer, Berlin, 1988.
112. ______, *Tata lectures on theta.* III (with the collaboration of M. Nori and P. Norman), Progr. Math., vol. 97, Birkhäuser, Boston, MA, 1991.
113. D. Mumford, J. Fogarty, and F. Kirwan, *Geometric invariant theory*, 3rd edition, Ergeb. Math. Grenzgeb. (2), vol. 34. Springer, Berlin, 1994.
114. J. P. Murre, *Algebraic cycles on abelian varieties: Application of abstract Fourier theory*, The Arithmetic and Geometry of Algebraic Cycles, NATO Sci. Ser. C Math. Phys. Sci., vol. 548, Kluwer Acad. Publ., Dordrecht, 2000, pp. 307–320.
115. V. K. Murty, *Introduction to abelian varieties*, CRM Monogr. Ser., vol. 3, Amer. Math. Soc., Providence, RI, 1993.
116. M. S. Narasimhan and M. V. Nori, *Polarisations on an abelian variety*, Geometry and Analysis, Indian Acad. Sci., Bangalore, 1980, pp. 125–128.
117. P. Norman, *An algorithm for computing local moduli of abelian varieties*, Ann. Math. (2) **101** (1975), 499–509.
118. P. Norman and F.Oort, *Moduli of abelian varieties*, Ann. of Math. (2) **112** (1980), no. 3, 413–439.
119. T. Oda, *The first de Rham cohomology group and Dieudonné modules*, Ann. Sci. École Norm. Sup. (4) **2** (1969), 63–135.
120. T. Oda and F. Oort, *Supersingular abelian varieties*, Proceedings of the International Symposium on Algebraic Geometry (Kyoto, 1977), Kinokuniya Book Store, Tokyo, 1978, pp. 595–621.
121. J. Oesterlé, *Travaux de Wiles (et Taylor, ...)*. II. Séminaire Bourbaki, vol. 1994/95., exp. no. 804, Astérisque, vol. 237, Soc. Math. France, Paris, 1996, pp. 333–355.
122. A. P. Ogg, *Modular forms and Dirichlet series*, W. A. Benjamin, Inc., New York–Amsterdam 1969.
123. ______, *Rational points on certain elliptic modular curves*, Analytic Number Theory (St. Louis, MO, 1972), Proc. Sympos. Pure Math., vol. 24, Amer. Math. Soc., Providence, RI, 1973, pp. 221–231.
124. ______, *Hyperelliptic modular curves*, Bull. Soc. Math. France **102** (1974), 449–462.
125. T. Ono, *Arithmetic of algebraic tori*, Ann. of Math. (2) **74** (1961), 101–139.
126. F. Oort, *Commutative group schemes*, Lecture Notes in Math. vol. 15, Springer, Berlin–New York, 1966.
127. ______, *Finite group schemes, local moduli for abelian varieties, and lifting problems*, Algebraic Geometry, (Oslo, 1970) (F. Oort, ed.), Wolters-Noordhoff Publishing, Groningen, 1972, pp. 223–254..
128. ______, *Which abelian surfaces are products of elliptic curves?* Math. Ann. **214** (1975), 35–47.
129. ______, *Endomorphism algebras of abelian varieties*, Algebraic Geometry and Commutative Algebra, vol. II, Kinokuniya, Tokyo, 1988, pp. 469–502.
130. ______, *Moduli of abelian varieties and Newton polygons*, C. R. Acad. Sci. Paris Sér. I Math. **312** (1991), no. 5, 385–389.
131. ______, *A stratification of a moduli space of abelian varieties*, Moduli of Abelian Varieties (Texel, 1999) (C. Faber, G. van der Geer and F. Oort, eds.) Progr. Math., vol. 195, Birkäuser, Boston, 2001, pp. 345–416.
132. G. Pappas, *Arithmetic models for Hilbert modular varieties*, Compositio Math. **98** (1995), no. 1, 43–76.
133. A. Pizer, *An algorithm for computing modular forms on* $\Gamma_0(N)$, J. Algebra **64** (1980), no. 2, 340–390.
134. R. Ramakrishna, *On a variation of Mazur's deformation functor*, Compositio Math. **87** (1993), no. 3, 269–286.
135. M. Rapoport, *Compactifications de l'espace de modules de Hilbert–Blumenthal*, Compositio Math. **36** (1978), no. 3, 255–335.

136. D. C. Ravenel, *Complex cobordism theory for number theorists*, Elliptic curves and modular forms in algebraic topology (Princeton, NJ, 1986), Lecture Notes in Math., vol. 1326, Springer, Berlin, 1988, pp. 123–133.
137. M. Raynaud, *Faisceaux amples sur les schémas en groupes et les espaces homogènes*, Lecture Notes in Math., vol. 119, Springer, Berlin–New York, 1970.
138. K. Ribet, *p-adic interpolation via Hilbert modular forms*, Algebraic Geometry (Arcata, CA, 1974), Proc. Sympos. Pure Math.,vol. 29, Amer. Math. Soc., Providence, RI, 1976, pp. 581–592.
139. K. Ribet and W. A. Stein, *Lectures on Serre's conjectures* (in preparation).
140. J. J. Rotman, *An introduction to the theory of groups*, 4th edition, Grad. Texts in Math., vol. 148, Springer, New York, 1995.
141. H.-G. Rück, *Abelian surfaces and Jacobian varieties over finite fields*, Compositio Math. **76** (1990), no.3, 351-36.
142. M. Schlessinger, *Functors of Artin rings*, Trans. Amer. Math. Soc. **130** (1968), 208–222.
143. B. Schoeneberg, *Elliptic modular functions: An introduction*, Grundlehren Math. Wiss., vol. 203, Springer, New York–Heidelberg, 1974.
144. J.-P. Serre, *Groupes p-divisibles (d'après J. Tate)*, Séminaire Bourbaki 1966/67, vol. 10, exp. no. 318, Soc. Math. France, Paris, 1995, pp. 73–86.
145. ———, Congruences et formes modulaires [d'après H. P. F. Swinnerton-Dyer], Séminaire Bourbaki, 24e année (1971/1972), exp. no. 416, Lecture Notes in Math., vol. 317, Springer, Berlin, 1973, pp. 319–338.
146. ———, *Formes modulaires et fonctions zêta p-adiques*, Modular Functions of One Variable. III (Antwerp, 1972), Lecture Notes in Math., vol. 350, Springer, Berlin, 1973, pp. 191–268.
147. ———, *A course in arithmetic*, Grad. Texts in Math., vol. 7, Springer, New York–Heidelberg, 1973.
148. ———, *Arithmetic groups*, Homological Group Theory (Durham, 1977), London Math. Soc. Lecture Note Ser., vol. 36, Cambridge Univ. Press, Cambridge–New York, 1979, pp. 105–136.
149. ———, *Local fields*, Grad. Texts in Math., vol. 67, Springer, New York–Berlin, 1979.
150. ———, *Sur les représentations modulaires de degré* 2 *de* $\mathrm{Gal}(\overline{Q}/Q)$, Duke Math. J. **54** (1987), no. 1, 179–230.
151. ———, *Travaux de Wiles (et Taylor, ...)*. I, Séminaire Bourbaki, vol. 1994/95, exp. no. 803, Astérisque, vol. 237, Soc. Math. France, Paris, 1996, pp. 319–332.
152. S. Shatz, *Group schemes, formal groups, and p-divisible groups*, Arithmetic Geometry (Storrs, CT, 1984), Springer, New York, 1986, pp. 29–78.
153. G. Shimura, *On analytic families of polarized abelian varieties and automorphic functions*, Ann. of Math. (2) **78** (1963), 149–192.
154. ———, *Introduction to the arithmetic theory of automorphic functions*, Publ. Math. Soc. Japan, vol. 11, Iwanami Shoten, Publishers, Tokyo; Princeton Univ. Press, Princeton, NJ, 1971.
155. ———, *The special values of the zeta functions associated with Hilbert modular forms*, Duke Math. J. **45** (1978), no. 3, 637–679.
156. J. H. Silverman, *The arithmetic of elliptic curves*, Grad. Texts in Math., vol. 106, Springer, New York, 1986.
157. ———, *Advanced topics in the arithmetic of elliptic curves*, Grad. Texts in Math., vol. 151, Springer, New York, 1994.
158. H. P. F. Swinnerton-Dyer, *On l-adic representations and congruences for coefficients of modular forms*, Modular Functions of One Variable. III (Antwerp, 1972), Lecture Notes in Math., vol. 350, Springer, 1973, Berlin, pp. 1–55.
159. ———, *Analytic theory of abelian varieties*, London Math. Soc. Lecture Note Ser., vol. 14, Cambridge Univ. Press, London–New York, 1974.
160. J. Tate, *Endomorphisms of abelian varieties over finite fields*, Invent. Math. **2** (1966), 134–144.
161. ———, *p-divisible groups*, Proc. Conf. Local Fields (Driebergen, 1966), Springer, Berlin, 1967, pp. 158–183.
162. ———, *On the conjectures of Birch and Swinnerton-Dyer and a geometric analogue*, Séminaire Bourbaki, vol. 9, exp. no. 306, Soc. Math. France, Paris, 1995, pp. 415–440.
163. ———, *Finite flat group schemes*, Modular Forms and Fermat's Last Theorem (Boston, MA, 1995), Springer, New York, 1997, pp. 121–154.

164. R. Taylor and A. Wiles, *Ring-theoretic properties of certain Hecke algebras*, Ann. of Math. (2) **141** (1995), no. 3, 553–572.

165. S. Tsuyumine, *On values of L-functions of totally real algebraic number fields at integers*, Acta Arith. **76** (1996), no. 4, 359–392.

166. D. L. Ulmer, *On universal elliptic curves over Igusa curves*, Invent. Math. **99** (1990), no. 2, 377–391.

167. L. C. Washington, *Introduction to cyclotomic fields*, 2nd edition, Grad. Texts in Math., vol. 83, Springer, New York, 1997.

168. W. C. Waterhouse, *Introduction to affine group schemes*, Grad. Texts in Math., vol. 66, Springer, New York-Berlin1979.

169. W. C. Waterhouse and J. S. Milne, *Abelian varieties over finite fields*, 1969 Number Theory Institue (Stony Brook, NY, 1969), Proc. Sympos. Pure Math., vol. 20, Amer. Math. Soc., Providence, RI, 1971, pp. 53-64.

170. T. Wedhorn, *Ordinariness in good reductions of Shimura varieties of PEL-type*, Ann. Sci. École Norm. Sup. (4) **32** (1999), no. 5, 575–618.

171. A. Weil, *Sur certains groupes d'opérateurs unitaires*, Acta Math. **111** (1964), 143–211.

172. ———, *Sur la formule de Siegel dans la théorie des groupes classiques*, Acta Math. **113** (1965), 1–87.

173. A. Wiles, *Modular elliptic curves and Fermat's last theorem*, Ann. of Math. (2) **141** (1995), no. 3, 443–551.

174. N. Yui, *On the Jacobian varieties of hyperelliptic curves over fields of characteristic* $p > 2$, J. Algebra **52** (1978), no. 2, 378–410.

175. ———, *Formal groups and some arithmetic properties of elliptic curves*, Algebraic Geometry (Copenhagen, 1978), Lecture Notes in Math., vol. 732, Springer, Berlin, 1979, pp. 630–658.

176. J. G. Zarhin, *Isogenies of abelian varieties over fields of finite characteristics*, Math. USSR-Sb. **24** (1974), 451–461.

177. ———, *A remark on endomorphisms of abelian varieties over function fields of finite characteristics*, Math. USSR-Izv. **8** (1974), 477–480.

178. Th. Zink, *The display of a formal p-divisible group*, Universität Bielefeld preprint 98-017, 1998.

Notation Index

Index